7/16/11
#58

GENETICS AND GENETIC ENGINEERING

ISSN 1546-6426

GENETICS AND GENETIC ENGINEERING

Barbara Wexler

INFORMATION PLUS® REFERENCE SERIES
Formerly Published by Information Plus, Wylie, Texas

GALE
CENGAGE Learning™

Detroit • New York • San Francisco • New Haven, Conn • Waterville, Maine • London

Genetics and Genetic Engineering

Barbara Wexler

Kepos Media, Inc., Paula Kepos and Janice Jorgensen, Series Editors

Project Editors: Kathleen J. Edgar, Elizabeth Manar, Kimberley A. McGrath

Rights Acquisition and Management: Robyn Young

Composition: Evi Abou-El-Seoud, Mary Beth Trimper

Manufacturing: Cynde Lentz

For product information and technology assistance, contact us at
Gale Customer Support, 1-800-877-4253.
For permission to use material from this text or product,
submit all requests online at **www.cengage.com/permissions.**
Further permissions questions can be e-mailed to
permissionrequest@cengage.com

Gale
27500 Drake Rd.
Farmington Hills, MI 48331-3535

ISBN-13: 978-0-7876-5103-9 (set)
ISBN-13: 978-1-4144-4860-2

ISBN-10: 0-7876-5103-6 (set)
ISBN-10: 1-4144-4860-0

ISSN 1546-6426

This title is also available as an e-book.
ISBN-13: 978-1-4144-7524-0 (set)
ISBN-10: 1-4144-7524-1 (set)
Contact your Gale sales representative for ordering information.

Printed in the United States of America
1 2 3 4 5 6 7 15 14 13 12 11

TABLE OF CONTENTS

PREFACE

Genetics and Genetic Engineering is part of the *Information Plus Reference Series.* The purpose of each volume of the series is to present the latest facts on a topic of pressing concern in modern American life. These topics include the most controversial and studied social issues of the 21st century: abortion, capital punishment, care of senior citizens, crime, the environment, health care, immigration, minorities, national security, social welfare, women, youth, and many more. Even though this series is written especially for high school and undergraduate students, it is an excellent resource for anyone in need of factual information on current affairs.

By presenting the facts, it is the intention of Gale, Cengage Learning to provide its readers with everything they need to reach an informed opinion on current issues. To that end, there is a particular emphasis in this series on the presentation of scientific studies, surveys, and statistics. These data are generally presented in the form of tables, charts, and other graphics placed within the text of each book. Every graphic is directly referred to and carefully explained in the text. The source of each graphic is presented within the graphic itself. The data used in these graphics are drawn from the most reputable and reliable sources, such as from the various branches of the U.S. government and from private organizations and associations. Every effort has been made to secure the most recent information available. Readers should bear in mind that many major studies take years to conduct and that additional years often pass before the data from these studies are made available to the public. Therefore, in many cases the most recent information available in 2011 is dated from 2008 or 2009. Older statistics are sometimes presented as well if they are landmark studies or of particular interest and no more-recent information exists.

Even though statistics are a major focus of the *Information Plus Reference Series*, they are by no means its only content. Each book also presents the widely held positions and important ideas that shape how the book's subject is discussed in the United States. These positions are explained in detail and, where possible, in the words of their proponents. Some of the other material to be found in these books includes historical background, descriptions of major events related to the subject, relevant laws and court cases, and examples of how these issues play out in American life. Some books also feature primary documents or have pro and con debate sections that provide the words and opinions of prominent Americans on both sides of a controversial topic. All material is presented in an even-handed and unbiased manner; readers will never be encouraged to accept one view of an issue over another.

HOW TO USE THIS BOOK

Genetics and genetic engineering is a hotly debated topic in the United States in the 21st century. Issues such as genetic testing to predict disease susceptibility and the implications of test results; whether to permit cloning of human organs and the moral dilemma of "playing God"; and the advisability of raising genetically modified crops are discussed in coffee shops and classrooms across the nation. These topics and more are covered in this volume.

Genetics and Genetic Engineering consists of 11 chapters and three appendixes. Each chapter is devoted to a particular aspect of genetics and genetic engineering in the United States. For a summary of the information covered in each chapter, please see the synopses provided in the Table of Contents. Chapters generally begin with an overview of the basic facts and background information on the chapter's topic, then proceed to examine subtopics of particular interest. For example, Chapter 6, Genetic Testing, begins by discussing the basic concepts of genetic testing, then goes on to examine the screening and diagnostic tests. The quality and utility of genetic tests depend on their reliability, validity, sensitivity, specificity, positive predictive value, and negative predictive value and the chapter details these issues as well as the laboratory techniques that are used to perform genetic testing. After a discussion of carrier identification, the chapter describes preimplantation genetic diagnosis, prenatal testing, and genetic testing of newborns, children, and adults. It concludes with discussions of the social and political issues surrounding genetic testing—including ethical considerations and legislation banning genetic discrimination—and genetic counseling. Readers can find their way through a chapter by looking for the section and subsection headings, which are clearly set off from the text. They can also refer to the book's extensive index if they already know what they are looking for.

Statistical Information

The tables and figures featured throughout *Genetics and Genetic Engineering* will be of particular use to readers in learning about this issue. These tables and figures represent an extensive collection of the most recent and important statistics on genetics and genetic engineering and related issues—for example, graphics cover the landmarks in the history of genetics, a diagram of deoxyribonucleic acid, public opinion on animal and human cloning, and public opinion on the moral acceptability of embryonic stem cell research. Gale, Cengage Learning believes that making this information available to readers is the most important way to fulfill the goal of this book: to help readers understand the issues and controversies surrounding genetics and genetic engineering in the United States and to reach their own conclusions.

Each table or figure has a unique identifier appearing above it for ease of identification and reference. Titles for the tables and figures explain their purpose. At the end of each table or figure, the original source of the data is provided.

To help readers understand these often complicated statistics, all tables and figures are explained in the text. References in the text direct readers to the relevant statistics. Furthermore, the contents of all tables and figures are fully indexed. Please see the opening section of the index at the back of this volume for a description of how to find tables and figures within it.

Appendixes

Besides the main body text and images, *Genetics and Genetic Engineering* has three appendixes. The first is the Important Names and Addresses directory. Here, readers will find contact information for a number of government and private organizations that can provide further information on genetics and genetic engineering. The second appendix is the Resources section, which can also assist readers in conducting their own research. In this section the author and editors of *Genetics and Genetic Engineering* describe some of the sources that were most useful during the compilation of this book. The final appendix is the detailed index. It has been greatly expanded from previous editions and should make it even easier to find specific topics in this book.

ADVISORY BOARD CONTRIBUTIONS

The staff of Information Plus would like to extend its heartfelt appreciation to the Information Plus Advisory Board. This dedicated group of media professionals provides feedback on the series on an ongoing basis. Their comments allow the editorial staff who work on the project to make the series better and more user-friendly. The staff's top priority is to produce the highest-quality and most useful books possible, and the Information Plus Advisory Board's contributions to this process are invaluable.

The members of the Information Plus Advisory Board are:

- Kathleen R. Bonn, Librarian, Newbury Park High School, Newbury Park, California
- Madelyn Garner, Librarian, San Jacinto College, North Campus, Houston, Texas
- Anne Oxenrider, Media Specialist, Dundee High School, Dundee, Michigan
- Charles R. Rodgers, Director of Libraries, Pasco-Hernando Community College, Dade City, Florida
- James N. Zitzelsberger, Library Media Department Chairman, Oshkosh West High School, Oshkosh, Wisconsin

COMMENTS AND SUGGESTIONS

The editors of the *Information Plus Reference Series* welcome your feedback on *Genetics and Genetic Engineering*. Please direct all correspondence to:

Editors
Information Plus Reference Series
27500 Drake Rd.
Farmington Hills, MI 48331-3535

CHAPTER 1
THE HISTORY OF GENETICS

Science seldom proceeds in the straightforward logical manner imagined by outsiders.

—James D. Watson, *The Double Helix: A Personal Account of the Discovery of the Structure of DNA* (1968)

Genetics is the biology of heredity, and geneticists are the scientists and researchers who study hereditary processes such as the inheritance of traits, distinctive characteristics, and diseases. Genetics considers the biochemical instructions that convey information from generation to generation.

Tremendous strides in science and technology have enabled geneticists to demonstrate that some genetic variation is related to disease and that the ability to vary genes improves the capacity of a species to survive changes in the environment. Even though some of the most important advances in genetics research—such as deciphering the genetic code, isolating the genes that cause or predict susceptibility to certain diseases, and successfully cloning plants and animals—have occurred since the mid-20th century, the history of genetics study spans a period of about 150 years. As the understanding of genetics progressed, scientific research became increasingly more specific. Genetics first considered populations, then individuals, then it advanced to explore the nature of inheritance at the molecular level.

By the end of the first decade of the 21st century geneticists had made several groundbreaking discoveries. For example, Raphael G. Satter reports in "Scientists: We've Cracked Wheat's Genetic Code" (Associated Press, August 27, 2010) that in 2010 British scientists uncovered the genetic code of wheat, a vital crop worldwide, which may help address pressing global concerns such as feeding the world's growing population and climate change. In "Clinical Assessment Incorporating a Personal Genome" (*Lancet*, vol. 375, no. 9725, May 1, 2010), Euan A. Ashley et al. note that by using an individual's genetic sequence scientists can predict future health risks and how the individual might respond to specific pharmaceutical drugs.

Even though genetics research, especially genetic engineering of synthetic organisms, offers the tantalizing promise of preventing and combating disease and global warming, it also raises a host of ethical, philosophical, and political issues. For example, many people have concerns about the morality of creating new or artificial life by using stem cells to develop medical treatments, the development of synthetic components of flu vaccines, and the creation of new food products such as genetically modified salmon. People are also concerned about the potential of genetic engineering to create biological weapons, the health consequences of consuming genetically engineered foods, and the environmental impact of genetic engineering. Another issue is self-regulation. According to Jeremy Hsu, in "One-Third of Americans Back Ban on Synthetic Biology" (September 9, 2010, http://www.livescience.com/health/synthetic-biology-manmade-life-100909.html), Americans are skeptical that the researchers and corporations involved in genetic engineering will be able to self-regulate.

EARLY BELIEFS ABOUT HEREDITY

From the earliest recorded history, ancient civilizations observed patterns in reproduction. Animals bore offspring of the same species, children resembled their parents, and plants gave rise to similar plants. Some of the earliest ideas about reproduction, heredity, and the transmission of information from parent to child were the particulate theories developed in ancient Greece during the fourth century BC. These theories posited that information from each part of the parent had to be communicated to create the corresponding body part in the offspring. For example, the particulate theories held that information from the parent's heart, lungs, and limbs was transmitted directly from these body parts to create the offspring's heart, lungs, and limbs.

Particulate theories were attempts to explain observed similarities between parents and their children. One reason these theories were inaccurate was that they relied on observations unaided by the microscope. Microscopy (the use of or investigation with the microscope) and the recognition of cells and microorganisms did not occur until the end of the 17th century, when the British physicist Robert Hooke (1635–1703) first observed cells through a microscope.

Until that time (and even for some time after) heredity remained poorly understood. During the Renaissance (from about the 14th to the 16th centuries) preformationist theories proposed that the parent's body carried highly specialized reproductive cells that contained whole, preformed offspring. Preformationist theories insisted that when these specialized cells containing the offspring were placed in suitable environments, they would spontaneously grow into new organisms with traits similar to the parent organism.

The Greek philosopher Aristotle (384–322 BC), who was such a keen observer of life that he is often referred to as the father of biology, noted that individuals sometimes resemble remote ancestors more closely than their immediate parents. He was a preformationist, positing that the male parent provided the miniature individual and the female provided the supportive environment in which it would grow. He also refuted the notion of a simple, direct transfer of body parts from parent to offspring by observing that animals and humans who had suffered mutilation or loss of body parts did not confer these losses to their offspring. Instead, he described a process that he called *epigenesist*, in which the offspring is gradually generated from an undifferentiated mass by the addition of parts.

Of Aristotle's many contributions to biology, one of the most important was his conclusion that inheritance involved the potential of producing certain characteristics rather than the absolute production of the characteristics themselves. This thinking was closer to the scientific reality of inheritance than any philosophy set forth by his predecessors. However, because Aristotle was developing his theories before the advent of microscopy, he mistakenly presumed that inheritance was conveyed via the blood. Regardless, his enduring influence is evident in the language and thinking about heredity. Even though blood is not the mode of transmission of heredity, people still refer to "blood relatives," "blood lines," and offspring as products of their own "flesh and blood."

One of the most important developments in the study of hereditary processes came in 1858, when the British naturalists Charles Darwin (1809–1882) and Alfred Russel Wallace (1823–1913) announced the theory of natural selection—the idea that members of a population who are better adapted to their environment will be the ones most likely to survive and pass their traits on to the next

generation. Darwin published his theories in *On the Origin of Species by Means of Natural Selection, or the Preservation of Favoured Races in the Struggle for Life* (1859). His work was not viewed favorably, especially by religious leaders who believed that it refuted the biblical interpretation of how life on the earth began. Even in the 21st century the idea that life evolves gradually through natural processes is not accepted by everyone, and the dispute over creationism and evolution continues.

CELL THEORY

In 1665, when Hooke used the microscope he had designed to examine a piece of cork, he saw a honeycomb pattern of rectangles that reminded him of cells, the chambers of monks in monasteries. His observations prompted scientists to speculate that living tissue as well as nonliving tissue was composed of cells. The French physiologist René-Joachim-Henri Dutrochet (1776–1847) performed microscopic studies and concluded in 1824 that both plant and animal tissue was composed of cells.

In 1838 the German botanist Matthias Jakob Schleiden (1804–1881) presented his theory that all plants were constructed of cells. The following year the German physiologist Theodor Schwann (1810–1882) suggested that animals were also composed of cells. Both Schleiden and Schwann theorized that cells were all created using the same process. Even though Schleiden's hypotheses about the process of cell formation were not entirely accurate, both he and Schwann are credited with developing cell theory. Describing cells as the basic units of life, they asserted that all living things are composed of cells, the simplest forms of life that can exist independently. Their pioneering work enabled other scientists to understand accurately how cells live, such as the German pathologist Rudolf Virchow (1821–1902), who launched theories of biogenesis when he posited in 1858 that cells reproduce themselves.

Improvements in microscopy and the increasing study of cytology (the formation, structure, and function of cells) enabled scientists to identify parts of the cell. Key cell components include the nucleus, which directs all cellular activities by controlling the synthesis of proteins, and the mitochondria, which are organelles (membrane-bound cell compartments) that catalyze reactions that produce energy for the cell. Figure 1.1 is a diagram of a typical animal cell that shows its component parts, including the contents of the nucleus, where chromosomes (which contain the genes) are located.

Germplasm Theory of Heredity

Studies of cellular components, processes, and functions produced insights that revealed the connection between cytology and inheritance. The German biologist August Weissmann (1834–1914) studied medicine, biology, and

FIGURE 1.1

A typical animal cell

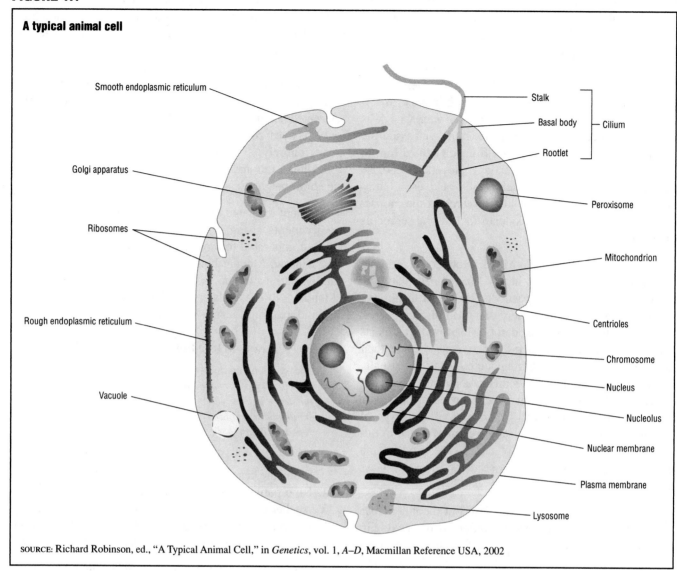

Smooth endoplasmic reticulum

Golgi apparatus

Ribosomes

Rough endoplasmic reticulum

Vacuole

Stalk

Basal body — Cilium

Rootlet

Peroxisome

Mitochondrion

Centrioles

Chromosome

Nucleus

Nucleolus

Nuclear membrane

Plasma membrane

Lysosome

SOURCE: Richard Robinson, ed., "A Typical Animal Cell," in *Genetics*, vol. 1, *A–D*, Macmillan Reference USA, 2002

zoology, and his contribution to genetics was an evolutionary theory known as the germplasm theory of heredity. Building on Darwin's idea that specific inherited characteristics are passed from one generation to the next, Weissmann asserted that the genetic code for each organism was contained in its germ cells (the cells that create sperm and eggs). The presence of genetic information in the germ cells explained how this information was conveyed and unchanged from one generation to the next.

In a series of essays about heredity published between 1889 and 1892, Weissmann observed that the amount of genetic material did not double when cells replicated, suggesting that there was some form of biological control of the chromosomes that occurred during the formation of the gametes (sperm and egg). His theory was essentially correct. Normal body growth is attributable to cell division, called mitosis, which produces cells that are genetically identical to the parent cells. The way to avoid giving

offspring a double dose of heredity information is through a cell division that reduces the amount of the genetic material in the gametes by one-half. Weissmann called this process reduction division; it is now known as meiosis.

Weissmann was also the first scientist to successfully refute the members of the scientific community who believed that physical characteristics acquired through environmental exposure were passed from generation to generation. He conducted experiments in which he cut the tails off several consecutive generations of mice and observed that none of their offspring were born tailless.

A FARMER'S SON BECOMES THE FATHER OF GENETICS

Gregor Mendel (1822–1884) was born into a peasant family in what is now Hyncice, Czech Republic, and spent much of his youth working in his family's orchards and gardens. At the age of 21 years he entered the Abbey of St. Thomas, a Roman Catholic monastery, where he studied theology, philosophy, and science. His interest in botany (the scientific study of plants) and an aptitude for natural science inspired his superiors to send him to the University of Vienna, where he studied to become a science teacher. However, Mendel was not destined to become an academic, despite his abiding interest in science and experimentation. In fact, the man who was eventually called the father of genetics never passed the qualifying examinations that would have enabled him to teach science at the highest academic level. Instead, he instructed students at a technical school. He also continued to study botany and conduct research at the monastery, and from 1868 until his death in 1884 he served as its abbot.

Between 1856 and 1863 Mendel conducted carefully designed experiments with nearly 30,000 pea plants he cultivated in the monastery garden. He chose to observe pea plants systematically because they had distinct, identifiable characteristics that could not be confused. Pea plants were also ideal subjects for his experiments because their reproductive organs were surrounded by petals and usually matured before the flower bloomed. As a result, the plants self-fertilized, and each plant variety tended to be a pure breed. Mendel raised several generations of each type of plant to be certain that his plants were pure breeds. In this way, he confirmed that tall plants always produce tall offspring, and plants with green seeds and leaves always produce offspring with green seeds and leaves.

His experiments were designed to test the inheritance of a specific trait from one generation to the next. For example, to test the inheritance of the characteristic of plant height, Mendel self-pollinated several short pea plants, and the seeds they produced grew into short plants. Similarly, self-pollinated tall plants and their resulting seeds, called the first or F1 generation, grew to be tall plants. These results seemed logical. When Mendel bred tall and short plants together and all their offspring in the F1 generation were tall, he concluded that the shortness trait had disappeared. However, when he self-pollinated the F1 generation, the offspring, called the F2 generation (second generation), contained both tall and short plants. After repeating this experiment many times, Mendel observed that in the F2 generation there were three tall plants for every short one—a 3:1 ratio.

Mendel's attention to rigorous scientific methods of observation, large sample sizes, and statistical analysis of the data he collected bolstered the credibility of his results. These experiments prompted him to theorize that characteristics, or traits, come in pairs—one from each parent—and that one trait will assume dominance over the other. The trait that appears more frequently is considered the stronger, or dominant, trait, whereas the one that appears less often is the recessive trait.

Focusing on plant height and other distinctive traits, such as the color of the pea pods, seed shape (smooth or wrinkled), and leaf color (green or yellow), enabled Mendel to record accurately and document the results of his plant breeding experiments. His observations about purebred plants and their consistent capacity to convey traits from one generation to the next represented a novel idea. The accepted belief of inheritance described a blending of traits, which, once combined, diluted or eliminated the original traits entirely. For example, it was believed that cross-breeding a tall and a short plant would produce a plant of medium height.

About the same time, Darwin was performing similar experiments using snapdragons, and his observations were comparable to those made by Mendel. Even though Darwin and Mendel both explained the units of heredity and variations in species in their published works, it was Mendel who was later credited with developing the groundbreaking theories of heredity.

Mendel's Laws of Heredity

The constant characters which appear in the several varieties of a group of plants may be obtained in all the associations which are possible according to the [mathematical] laws of combination, by means of repeated artificial fertilization.

—Gregor Mendel, "Versuche über Pflanzen-Hybriden" (1865)

From the results of his experiments, Mendel formulated and published three interrelated theories in the paper "Versuche über Pflanzen-Hybriden" ("Experiments in Plant Hybridization" [1865; translated into English in 1901]). This work established the basic tenets of heredity:

- Two heredity factors exist for each characteristic or trait.

- Heredity factors are contained in equal numbers in the gametes.

- The gametes contain only one factor for each characteristic or trait.

- Gametes combine randomly, no matter which hereditary factors they carry.

- When gametes are formed, different hereditary factors sort independently.

When Mendel presented his paper, it was virtually ignored by the scientific community, which was otherwise engaged in a heated debate about Darwin's theory of evolution. Years later, well after Mendel's death in 1884, his observations and assumptions were revisited and became known as Mendel's laws of heredity. His first principal of heredity, the law of segregation, states that hereditary units, now known as genes, are always paired and that genes in a pair separate during cell division, with the sperm and egg each receiving one gene of the pair. As

FIGURE 1.2

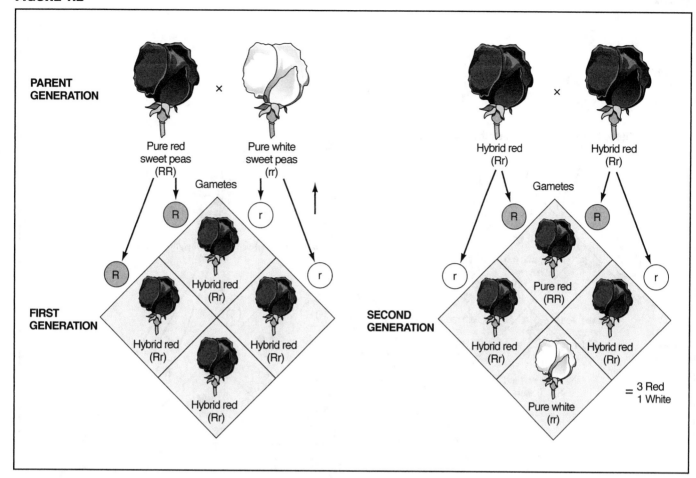

Mendel's law of segregation. *Hans & Cassidy, Cengage Gale.*

a result, each gene in a pair will be present in half the sperm or egg cells. In other words, each gamete receives from a parent cell only one-half of the pair of genes it carries. Because two gametes (male and female) unite to reproduce and form a new cell, the new cell will have a unique pair of genes of its own, half from one parent and half from the other.

Diagrams of genetic traits conventionally use capital letters to represent the dominant traits and lowercase letters to represent recessive traits. Figure 1.2 uses this system to demonstrate Mendel's law of segregation. The pure red sweet pea and the pure white sweet pea each have two genes—RR for the red and rr for the white. The possible outcomes of this mating in the first generation are all hybrid (a combination of two different types) red plants (Rr)—plants that all have the same outward appearance (or phenotype) as the pure red parent but that also carry the white gene. As a result, when two of the hybrid F1 generation plants are bred, there is a 50% chance that the resulting offspring will be hybrid red, a 25% chance that the offspring will be pure red, and a 25% chance that the offspring will be pure white.

Mendel also provided compelling evidence from his experiments for the law of independent assortment. This law establishes that each pair of genes is inherited independently of all other pairs. Figure 1.3 shows the chance distribution of any possible combination of traits. The F1 generation of tall red and dwarf white sweet pea plants produced four tall hybrid red plants with the identical phenotype. However, each one has a combination of genetic information different from that of the original parent plants. The unique combination of genetic information is known as a genotype. The F2 generation, bred from two tall red hybrid flowers, produced four different phenotypes: tall with red flowers, tall with white flowers, dwarf with red flowers, and dwarf with white flowers. Both Figure 1.2 and Figure 1.3 demonstrate that recessive traits that disappear in the F1 generation may reappear in future generations in definite, predictable percentages.

The law of dominance, the third tenet of inheritance identified by Mendel, asserts that heredity factors (genes) act together as pairs. When a cross occurs between organisms that are pure for contrasting traits, only one trait, the

FIGURE 1.3

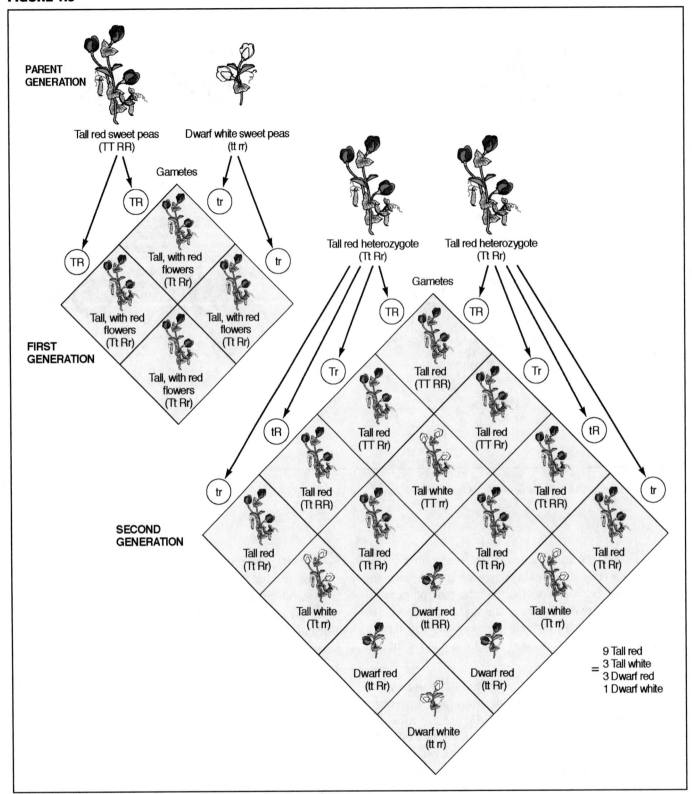

Mendel's law of independent assortment. *Hans & Cassidy, Cengage Gale.*

dominant one, appears in the hybrid offspring. In Figure 1.2 all the F1 generation offspring are red—an identical phenotype to the parent plant—though they also carry the recessive white gene.

Mendel's contributions to the understanding of heredity were not acknowledged during his lifetime. When he applied the knowledge gained from his pea plant studies to hawkweed plants and honeybees and failed to replicate the results, Mendel became dispirited. He set aside his botany research and returned to monastic life until his death. It was not until the early 20th century, nearly 40 years after Mendel published his findings, that the scientific community resurrected his work and affirmed the importance of his ideas.

GENETICS AT THE DAWN OF THE 20TH CENTURY

During the years following Mendel's work, understanding of cell division and fertilization increased, as did insight into the component parts of cells known as subcellular structures. For example, in 1869 the Swiss physiologist Johann Friedrich Miescher (1844–1895) looked at pus he had scraped from the dressings of soldiers wounded in the Crimean War (1853–1856). In the white blood cells from the pus, and later in salmon sperm, he identified a substance he called nuclein. In 1874 Miescher separated nuclein into a protein and an acid, and it was renamed nucleic acid. He proposed that it was the "chemical agent of fertilization."

In 1900 three scientists—Karl Erich Correns (1864–1933), Hugo Marie de Vries (1848–1935), and Eric von Tschermak-Seysenegg (1871–1962)—independently rediscovered and verified Mendel's principles of heredity, and Mendel's contributions to modern genetics were finally acknowledged. In 1902 Sir Archibald E. Garrod (1857–1936), a British physician and chemist, applied Mendel's principles and identified the first human disease attributable to genetic causes, which he called "inborn errors of metabolism." The disease was alkaptonuria, a condition in which an abnormal buildup of an acid (homogentisic acid or alkapton) accumulates.

Seven years later Garrod published the textbook *Inborn Errors of Metabolism* (1909), which described various disorders that he believed were caused by these inborn metabolic errors. These included albinism (a pigment disorder in which affected individuals have abnormally pale skin, hair, and eyes) and porphyria (a group of disorders resulting from abnormalities in the production of heme, a vitally important substance that carries oxygen in the blood, bone, liver, and other tissues). Garrod's was the first effort to distinguish diseases caused by bacteria from those attributable to genetically programmed enzyme deficiencies that interfered with normal metabolism.

In 1905 the British embryologist William Bateson (1861–1926) coined the term *genetics*, along with other descriptive terms used in modern genetics, including *allele* (a particular form of a gene), *zygote* (a fertilized egg), *homozygote* (an organism with genetic information that contains two identical forms of a gene), and *heterozygote* (an organism with two different forms of a particular gene). Arguably, his most important contributions to the progress of genetics were his translations of Mendel's work from German to English and his vigorous endorsement and promotion of Mendel's principles.

In 1908 the British mathematician Godfrey Harold Hardy (1877–1947) and the German geneticist Wilhelm Weinberg (1862–1937) independently developed a mathematical formula that describes the actions of genes in populations. Their assumptions that algebraic formulas could be used to analyze the occurrence of, and reasons for, genetic variation became known as the Hardy-Weinberg equilibrium. It advanced the application of Mendel's laws of heredity from individuals to populations, and by applying Mendelian genetics to Darwin's theory of evolution, it improved geneticists' understanding of the origin of mutations and how natural selection gives rise to hereditary adaptations. The Hardy-Weinberg equilibrium enables geneticists to determine whether evolution is occurring in populations.

Chromosome Theory of Inheritance

Bateson is often cited for having said, "Treasure your exceptions." I believe Sturtevant's admonition would be, "Analyze your exceptions."

—Edward B. Lewis, "Remembering Sturtevant," *Genetics*, 1995

The American geneticist Walter Stanborough Sutton (1877–1916) conducted studies using grasshoppers (*Brachystola magna*) he collected at his family's farm in Kansas. Sutton was strongly influenced by reading William Bateson's work and sought to clarify the role of the chromosomes in sexual reproduction. The results of his research, which were published in 1902, demonstrated that chromosomes exist in pairs that are structurally similar and proved that sperm and egg cells each have one pair of chromosomes. Sutton's work advanced genetics by identifying the relationship between Mendel's laws of heredity and the role of the chromosome in meiosis.

Along with Bateson, the American geneticist Thomas Hunt Morgan (1866–1945) is often referred to as the father of classical genetics. In 1907 Morgan performed laboratory research using the fruit fly *Drosophila melanogaster*. He chose to study fruit flies because they bred quickly, had distinctive characteristics, and had just four chromosomes. The aim of his research was to replicate the genetic variation de Vries had reported from his experiments with plants and animals.

Working in a laboratory they called the "Fly Room," Morgan and his students Calvin Blackman Bridges (1889–1938), Hermann Joseph Muller (1890–1967), and Alfred H. Sturtevant (1891–1970) conducted research that unequivocally confirmed the findings and conclusions of Mendel, Bateson, and Sutton. Breeding both red- and white-eyed fruit flies, they demonstrated that all the offspring were red eyed, indicating that the red-eye gene was dominant and the white-eye gene was recessive. The offspring carried the white-eye gene but it did not appear in the F1 generation. When, however, the F1 offspring were crossbred, the ratio of red-eyed to white-eyed flies was 3:1 in the F2 generation. (A similar pattern is shown for red and white flowers in Figure 1.2.)

The investigators also observed that all the white-eyed flies were male, prompting them to investigate sex chromosomes and hypothesize about sex-linked inheritance. The synthesis of their research with earlier work produced the chromosomal theory of inheritance, the premise that genes are the fundamental units of heredity and are found in the chromosomes. It also confirmed that specific genes are found on specific chromosomes, that traits found on the same chromosome are not always inherited together, and that genes are actual physical objects. In 1915 the four researchers published *The Mechanism of Mendelian Heredity*, which detailed the results of their research, conclusions, and directions for future research.

In *The Theory of the Gene* (1926), Morgan asserted that the ability to quantify or number genes enables researchers to predict accurately the distribution of specific traits and characteristics. He contended that the mathematical principles governing genetics qualify it as science: "That the characters of the individual are referable to paired elements (genes) in the germinal matter that are held together in a definite number of linkage groups. ... The members of each pair of genes separate when germ-cells mature. ... Each germ-cell comes to contain only one set. ... These principles ... enable us to handle problems of genetics in a strictly numerical basis, and allow us to predict ... what will occur. ... In these respects the theory [of the gene] fulfills the requirements of a scientific theory in the fullest sense."

In 1933 Morgan was awarded the Nobel Prize in Physiology or Medicine for his groundbreaking contributions to the understanding of inheritance. Muller also became a distinguished geneticist, and after pursuing research on flies to determine if he could induce genetic changes using radiation, he turned his attention to studies of twins to gain a better understanding of human genetics.

Bridges eventually discovered the first chromosomal deficiency as well as chromosomal duplication in fruit flies. He served in various academic capacities at Columbia University, the Carnegie Institution, and the California Institute of Technology and was a member of the National Academy of Sciences and a fellow of the American Association for the Advancement of Science.

Sturtevant was awarded the National Medal of Science in 1968. His most notable contribution to genetics was the detailed outline and instruction he provided about gene mapping (the process of determining the linear sequence of genes in genetic material). In 1913 he began construction of a chromosome map of the fruit fly that was completed in 1951. Because of his work in gene mapping, he is often referred to as the father of the Human Genome Project, the comprehensive map of humanity's 20,000 to 25,000 genes. His book *A History of Genetics* (1965) recounts the ideas, events, scientists, and philosophies that shaped the development of genetics.

CLASSICAL GENETICS

Another American geneticist awarded a Nobel Prize was Barbara McClintock (1902–1992), who described key methods of exchange of genetic information. Performing chromosomal studies of maize in the botany department at Cornell University, she observed colored kernels on an ear of corn that should have been clear. McClintock hypothesized that the genetic information that normally would have been conveyed to repress color had somehow been lost. She explained this loss by seeking and ultimately producing cytological proof of jumping genes, which could be released from their original position and inserted, or transposed, into a new position. This genetic phenomenon of chromosomes exchanging pieces became known as crossing over, or recombination.

With another pioneering female researcher, Harriet Baldwin Creighton (1909–2004), McClintock published a series of research studies, including a 1931 paper that offered tangible evidence that genetic information crosses over during the early stages of meiosis. Along with the 1983 Nobel Prize in Physiology or Medicine, McClintock received the prestigious Albert Lasker Basic Medical Research Award in 1981, making her the most celebrated female geneticist in history.

During the same period the British microbiologist Frederick Griffith (1879–1941) performed experiments with *Streptococcus pneumoniae*, demonstrating that the ability to cause deadly pneumonia in mice could be transferred from one strain of bacteria to another. Griffith observed that the hereditary ability of bacteria to cause pneumonia could be altered by a transforming principle. Even though Griffith mistakenly believed the transforming factor was a protein, his observation offered the first tangible evidence linking deoxyribonucleic acid (DNA, the molecule that carries the genetic code) to heredity in cells. His experiment provided a framework for researching the biochemical basis of heredity in bacteria. In 1944 the American biologist Oswald Theodore Avery (c. 1877–c. 1955), along with the American bacteriologists Colin Munro Macleod (1909–1972) and Maclyn McCarty (1911–2005), performed studies demonstrating that Griffith's transforming factor was DNA rather than simply a protein. Among the experiments Avery, Macleod, and McCarty performed was one similar to Griffith's, which confirmed that DNA from one strain of bacteria could transform a harmless strain of bacteria into a deadly strain. (See Figure 1.4.) Their findings gave credence to the

FIGURE 1.4

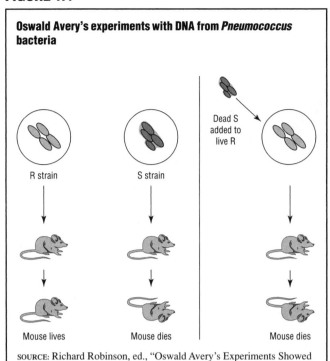

Oswald Avery's experiments with DNA from *Pneumococcus* bacteria

R strain

S strain

Dead S added to live R

Mouse lives

Mouse dies

Mouse dies

SOURCE: Richard Robinson, ed., "Oswald Avery's Experiments Showed That DNA from Dead S Strain of *Pneumococcus* Bacteria Could Transform a Harmless Strain into a Deadly Strain," in *Genetics*, vol. 1, A–D, Macmillan Reference USA, 2002

premise that DNA was the molecular basis for genetic information.

The first half of the 20th century was devoted to classical genetics research and the development of increasingly detailed and accurate descriptions of genes and their transmission. In 1929 the American organic chemist Phoebus A. Levene (1869–1940) isolated and discovered the structure of the individual units of DNA. Called nucleotides, the molecular building blocks of DNA are composed of deoxyribose (a sugar molecule), a phosphate molecule, and four types of nucleic acid bases. (Figure 1.5 uses letters to represent the four bases—adenine [A], cytosine [C], guanine [G], and thymine [T].) Also in 1929 Theophilus Shickel Painter (1889–1969), an American cytologist, made the first estimate of the number of human chromosomes. His count of 48 was off by only 2—25 years later researchers were able to stain and view human chromosomes microscopically to determine that they number 46. Analysis of chromosome number and structure would become pivotal to medical diagnosis of diseases and disorders associated with altered chromosomal numbers or structure.

Another milestone during the first half of the 20th century was the determination by the American chemist Linus Pauling (1901–1994) that sickle-cell anemia (the presence of oxygen-deficient, abnormal red blood cells that cause affected individuals to suffer from obstruction of capillaries, resulting in pain and potential organ damage) was caused by the change in a single amino acid (a building block of protein) of hemoglobin (the oxygen-bearing, iron-containing protein in red blood cells). Pauling's work paved the way for research showing that genetic information is used by cells to direct the synthesis of protein and that mutation (a change in genetic information) can directly cause a change in a protein. This explains hereditary genetic disorders such as sickle-cell anemia.

From 1950 to 1952 the American geneticists Martha Cowles Chase (1927–2003) and Alfred Day Hershey (1908–1997) conducted experiments that provided definitive proof that DNA was genetic material. In research that would be widely recounted as the "Waring blender experiment," the investigators dislodged virus particles that infect bacteria by spinning them in a blender and found that the viral DNA, and not the viral protein, that remains inside the bacteria directed the growth and multiplication of new viruses.

MODERN GENETICS EMERGES

The period of classical genetics focused on refining and improving the structural understanding of DNA. In contrast, modern genetics seeks to understand the processes of heredity and how genes work.

Many historians consider 1953—the year that the American geneticist James D. Watson (1928–) and the British biophysicist Francis Crick (1916–2004) famously described the structure of DNA—as the birth of modern genetics. It is important, however, to remember that Watson and Crick's historic accomplishment was not the discovery of DNA—Miescher had identified nucleic acid in cells nearly a century earlier. Similarly, even though Watson and Crick earned recognition and public acclaim for their landmark research, it would not have been possible without the efforts of their predecessors and colleagues such as the British biophysicist Maurice Wilkins (1916–2004) and the British molecular biologist Rosalind Franklin (1920–1958). Wilkins and Franklin were the molecular biologists who in 1951 obtained sharp X-ray diffraction photographs of DNA crystals, revealing a regular, repeating pattern of molecular building blocks that correspond to the components of DNA. (Wilkins shared the Nobel Prize with Watson and Crick, but Franklin was ineligible to share the prize because she died in 1958, four years before it was awarded.)

Another pioneer in biochemistry, the Austrian Erwin Chargaff (1905–2002), also provided information about DNA that paved the way for Watson and Crick. Chargaff suggested that DNA contained equal amounts of the four nucleotides: the nitrogenous (containing nitrogen, a non-metallic element that constitutes almost four-fifths of the air by volume) bases adenine (A), thymine (T), guanine (G), and cytosine (C). In DNA there is always one A for

FIGURE 1.5

Building blocks of deoxyribonucleic acid (DNA)

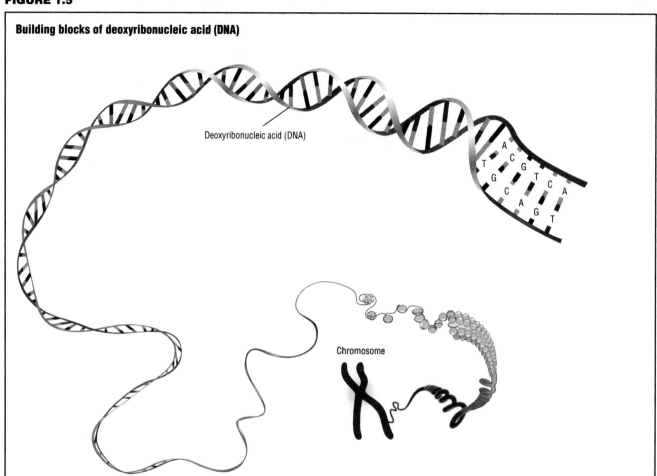

SOURCE: "DNA (Deoxyribonucleic Acid)," in *Talking Glossary of Genetic Terms*, U.S. Department of Health and Human Services, National Institutes of Health, National Human Genome Research Institute, Division of Intramural Research, undated, http://www.genome.gov/Glossary/index.cfm?id=48 (accessed September 5, 2010)

each T, and one G for each C. This relationship became known as base pairing or Chargaff's rules, which also includes the observation that the ratio of AT to GC varies from species to species but remains consistent across different cell types within each species. (See Figure 1.6.)

Watson and Crick

Watson is an American geneticist known for his willingness to grapple with big scientific challenges and his expansive view of science. In *The Double Helix: A Personal Account of the Discovery of the Structure of DNA* (1968), he chronicles his collaboration with Crick to create an accurate model of DNA. He credits his inclination to take intellectual risks and venture into uncharted territory as his motivation for this ambitious undertaking.

Watson was just 25 years old when he announced the triumph that was hailed as one of the greatest scientific achievements of the 20th century. Following this remarkable accomplishment, Watson served on the faculty of Harvard University for two decades and assumed the directorship and

FIGURE 1.6

$$A + G = C + T$$

$$\text{and}$$

$$A = T \qquad G = C$$

Chargaff's rules. *Argosy Publishing, Cengage Gale.*

then the presidency of the Cold Spring Harbor Laboratory in Long Island, New York. From 1989 to 1992 he headed the National Institutes of Health's (NIH) Human Genome Project, the effort to sequence (or discover the order of) the entire human genome.

Crick was a British scientist who had studied physics before turning his attention to biochemistry and biophysics. He became interested in discovering the structure of DNA, and when in 1951 Watson came to work at the Cavendish Laboratory in Cambridge, England, the two scientists decided to work together to unravel the structure of DNA.

After his landmark accomplishment with Watson, Crick continued to study the relationship between DNA and genetic coding. He is credited with predicting the ways in which proteins are created and formed, a process known as protein synthesis. During the mid-1970s Crick turned his attention to the study of brain functions, including vision and consciousness, and assumed a professorship at the Salk Institute for Biological Studies in San Diego, California. Like Watson, he received many professional awards and accolades for his work, and besides the scientific papers he and Watson coauthored, he wrote four books. Published a decade before his death in 2004, Crick's last book, *The Astonishing Hypothesis: The Scientific Search for the Soul* (1994), detailed his ideas and insights about human consciousness.

WATSON AND CRICK MODEL OF DNA. Using the X-ray images of DNA created by Franklin and Wilkins, who also worked in the Cavendish Laboratory, Watson and Crick worked out and then began to build models of DNA. Crick contributed his understanding of X-ray diffraction techniques and imaging and relied on Watson's expertise in genetics. In 1953 Watson and Crick published the paper "Molecular Structure of Nucleic Acids: A Structure for Deoxyribose Nucleic Acid" (*Nature*, vol. 171, no. 4356, April 25, 1953), which contained the famously understated first lines, "We wish to suggest a structure for the salt of deoxyribose nucleic acid (D.N.A.). This structure has novel features which are of considerable biological interest." Watson and Crick then described the shape of a double helix, an elegant structure that resembles a latticework spiral staircase. (See Figure 1.7.)

Their model enabled scientists to better understand functions such as carrying hereditary information to direct protein synthesis, replication, and mutation at the molecular level. The three-dimensional Watson and Crick model consists of two strings of nucleotides connected across like a ladder. Each rung of the ladder contains an A-T pair or a G-C pair, which is consistent with Chargaff's rule that there is an A for every T and a G for every C in DNA. (See Figure 1.7.) Watson and Crick posited that changes in the sequence of nucleotide pairs in the double helix would produce mutations.

FIGURE 1.7

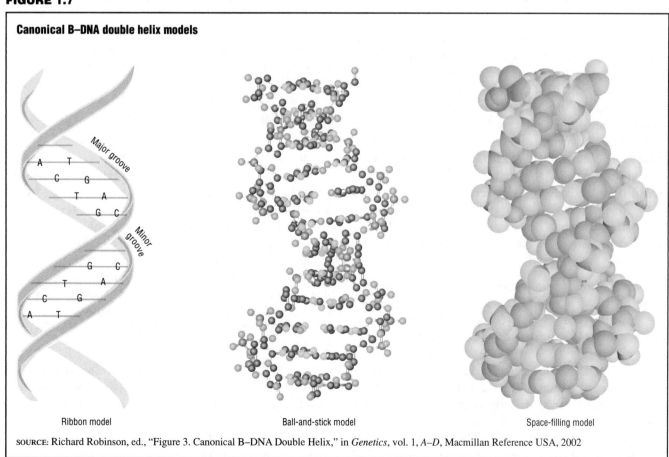

Canonical B–DNA double helix models

Ribbon model   Ball-and-stick model   Space-filling model

SOURCE: Richard Robinson, ed., "Figure 3. Canonical B–DNA Double Helix," in *Genetics*, vol. 1, *A–D*, Macmillan Reference USA, 2002

MILESTONES IN MODERN GENETICS

Geneticists and other researchers made remarkable strides during the second half of the 20th century. In 1956 the American biochemist Vernon M. Ingram (1924–2006), who would soon be recognized as the father of molecular medicine, identified the single base difference between normal and sickle-cell hemoglobin. The implications of his finding that the mutation of a single letter in the DNA genetic code was sufficient to cause a hereditary medical disorder were far reaching. This greater insight into the mechanisms of sickle-cell disease suggested directions for research into prevention and treatment. It also prompted research that uncovered other diseases with similar causes, such as hemophilia (an inherited blood disease that is associated with insufficient clotting factors and excessive bleeding) and cystic fibrosis (an inherited disease of the mucous glands that produces problems associated with the lungs and pancreas). Just three years later the first human chromosome abnormality was identified: people with Down syndrome were found to have an extra chromosome, demonstrating that it is a genetic disorder that may be diagnosed by direct examination of the chromosomes.

Ingram's work became the foundation for current research to map genetic variations that affect human health. For example, in 1989, more than 30 years after Ingram's initial work, the gene for cystic fibrosis was identified and a genetic test for the gene mutation was developed.

Using radioactive labeling to track each strand of the DNA in bacteria, the American molecular biologist Matthew Meselson (1930–) and the American geneticist Franklin W. Stahl (1929–) demonstrated with an experiment in 1958 that the replication of DNA in bacteria is semiconservative. Semiconservative replication occurs as the double helix unwinds at several points and knits a new strand along each of the old strands. Meselson and Stahl's experiment revealed that one strand remained intact and combined with a newly synthesized strand when DNA replicated, precisely as Watson and Crick's model predicted. In other words, each of the two new molecules created contains one of the two parent strands and one new strand.

During the early 1960s Crick, the American biochemist Marshall Nirenberg (1927–2010), the Russian-born American physicist George Gamow (1904–1968), and other researchers performed experiments that detected a direct relationship between DNA nucleotide sequences and the sequence of the amino acid building blocks of proteins. They determined that the four nucleotide letters (A, T, C, and G) may be combined into 64 different triplets. The triplets are code for instructions that determine the amino acid structure of proteins. Ribosomes are cellular organelles (membrane-bound cell compartments) that interpret a sequence of genetic code three letters at a time and link together amino acid building blocks of proteins specified by the triplets to construct a specific protein. The 64 triplets of nucleotides that can be coded in the DNA—which are copied during cell division, infrequently mutate, and are read by the cell to direct protein synthesis—make up the universal genetic code for all cells and viruses.

Origins of Genetic Engineering

The late 1960s and early 1970s were marked by research that would lay the groundwork for modern genetic engineering technology. In 1966 DNA was found to be present not only in chromosomes but also in the mitochondria. The first single gene was isolated in 1969, and the following year the first artificial gene was created. In 1972 the American biochemist Paul Berg (1926–) developed a technique to splice DNA fragments from different organisms and created the first recombinant DNA, or DNA molecules formed by combining segments of DNA, usually from different types of organisms. In 1980 Berg was awarded the Nobel Prize in Chemistry for this achievement, which is now referred to as recombinant DNA technology.

In 1976 an artificial gene was inserted into a bacterium and functioned normally. The following year DNA from a virus was fully decoded, and three researchers, working independently, developed methods to sequence DNA—in other words, to determine how the building blocks of DNA (the nucleotides A, C, G, and T) are ordered along the DNA strand. In 1978 bacteria were engineered to produce insulin, a pancreatic hormone that regulates carbohydrate metabolism by controlling blood glucose levels. Just four years later the Eli Lilly pharmaceutical company marketed the first genetically engineered drug: a type of human insulin grown in genetically modified bacteria.

In 1980 the U.S. Supreme Court decision in *Diamond v. Chakrabarty* (447 U.S. 303) permitted patents for genetically modified organisms; the first one was awarded to the General Electric Company for bacteria to assist in clearing oil spills. The following year a gene was transferred from one animal species to another. In 1983 the first artificial chromosome was created. That same year the marker (the usually dominant gene or trait that serves to identify genes or traits linked with it) for Huntington's disease (an inherited neuropsychiatric disease that affects the body and mind) was identified; in 1993 the disease gene was identified.

In 1984 the observation that some nonfunctioning DNA is different in each individual launched research to refine tools and techniques developed by the British geneticist Alec Jeffreys (1950–) that perform genetic fingerprinting. Initially, the technique was used to determine the paternity of children, but it rapidly gained acceptance among forensic medicine specialists, who are often called on to assist in the investigation of crimes and interpret medicolegal issues.

The 1985 invention of the polymerase chain reaction (PCR), which amplifies (or produces many copies of) DNA, enabled geneticists, medical researchers, and forensic specialists to analyze and manipulate DNA from the smallest samples. PCR allowed biochemical analysis of even trace amounts of DNA. In *A Short History of Genetics and Genomics* (2003, http://www.dna50.com/dna50.swf), Ricki Lewis and Bernard Possidente describe the American biochemist Kary B. Mullis's (1944–) development of PCR as the "genetic equivalent of a printing press," with the potential to revolutionize genetics in the same way that the printing press had revolutionized mass communications.

In 1990 the first gene therapy (an approach that aims to prevent, treat, or cure disease by changing the pattern of gene expression) was administered. The patient was a four-year-old girl with the inherited immunodeficiency disorder adenosine deaminase deficiency. If left untreated, the deficiency is fatal. When given along with conventional medical therapy, the gene therapy treatment was considered effective. The 1999 death of another gene therapy patient, as a result of an immune reaction to the treatment, tempered enthusiasm for gene therapy and prompted medical researchers to reconsider its safety and effectiveness. In January 2003 the U.S. Food and Drug Administration temporarily halted gene therapy trials; however, the following April it allowed selected trials to resume.

In "Effect of Gene Therapy on Visual Function in Leber's Congenital Amaurosis" (*New England Journal of Medicine*, vol. 358, no. 21, May 22, 2008), James W. Bainbridge et al. report successful gene therapy for Leber's congenital amaurosis, a type of inherited blindness. Effective gene therapy has also been developed for some inherited immunodeficiency diseases. For example, Barbara Cappelli and Alessandro Aiuti indicate in "Gene Therapy for Adenosine Deaminase Deficiency" (*Immunology and Allergy Clinics of North America*, vol. 30, no. 2, May 2010) that adenosine deaminase deficiency has historically been treated using bone marrow transplant (a procedure that is often complicated by the challenge of identifying a suitable donor and carries a high risk of rejection and infection) and replacing the deficient enzyme, which is a lifesaving but not curative treatment. Gene therapy has proven to be an effective alternative to other therapies, sparing patients the risks that are associated with transplant and the need for lifetime enzyme replacement.

Cloning (the production of genetically identical organisms) was performed first with carrots. A cell from the root of a carrot plant was used to generate a new plant. By the early 1950s scientists had cloned tadpoles, and during the 1970s attempts were under way to clone mice, cows, and sheep. These clones were created using embryos, and many did not produce healthy offspring, offspring with normal life spans, or offspring with the ability to reproduce. In 1993 researchers at George Washington University cloned nearly 50 human embryos, but their experiment was terminated after just six days.

In 1996 the British embryologist Ian Wilmut (1944–) and his colleagues at the Roslin Institute in Edinburgh, Scotland, successfully cloned the first adult mammal that was able to reproduce. Dolly the cloned sheep, named for the country singer Dolly Parton (1946–), focused public attention on the practical and ethical considerations of cloning.

Human Genome Project and More

The term *genetics* refers to the study of a single gene at a time, and *genomics* is the study of all genetic information contained in a cell. The Human Genome Project (HGP) set as one of its goals to map the entire nucleotide sequence of the more than 3 billion bases of DNA contained in the nucleus of a human cell. Initial discussions about the feasibility and value of conducting the HGP began in 1986. The following year the first automated DNA sequencer was produced commercially. Automated sequencing, which enabled researchers to decode millions, as opposed to thousands, of letters of genetic code per day, was a pivotal technological advance for the HGP, which began in 1987 under the auspices of the U.S. Department of Energy (DOE).

In 1988 the HGP was relocated to the NIH, and Watson was recruited to direct the project. The following year the NIH opened the National Center for Human Genome Research, and a committee composed of professionals from the NIH and the DOE was named to consider ethical, legal, and social issues that might arise from the project. In 1990 the project began in earnest, with work on preliminary genetic maps of the human genome and four other organisms believed to share many genes with humans.

During the early 1990s several new technologies were developed that further accelerated progress in analyzing, sequencing, and mapping sections of the genome. The advisability of granting private biotechnology firms the right to patent specific genes and DNA sequences was hotly debated. In April 1992 Watson resigned as director of the project to express his vehement disapproval of the NIH decision to patent human gene sequences. Later that year preliminary physical and genetic maps of the human genome were published.

The American geneticist Francis S. Collins (1950–) was named as the director of the HGP in April 1993, and international efforts to assist were under way in England,

France, Germany, Japan, and other countries. The 1995 release of DNA chip technology that simultaneously analyzes genetic information representing thousands of genes also helped speed the project to completion.

In 1995 investigators at the Institute for Genomic Research published the first complete genome sequence for any organism: the bacterium *Haemophilus influenzae*, with nearly 2 million genetic letters and 1,000 recognizable genes. In 1997 a yeast genome, *Saccharomyces cerevisiae*, composed of about 6,000 genes, was sequenced, and later that year the genome of the bacterium *Escherichia coli*, which contains approximately 4,600 genes, was sequenced.

In 1998 the genome of the first multicelled animal, the nematode worm *Caenorhabditis elegans*, was sequenced, containing approximately 18,000 genes. The following year the first complete sequence of a human chromosome (number 22) was published by the HGP. In 2000 the genome of the fruit fly *Drosophila melanogaster*, which Morgan and his colleagues had used to study genetics nearly a century earlier, was sequenced by the private firm Celera Genomics. The fruit fly sequence contains about 13,000 genes, with many sequences matching already identified human genes that are associated with inherited disorders.

In 2000 the first draft of the human genome was announced, and it was published in 2001. Also in 2000 the first plant genome, *Arabidopsis thaliana*, was sequenced. This feat spurred research in plant biology and agriculture. Even though tomatoes that had been genetically engineered for longer shelf life had been marketed during the mid-1990s, agricultural researchers began to see new possibilities to enhance crops and food products. For example, in 2000 plant geneticists developed genetically engineered rice that manufactured its own vitamin A. Many researchers believe the genetically enhanced strain of rice holds great promise in terms of preventing vitamin A deficiency in developing countries.

The 2001 publication of the human genome estimated that humans have between 30,000 and 35,000 genes. The HGP was completed in 2003, the same year that Cold Spring Harbor Laboratory held educational events to commemorate and celebrate the 50th anniversary of the discovery of the double helical structure of DNA. In October 2004 human gene count estimates were revised downward to between 20,000 and 25,000. During 2005 and 2006 sequencing of more than 10 human chromosomes was completed, including the human X chromosome, which is one of the two sex chromosomes; the other is the Y chromosome.

In 2006 the American structural biologist Roger D. Kornberg (1947–) was awarded the Nobel Prize in Chemistry for determining the intricate way in which information in the DNA of a gene is copied to provide the instructions for building and running a living cell. In 2007 the Nobel Prize in Physiology or Medicine was awarded jointly to Mario R. Capecchi (1937–), Martin J. Evans (1941–), and Oliver Smithies (1925–), acknowledging their landmark discoveries about the use of embryonic stem cells (undifferentiated cells from the embryo that have the potential to become a wide variety of specialized cell types) to introduce gene modifications in mice, and their development of a technique known as gene targeting that is used to inactivate single genes.

In 2008 Jeffrey M. Kidd et al. published in "Mapping and Sequencing of Structural Variation from Eight Human Genomes" (*Nature*, vol. 453, no. 7191, May 1, 2008) the first high-resolution genomic maps that detailed genetic structural variations among eight individuals of varied ancestry. These variations are important in terms of health and understanding the risk of developing genetic diseases. Furthermore, these eight genomes may be used as benchmarks for comparison with other human genomes that are sequenced.

The HGP is one of the principal achievements in recent biomedical history, but because it involved the time, energy, and efforts of thousands of researchers around the world, some observers question whether any individual researchers should be awarded the Nobel Prize for the completion of the human genome sequencing effort. In "The Nobel Prize as a Reward Mechanism in the Genomics Era: Anonymous Researchers, Visible Managers, and the Ethics of Excellence" (*Journal of Bioethical Inquiry*, vol. 7, no. 3, September 2010), Hub Zwart of Radboud University Nijmegen observes that it is difficult to ignore the magnitude of the accomplishment of the HGP but opines, "If a Nobel Prize is to be awarded for deciphering the human genome, therefore, it is difficult to see how this can be done in a manner that is both meaningful and fair." Zwart suggests that the Nobel Prize should be awarded to the researchers Francis S. Collins; J.Craig Venter (1946–), a biologist and entrepreneur involved in the HGP as well as the lead researcher responsible for the 2010 creation of a cell with a synthetic genome; and Eric Lander (1957–), a professor of biology at the Massachusetts Institute of Technology, a cochair of the President's Council of Advisers on Science and Technology, and the lead author of the first article presenting the genome sequence.

The Postgenomic Era and Epigenetics

Epigenetic changes are heritable changes in gene function that occur without a change in the sequence of DNA. Many of these changes occur in response to environmental influences such as diet, exposure to chemicals and toxins, or psychological stress. One example is DNA methylation (the process in which methyl groups are added to nucleotides

FIGURE 1.8

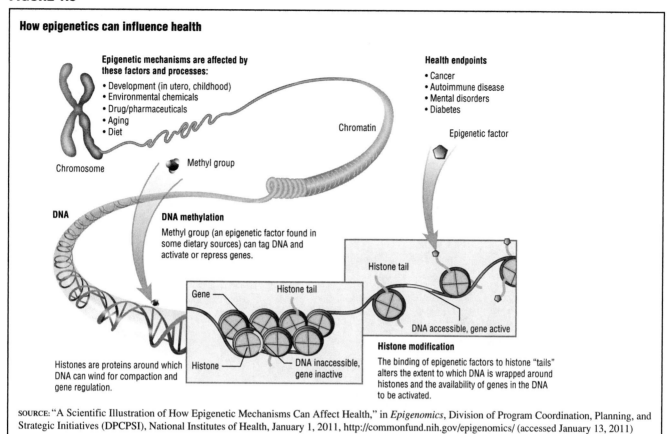

How epigenetics can influence health

Epigenetic mechanisms are affected by these factors and processes:
- Development (in utero, childhood)
- Environmental chemicals
- Drug/pharmaceuticals
- Aging
- Diet

Health endpoints
- Cancer
- Autoimmune disease
- Mental disorders
- Diabetes

Chromosome

Chromatin

Methyl group

Epigenetic factor

DNA

DNA methylation
Methyl group (an epigenetic factor found in some dietary sources) can tag DNA and activate or repress genes.

Histone tail

Gene

Histone tail

DNA accessible, gene active

Histones are proteins around which DNA can wind for compaction and gene regulation.

Histone

DNA inaccessible, gene inactive

Histone modification
The binding of epigenetic factors to histone "tails" alters the extent to which DNA is wrapped around histones and the availability of genes in the DNA to be activated.

SOURCE: "A Scientific Illustration of How Epigenetic Mechanisms Can Affect Health," in *Epigenomics*, Division of Program Coordination, Planning, and Strategic Initiatives (DPCPSI), National Institutes of Health, January 1, 2011, http://commonfund.nih.gov/epigenomics/ (accessed January 13, 2011)

in genomic DNA). (See Figure 1.8.) DNA methylation governs many important functions including decreasing gene expression, especially viral genes and other microbial threats that may have entered the host's genome. DNA methylation is vital for normal growth and development. Alteration of DNA methylation also plays a key role in the development of diseases, including cancer.

In late 2008 the NIH's Roadmap Epigenomics Program (http://nihroadmap.nih.gov/epigenomics/) began mapping the epigenetic markers that constitute the epigenome, which is a code of chemicals that modify DNA by tightly wrapping certain genes to inactivate them and relaxing others to enable their expression. Epigenetics is the study of these markers, which switch parts of the genome on and off at specific times and locations, and the factors that influence them, which in some cases may be passed down to successive generations. Epigenetic modifications to DNA exert powerful influences on gene activity; they do not alter genes but instead affect how they are expressed. Epigenetics is thought to explain

much of the diversity seen in nature and promises to deepen the understanding of health and disease. Epigenetics considers single genes or sets of genes, whereas epigenomics is the global analysis of epigenetic changes across the entire genome. Charlie McDermott reports in "Human Epigenome Project Launched" (*BioNews*, February 15, 2010) that the International Human Epigenome Consortium (IHEC) was launched in February 2010 in Paris, France. The IHEC aims to map 1,000 epigenomes of different tissues, which "will allow researchers to compare their findings with a robust and trustworthy template" that can be used as a reference point when examining diseased tissue.

The postgenomic era began with a firestorm of controversies about the direction of genetic research, human cloning, stem cell research, and genetically modified food and crops. It also ushered in the epigenomic era, which promises new ways to prevent, diagnose, and treat many of the chronic diseases, such as heart disease and cancer, that threaten human health.

CHAPTER 2

UNDERSTANDING GENETICS

The study of genetics (the inheritance of traits in living organisms) is a basic concept in biology. The same processes that provide the mechanism for organisms to pass genetic information to their offspring lead to the gradual change of species over time, which in turn produces biodiversity (the variety of life and the genetic differences among living organisms) and the evolution of new species. An understanding of genetics is becoming increasingly important as genetics research and technology—and the controversies surrounding them—gain greater influence on social trends and individual lives.

Rapid and revolutionary advances in genetics, medicine, and biotechnology have created new opportunities as well as a complex and far-reaching range of ethical, legal, and social issues. Genetic testing to screen for disease susceptibility, moral questions about the cloning of human organs, and the debate over genetically modified foods affect many lives. To fully appreciate the benefits, risks, and ramifications of genetics research and to critically evaluate the related issues raised by ethicists, scientists, and other stakeholders, it is vital to understand the basics of genetics, a discipline that integrates biology, mathematics, sociology, medicine, and public health.

To understand genetics, the development of organisms, and the diversity of species, it is important to learn how deoxyribonucleic acid (DNA) functions as the information-bearing molecule of living organisms. This chapter contains definitions of basic genetics terms such as *DNA*, *chromosomes*, and *proteins*, as well as an introduction to genetic concepts and processes, including reproduction and inheritance.

DNA CODES FOR PROTEINS

From the perspective of genetics, the DNA molecule has two major attributes. The first is that it is able to replicate—that is, to make an exact copy of itself that can be passed to another cell, thereby conveying its precise genetic characteristics. Figure 2.1 is a diagram that shows how DNA replication produces two completely new and identical daughter strands of DNA. The second critical attribute is that it stores detailed instructions to manufacture specific proteins—molecules that are essential to every aspect of life. DNA is a blueprint or template for making proteins, and much of the behavior and physiology (life processes and functions) of a living organism depends on the repertoire of proteins its DNA molecules know how to manufacture.

The function of DNA depends on its structure. The double strand of DNA is composed of individual building blocks called nucleotides that are paired and connected by chemical bonds. A nucleotide contains one of four nitrogenous (containing nitrogen, a nonmetallic element that constitutes almost four-fifths of the air by volume) bases: the purines (nitrogenous bases with two rings) adenine (A) and guanine (G), or the pyrimidines (nitrogenous bases with one ring; pyrimidines are smaller than purines) cytosine (C) and thymine (T). The two strands of DNA lie side by side to create a predictable sequence of nitrogenous base pairs. (See Figure 2.2.) A stable DNA structure is formed when the two strands are a constant distance apart, which can occur only when a purine (A or G) on one strand is paired with a pyrimidine (T or C) on the other strand. A generally pairs with T, and G generally pairs with C.

Proteins are molecules that perform all the chemical reactions necessary for life and provide structure and shape to cells. The properties of each protein depend primarily on its shape, which is determined by the sequence of its building blocks, known as amino acids. Proteins may be tough like collagen, the most abundant protein in the human body, or they may be stretchy like elastin, a protein that mixes with collagen to make softer, more flexible tissues such as skin. Figure 2.3 shows the shapes of four types of protein structures.

FIGURE 2.1

DNA Replication

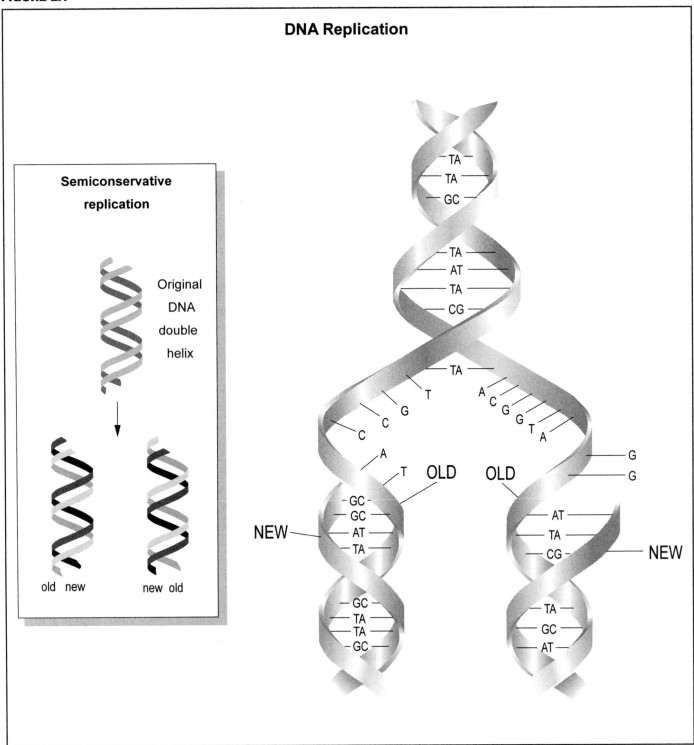

DNA replication *Argosy Publishing, Cengage Gale.*

Proteins can act as structural components by building the tissues of the body. For example, some of the proteins in an egg include a bond that acts like an axle, allowing other parts of the molecule to spin around like wheels. However, when the egg is heated, these bonds break or denature, locking the "wheels" of the molecule in place. This is why an egg gets hard when it is cooked.

Enzymes such as lactase, which helps in the digestion of lactose (milk sugar), and hormones such as insulin (a pancreatic hormone that regulates carbohydrate metabolism by controlling blood glucose levels) are proteins that act to facilitate and direct chemical reactions. Defense proteins, which are able to combat invasion by bacteria or viruses, are embedded in the walls of cells and act as

FIGURE 2.2

DNA

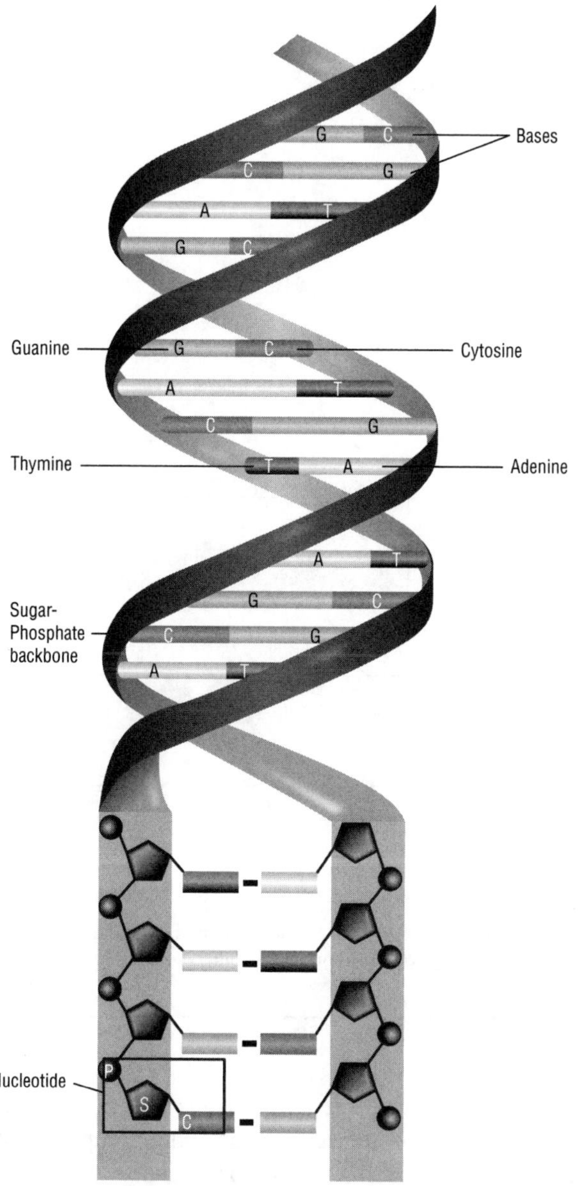

Bases

Guanine — Cytosine

Thymine — Adenine

Sugar-
Phosphate
backbone

Nucleotide

SOURCE: "DNA," in *The New Genetics*, National Institute of General Medical Sciences, National Institutes of Health, 2006, http://publications.nigms.nih.gov/thenewgenetics/chapter1.html (accessed September 6, 2010)

FIGURE 2.3

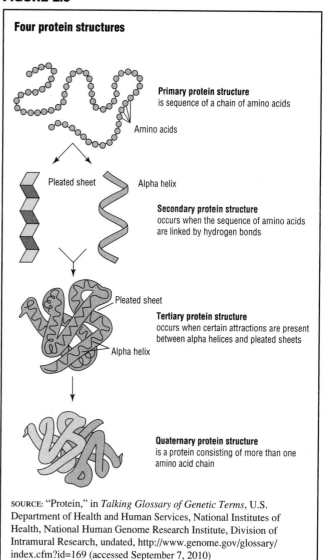

Four protein structures

Primary protein structure
is sequence of a chain of amino acids

Amino acids

Pleated sheet — Alpha helix

Secondary protein structure
occurs when the sequence of amino acids are linked by hydrogen bonds

Pleated sheet

Tertiary protein structure
occurs when certain attractions are present between alpha helices and pleated sheets

Alpha helix

Quaternary protein structure
is a protein consisting of more than one amino acid chain

SOURCE: "Protein," in *Talking Glossary of Genetic Terms*, U.S. Department of Health and Human Services, National Institutes of Health, National Human Genome Research Institute, Division of Intramural Research, undated, http://www.genome.gov/glossary/index.cfm?id=169 (accessed September 7, 2010)

Proteins and Amino Acids

All proteins are composed of building blocks called amino acids. (See Figure 2.4.) There are 20 different kinds of amino acids, and each has a slightly different chemical composition. The structure and function of each protein depends on its amino acid sequence—in a protein containing a hundred amino acids, a change in a single one may dramatically affect the function of the protein. Amino acids are small molecular groups that act like jigsaw puzzle pieces, linking together in a chain to make up the protein. Each amino acid links to its neighbor with a special kind of covalent bond (covalent bonds hold atoms together) called a peptide bond. Many amino acids link together, side by side, to make a protein. Figure 2.5 is a diagram of a protein structure. Proteins range in length from 50 to 500 amino acids that are linked head to tail.

Besides their ability to form peptide bonds to their neighbors, amino acids also contain molecular appendages

channels, determining which substances to let into the cell and which to block. Some bacteria know how to make proteins that protect them from antibiotics (substances such as penicillin and streptomycin that inhibit the growth of or destroy microorganisms), while the human immune system can make proteins that target bacteria or other germs for destruction. Many essential biological processes depend on the highly specific functions of proteins.

FIGURE 2.4

Amino acids

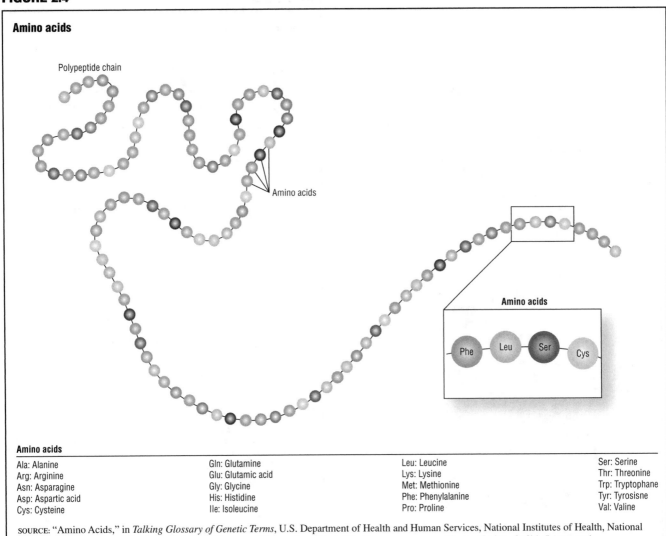

Polypeptide chain

Amino acids

Amino acids

Phe Leu Ser Cys

Amino acids

Ala: Alanine	Gln: Glutamine	Leu: Leucine	Ser: Serine
Arg: Arginine	Glu: Glutamic acid	Lys: Lysine	Thr: Threonine
Asn: Asparagine	Gly: Glycine	Met: Methionine	Trp: Tryptophane
Asp: Aspartic acid	His: Histidine	Phe: Phenylalanine	Tyr: Tyrosisne
Cys: Cysteine	Ile: Isoleucine	Pro: Proline	Val: Valine

SOURCE: "Amino Acids," in *Talking Glossary of Genetic Terms*, U.S. Department of Health and Human Services, National Institutes of Health, National Human Genome Research Institute, Division of Intramural Research, undated, http://www.genome.gov/glossary/index.cfm?id=5 (accessed September 7, 2010)

FIGURE 2.5

A primary protein structure

Amino acids are linked head to tail, so that at one end there is a free amino group, and at the other a free carboxyl group. Proteins are typically 50–500 amino acids in length.

H_2N — AA$_1$ — AA$_2$ — AA$_3$ — AA$_4$ — AA$_5$ — COOH

Amino end   Carboxyl end

SOURCE: Adapted from Richard Robinson, ed., "Schematic Diagram of a Primary Protein Structure," in *Genetics*, vol. 3, *K–P*, Macmillan Reference USA, 2002

peptide bond linking the amino acids into a chain, along with the extra influences of the side groups, twists a protein into a specific shape. This shape is called the protein's conformation, which determines how the protein interacts with other molecules.

Genes and Proteins

Along the length of a DNA molecule there are regions that hold the instructions to manufacture specific proteins. These regions are called protein-encoding genes and are an essential element of the modern understanding of genetics. Like an MP3 player that holds thousands of songs but only plays the one that is selected at any given moment, a DNA molecule can contain anywhere from a dozen to several thousand of these protein-encoding genes. However, as with the MP3 player, at any given time only some of these genes will be expressed—that is, switched on to actively produce the protein they know how to make.

called side groups. Depending on the particular atomic arrangement of the side group, neighboring amino acids experience different pushes and pulls as they attract or repel one another. The combination of the side-by-side

FIGURE 2.6

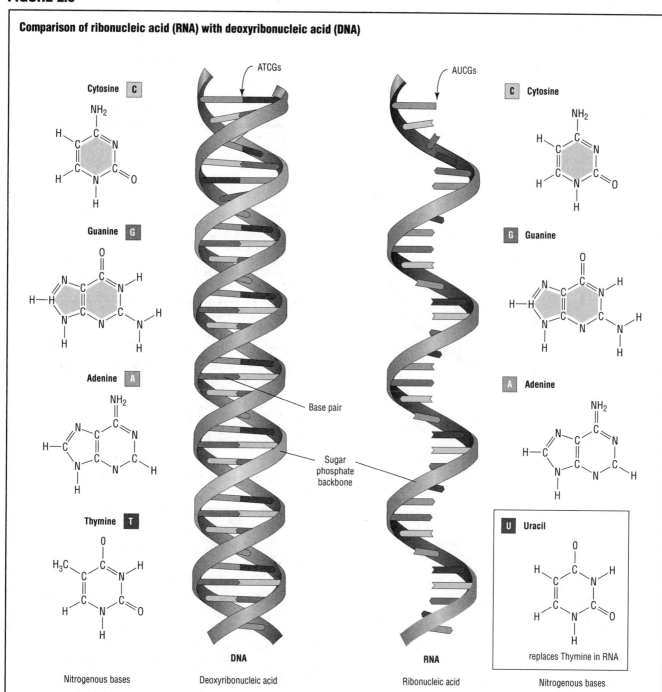

Comparison of ribonucleic acid (RNA) with deoxyribonucleic acid (DNA)

Cytosine **C**

Guanine **G**

Adenine **A**

Thymine **T**

Nitrogenous bases

DNA
Deoxyribonucleic acid

ATCGs

Base pair

Sugar
phosphate
backbone

AUCGs

RNA
Ribonucleic acid

C Cytosine

G Guanine

A Adenine

U Uracil

replaces Thymine in RNA

Nitrogenous bases

SOURCE: Adapted from "Ribonucleic Acid (RNA)," in *Talking Glossary of Genetic Terms*, U.S. Department of Health and Human Services, National Institutes of Health, National Human Genome Research Institute, Division of Intramural Research, undated, http://www.genome.gov/Glossary/index.cfm?id=180 (accessed September 7, 2010)

The protein-synthesizing instructions present in DNA are interpreted and acted on by ribonucleic acid (RNA). RNA, as its name suggests, is similar to DNA, except that the sugar in RNA is ribose (instead of deoxyribose), the base uracil (U) replaces thymine (T), and RNA molecules are usually single stranded and shorter than DNA molecules. (See Figure 2.6.) RNA is used to transcribe and translate the genetic code contained in DNA. Transcription is the process by which a molecule of messenger RNA (mRNA) is made, and translation is the synthesis of a protein using mRNA code. Figure 2.7 shows the transcription of mRNA and how mRNA is involved in protein synthesis (translation).

Genetic Synthesis of Proteins

How does a gene make a protein? The process of protein synthesis is quite complex. This overview describes

FIGURE 2.7

Messenger RNA (mRNA)

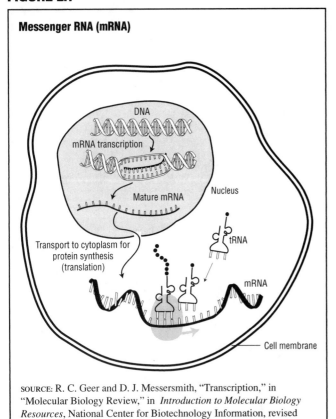

SOURCE: R. C. Geer and D. J. Messersmith, "Transcription," in "Molecular Biology Review," in *Introduction to Molecular Biology Resources*, National Center for Biotechnology Information, revised 2007, http://www.ncbi.nlm.nih.gov/Class/MLACourse/Modules/MolBioReview/transcription.html (accessed September 7, 2010)

FIGURE 2.8

Codon

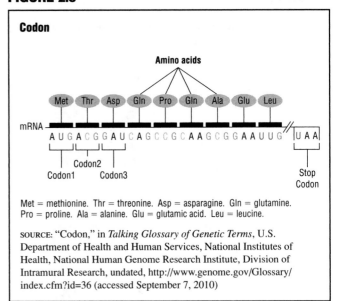

Met = methionine. Thr = threonine. Asp = asparagine. Gln = glutamine. Pro = proline. Ala = alanine. Glu = glutamic acid. Leu = leucine.

SOURCE: "Codon," in *Talking Glossary of Genetic Terms*, U.S. Department of Health and Human Services, National Institutes of Health, National Human Genome Research Institute, Division of Intramural Research, undated, http://www.genome.gov/Glossary/index.cfm?id=36 (accessed September 7, 2010)

the basic sequence of protein synthesis, which includes the following steps:

1. A gene is triggered for expression—to synthesize a protein.

2. Half of the gene is copied into a single strand of mRNA in a process called transcription.

3. The mRNA anchors to a ribosome, an organelle (or membrane-bound cell compartment) where protein synthesis occurs.

4. Each sequence of three bases on the mRNA, called a codon, uses the right type of transfer RNA (tRNA) to pick up a corresponding amino acid from the cell. (See Figure 2.8.)

5. A string of amino acids is assembled on the ribosome, side by side, in the same order as the codons of the mRNA, which, in turn, correspond to the sequence of bases from the original DNA molecule in a process called translation.

6. When the codons have all been read and the entire sequence of amino acids has been assembled, the protein is released to twist into its final form.

First, the DNA molecule receives a trigger telling it to express a particular gene. Many influences may trigger

gene expression, including chemical signals from hormones and energetic signals from light or other electromagnetic energy. For example, the spiral backbone of the DNA molecule can actually carry pulsed electrical signals that participate in activating gene expression.

Once a gene has been triggered for expression, a special enzyme system causes the DNA's double spiral to spring apart between the beginning and end of the gene sequence. The process is similar to a zipper with teeth that remain connected above and below a certain area, but pop open to create a gap along part of the zipper's length. At this point, the DNA base pairs that make up the gene sequence are separated. DNA replication is based on the understanding that any exposed nucleotide thymine (T) will pick up an adenine (A) and vice versa, whereas an exposed cytosine (C) will connect to an available guanine (G) and vice versa. Here the same process takes place except that instead of unzipping and copying the entire length of the DNA molecule, only the region between the beginning and end of the gene is copied. Instead of making a new double spiral, only one side of the gene is replicated, creating a special, single strand of bases called mRNA.

Once the mRNA has made a copy of one side of the gene, it separates from the DNA molecule. The DNA returns to its original state and the mRNA molecule breaks away, eventually connecting to an organelle in the cell called a ribosome. (Figure 2.9 is a drawing of a ribosome, which is a tiny particle of RNA and protein found in the cell's cytoplasm.) Here it anchors to another type of RNA called ribosomal RNA (rRNA). This is where the actual protein synthesis takes place. Using a special genetic coding system, each sequence of three bases on the mRNA copied from the gene is used to catch

FIGURE 2.9

Ribosome

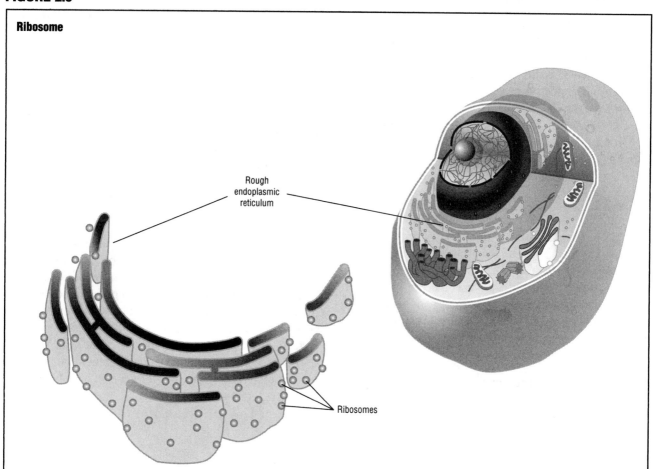

Rough
endoplasmic
reticulum

Ribosomes

SOURCE: "Ribosome," in *Talking Glossary of Genetic Terms*, U.S. Department of Health and Human Services, National Institutes of Health, National Human Genome Research Institute, Division of Intramural Research, undated, http://www.genome.gov/Glossary/index.cfm?id=178 (accessed September 7, 2010)

a corresponding amino acid floating inside the cell. These sequences of three bases are the codons, and they link to tRNA. A different form of tRNA is used to catch each different type of amino acid. The tRNA then catches the appropriate amino acid.

One after another, the mRNA's codons cause the corresponding tRNAs to be captured and linked, side by side, using the peptide bonds to connect them. When the last codon has been read and the entire peptide sequence is complete, the newly formed protein molecule is released from the ribosome. When this happens, all the side groups are able to interact, twisting the protein into its final shape. In this way, a protein-encoding gene is able to manufacture a protein from a series of DNA bases.

THE CELL IS THE BASIC UNIT OF LIFE

Ever since Matthias Jakob Schleiden (1804–1881) and Theodor Schwann (1810–1882) put forth their theories—Schleiden in 1838 and Schwann in 1839—that all plants and animals are composed of cells, there has been continuous refinement of cell theory. The early view that cells were made up of protoplasm (a jellylike substance) has given way to the more sophisticated understanding that cells are highly complex organizations of even smaller molecules and substructures. Cytology (the study of the formation, structure, and function of cells) has benefited from ever-improving technology, including powerful microscopes that enable researchers to identify the organelles (component parts of the cell) and determine their roles in inheritance.

Cells are the basic units and building blocks of nearly every organism. (One exception is viruses, which are simple organisms that are not composed of cells.) Each cell of an organism contains the same genetic information, which is passed on faithfully when cells divide. Different types of cells arise because they use different parts of the information, as determined by the cell's history and the immediate environment. Different cell types may be organized into tissues and organs.

Cell Structure and Function

Plants and animals, as well as other organisms such as fungi, are composed of eukaryotic cells, or eukaryotes, because they have nuclei and membrane-bound structures known as organelles. In eukaryotic cells the organelles within the cell sustain, support, and protect it by creating a barrier between the cell and its environment, acting to build and repair cell parts, storing and releasing energy, transporting material, disposing of waste, and increasing in number.

Each organelle functions like an organ system for the cell. For example, the nucleus is the command center, masterminding protein synthesis within the cell. (See Figure 1.1 in Chapter 1.) The ribosomes work as protein factories, the Golgi apparatus is a protein sorter, and the endoplasmic reticulum operates as a protein processor. Lysosomes and peroxisomes serve as the cell's digestive system, and mitochondria convert energy in the cell. The surface membrane of the cell acts like skin by selectively permitting molecules in and out of the cell.

The nuclei of eukaryotes contain the chromosomes, which are chains of genetic material coded in DNA. The threadlike chromosomes are contained in the nucleus of a typical animal cell. Genes are segments of DNA that carry a basic unit of hereditary information in coded form. (See Figure 2.10.) They contain instructions for making proteins.

The eukaryotic chromosome is composed of chromatin (a combination of nuclear DNA and protein) and contains a linear array of genes. It is visible just before and during cell division. (See Figure 2.11.) Human cells normally contain 23 pairs of chromosomes, or a total of 46 chromosomes, that may be examined using a process known as karyotyping (the organization of a standard picture of the chromosomes). Figure 2.12 is a karyotype—that is, a photo of an individual's chromosomes.

Cells without nuclei, such as bacteria and blue-green algae, are called prokaryotic cells or prokaryotes. Prokaryotic cells are smaller than eukaryotic cells, contain less genetic information, and are able to grow and divide more quickly. They perform these functions without organelles. A prokaryotic cell's DNA is not contained in one location; instead, it floats in different regions of the cell. The DNA of prokaryotes also contains jumping genes that are able to bind to other genes and transfer gene sequences from one site of a chromosome to another.

FIGURE 2.10

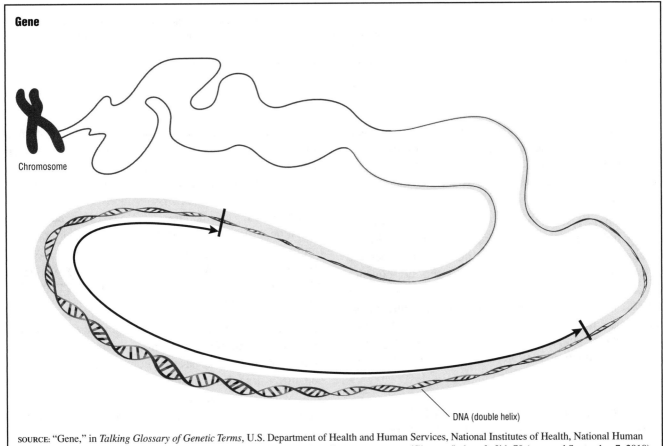

SOURCE: "Gene," in *Talking Glossary of Genetic Terms*, U.S. Department of Health and Human Services, National Institutes of Health, National Human Genome Research Institute, Division of Intramural Research, undated, http://www.genome.gov/Glossary/index.cfm?id=70 (accessed September 7, 2010)

FIGURE 2.11

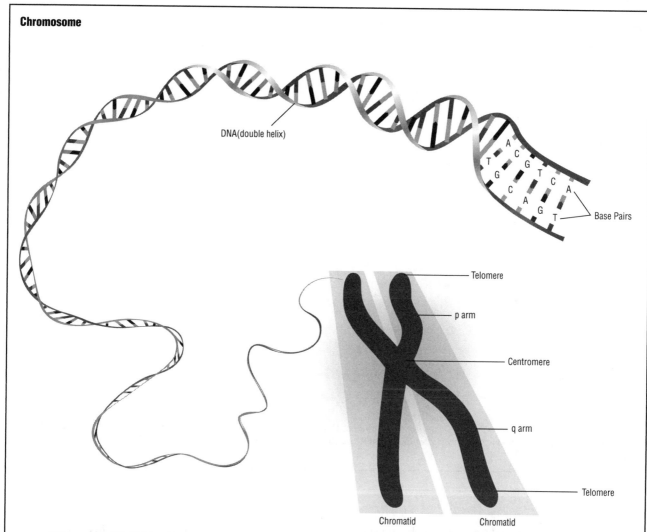

Chromosome

DNA(double helix)

Base Pairs

Telomere

p arm

Centromere

q arm

Telomere

Chromatid   Chromatid

SOURCE: "Chromosome," in *Talking Glossary of Genetic Terms*, U.S. Department of Health and Human Services, National Institutes of Health, National Human Genome Research Institute, Division of Intramural Research, undated, http://www.genome.gov/Glossary/index.cfm?id=33 (accessed September 7, 2010)

GROWTH AND REPRODUCTION

The growth of an organism occurs as a result of cell division in a process known as mitosis. Many cells are relatively short-lived, and mitosis allows for regular renewal of these cells. It is also the process that generates the millions of cells needed to grow an organism, or in the case of a human being, the trillions of cells needed to grow from birth to adulthood.

Mitosis is a continuous process that occurs in several stages. Between cell divisions, the cells are in interphase, during which there is cell growth, and the genetic material (DNA) contained in the chromosomes is duplicated so that when the cell divides, each new cell has a full-scale version of the same genetic material. The process of mitosis involves exact duplication—gene by gene—of the cell's chromosomal material and a systematic method for evenly distributing this material. It concludes with the physical division known as cytokinesis, when the identical chromosomes pull apart and each heads for the nucleus of one of the new daughter cells. Mitosis occurs in all eukaryotic cells, except the gametes (sperm and egg), and always produces genetically identical daughter cells with a complete set of chromosomes.

If, however, mitosis occurred in the gametes, then when fertilization (the joining of sperm and egg) took place, the offspring would receive a double dose of hereditary information. To prevent this from occurring, the gametes undergo a process of reduction division known as meiosis. Meiosis reduces the number of chromosomes in the gametes by half, so that when fertilization occurs the normal number of chromosomes is restored. For example, in humans the gametes produced by meiosis are haploid—they have just one copy of each of the 23 chromosomes. Besides preventing the number

FIGURE 2.12

Karyotype

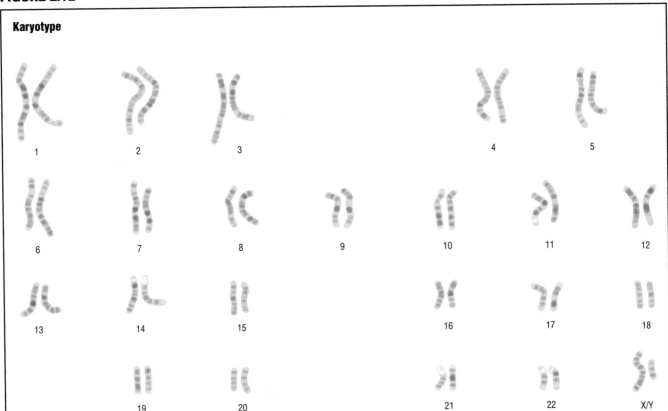

SOURCE: Adapted from "Karyotype," in *Talking Glossary of Genetic Terms*, U.S. Department of Health and Human Services, National Institutes of Health, National Human Genome Research Institute, Division of Intramural Research, undated, http://www.genome.gov/Glossary/index.cfm?id=114 (accessed September 7, 2010)

of chromosomes from doubling with each successive generation, meiosis also provides genetic diversity in offspring.

During meiosis the chromosomal material replicates and concentrates itself into homologous chromosomes (doubled chromosomes), each of which is joined at a central spot called the centromere. Figure 2.11 shows chromosomes joined at the centromere, and Figure 2.13 shows the location of the centromere in the chromosome. Pairing up along their entire lengths, they are able to exchange genetic material in a process known as crossing over. Figure 2.14 shows homologous chromosomes crossing over during meiosis to create new gene combinations. Crossing over results in much of the genetic variation that is observed among parents and their offspring. The pairing of homologous chromosomes and crossing over occur only in meiosis.

The process of meiosis also creates another opportunity to generate genetic diversity. During one phase of meiosis, called metaphase, the arrangement of each pair of homologous chromosomes is random, and different combinations of maternal and paternal chromosomes line up with varying orientations to create new gene combi-

nations on different chromosomes. This action is called independent assortment. Figure 2.15 shows the process of meiosis in an organism with six chromosomes. Because recombination and independent assortment of parental chromosomes takes place during meiosis, the daughter cells are not genetically identical to one another.

Gamete Formation

In male animals gamete formation, known as spermatogenesis, begins at puberty and takes place in the testes. Spermatogenesis involves a sequence of events that begins with the mitosis of primary germ cells to produce primary spermatocytes. (See Figure 2.16.) Four primary spermatocytes undergo two meiotic divisions, and as they undergo spermatogenesis they lose much of their cytoplasm and develop the spermatozoa's characteristic tail, the motor apparatus that provides the propulsion necessary to reach the egg cell. Like oocytes (egg cells) produced by the female, the mature sperm cells will be haploid (possessing just one copy of each chromosome).

In female animals gamete formation, known as oogenesis, takes place in the ovaries. Primary oocytes are produced by mitosis in the fetus before birth. Unlike

FIGURE 2.13

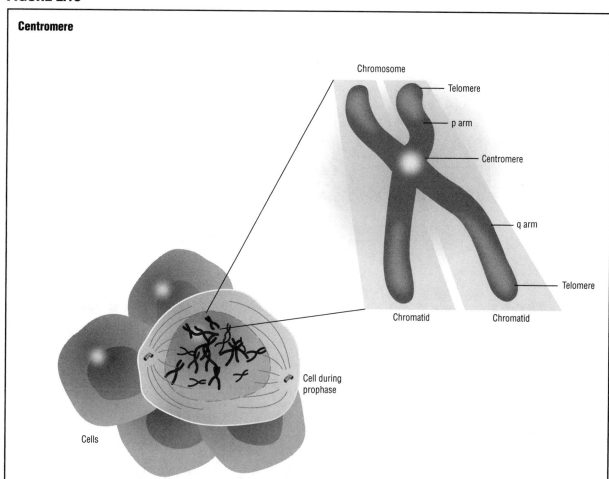

Centromere

SOURCE: "Centromere," in *Talking Glossary of Genetic Terms*, U.S. Department of Health and Human Services, National Institutes of Health, National Human Genome Research Institute, Division of Intramural Research, undated, http://www.genome.gov/Glossary/index.cfm?id=29 (accessed September 7, 2010)

the male, which continues to produce sperm cells throughout life, in the female the total number of eggs ever to be produced is present at birth. In April 2009, however, the online edition of the journal *Nature Cell Biology* reported the results of experiments by Chinese researchers on female mice showing evidence that mammalian ovaries might actually hold primitive cells that could be manipulated into producing new eggs ("Making New Eggs in Old Mice," April 11, 2009, http://www.nature.com/news/2009/090411/full/news.2009.362.html). At birth, or shortly before, meiosis begins and primary oocytes remain in the prophase of meiosis until puberty. At puberty the first meiotic division is completed, and a diploid cell becomes two haploid daughter cells; one large cell becomes the secondary oocyte and the other the first polar body. The secondary oocyte undergoes meiosis a second time but the meiosis does not continue to completion without fertilization. Figure 2.17 shows the sequence of events leading to the production of a mature ovum.

Fertilization

Like gamete formation, fertilization is a process, as opposed to a single event. It begins when the sperm and egg first come into contact and fuse together and culminates with the intermingling of two sets of haploid genes to reconstitute a diploid cell with the potential to become a new organism.

Of the millions of sperm released during an ejaculation, less than 1% survive to reach the egg. Of the few hundred sperm that reach the egg, only one will successfully fertilize it. While the sperm are in the female reproductive tract, swimming toward the egg, they undergo a process known as capacitation, during which they acquire the capacity to fertilize the egg. As the sperm approach the egg they become hyperactivated, and in a frenzy of mechanical energy the sperm attempt to burrow their way through the outer shell of the egg called the zona pellucida.

The cap of the sperm, known as the acrosome, contains enzymes that are crucial for fertilization. These acrosomal

FIGURE 2.14

Gene crossovers

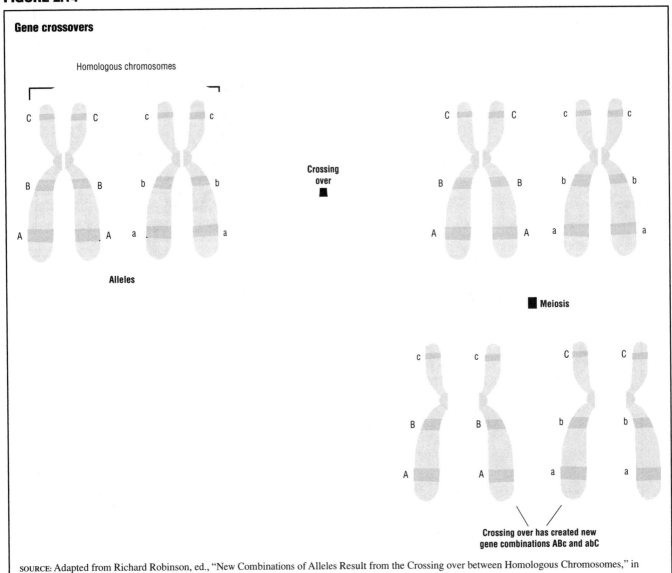

SOURCE: Adapted from Richard Robinson, ed., "New Combinations of Alleles Result from the Crossing over between Homologous Chromosomes," in *Genetics*, vol. 3, *K–P*, Macmillan Reference USA, 2002

enzymes dissolve the zona pellucida by making a tiny hole in it, so that one sperm can swim through and reach the surface of the egg. At this time, the egg transforms the zona pellucida by creating an impenetrable barrier, so that no other sperm may enter.

Sperm penetration triggers the second meiotic division of the egg. With this division the chromosomes of the sperm and egg form a single nucleus. The resulting cell—the first cell of an entirely new organism—is called a zygote. The zygote then divides into two cells, which, in turn, continue to divide rapidly, producing a ball of cells now called the blastocyst. The blastocyst is an early stage of embryogenesis, the process that describes the development of the fertilized egg as it becomes an embryo. In humans the developing baby is considered an embryo until the end of the eighth week of pregnancy.

GENETIC INHERITANCE

For inheritance of simple genetic traits, the two inherited copies of a gene determine the phenotype (the observable characteristic) for that trait. When genes for a particular trait exist in two or more different forms that may differ among individuals and populations, they are called alleles. For example, brown and blue eye colors are different alleles for eye color. For every gene, the offspring receives two alleles, one from each parent. The combination of inherited alleles is the genotype of the organism, and its expression (the observable characteristic) is its phenotype. Figure 2.18 is a graphic example of phenotype.

For many traits the phenotype is a result of an interaction between the genotype and the environment. Some of the most readily apparent traits in humans, such as

FIGURE 2.15

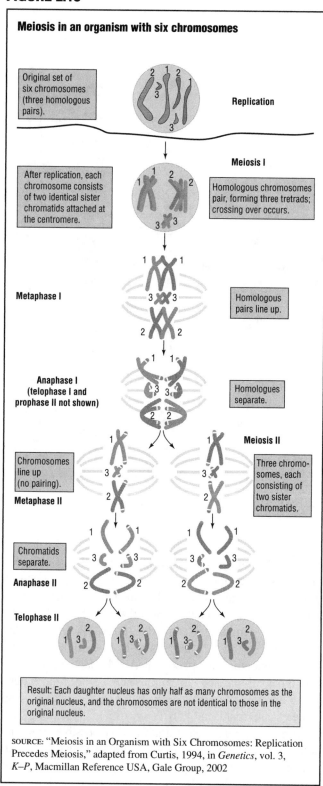

Meiosis in an organism with six chromosomes

Original set of six chromosomes (three homologous pairs).

Replication

After replication, each chromosome consists of two identical sister chromatids attached at the centromere.

Meiosis I

Homologous chromosomes pair, forming three tretrads; crossing over occurs.

Metaphase I

Homologous pairs line up.

Anaphase I (telophase I and prophase II not shown)

Homologues separate.

Meiosis II

Chromosomes line up (no pairing).

Metaphase II

Three chromosomes, each consisting of two sister chromatids.

Chromatids separate.

Anaphase II

Telophase II

Result: Each daughter nucleus has only half as many chromosomes as the original nucleus, and the chromosomes are not identical to those in the original nucleus.

SOURCE: "Meiosis in an Organism with Six Chromosomes: Replication Precedes Meiosis," adapted from Curtis, 1994, in *Genetics*, vol. 3, *K–P*, Macmillan Reference USA, Gale Group, 2002

indicates susceptibility or likelihood to develop a certain trait, it does not guarantee expression of the trait.

For a specific trait some alleles may be dominant whereas others are recessive. The phenotype of a dominant allele is always expressed, whereas the phenotype of a recessive allele is expressed only when both alleles are recessive. Recessive genes continue to pass from generation to generation, but they are only expressed in individuals who do not inherit a copy of the dominant gene for the specific trait. Figure 2.19 shows the inheritance of a recessive trait, in this example a recessive mutation that produces offspring with sickle-cell anemia (the presence of oxygen-deficient, abnormal red blood cells that cause affected individuals to suffer from obstruction of capillaries, resulting in pain and potential organ damage) or cystic fibrosis (an inherited disease of the mucous glands that produces problems associated with the lungs and pancreas).

There are also some instances, known as incomplete dominance, when one allele is not completely dominant over the other, and the resulting phenotype is a blend of both traits. Skin color in humans is an example of a trait often governed by incomplete dominance, with offspring appearing to be a blend of the skin tones of each parent. Furthermore, some traits are determined by a combination of several genes (multigenic or polygenic), and the resulting phenotype is determined by the final combination of alleles of all the genes that govern the particular trait.

Some multigenic traits are governed by many genes, each contributing equally to the expression of the trait. In such instances a defect in a single gene pair may not have a significant impact on expression of the trait. Other multigenic traits are predominantly directed by one major gene pair and only mildly influenced by the effects of other gene pairs. For these traits the impact of a defective gene pair depends on whether it is the major pair governing expression of the trait or one of the minor pairs influencing its expression.

A range of other factors enters into whether a trait will be evidenced and the extent to which it is expressed. For example, different individuals may express a trait with different levels of severity. This phenomenon is known as variable expressivity.

Determining Genetic Probabilities of Inheritance

Conventionally, geneticists use uppercase letters to represent dominant alleles and lowercase letters to stand for recessive alleles. An organism with a pair of identical alleles for a trait is described as a homozygote, or homozygous for that particular trait. When organisms are homozygous for a dominant trait, all uppercase letters symbolize the trait, whereas those that are homozygous for a recessive trait are represented by all lowercase letters. A heterozygote is an organism with different alleles for a trait, one donated

height, weight, and skin color, result from interactions between genetic and environmental factors. In addition, there are complex phenotypes that involve multiple gene-encoded proteins; the alleles of these particular genes are influenced by other factors, either genetic or environmental. So even though the presence of certain genes

FIGURE 2.16

Process of spermatogenesis

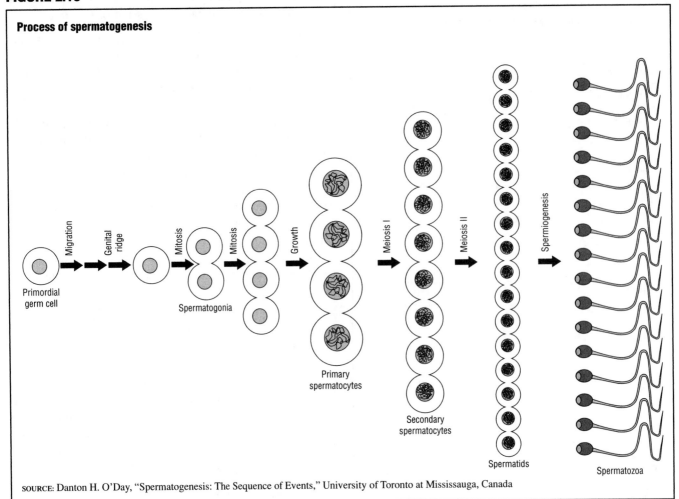

SOURCE: Danton H. O'Day, "Spermatogenesis: The Sequence of Events," University of Toronto at Mississauga, Canada

FIGURE 2.17

Process of oogenesis

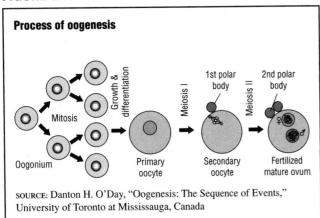

SOURCE: Danton H. O'Day, "Oogenesis: The Sequence of Events," University of Toronto at Mississauga, Canada

from each parent, and when one is dominant and the other is recessive the trait is shown using a combination of uppercase and lowercase letters.

Even though the combination of alleles is a random event, it is possible to predict the probability that an offspring will have the same or a different phenotype from its parents when the genotypes with respect to the specific trait of both parents and the phenotype associated with each possible combination of alleles are known. The formula used to determine genetic possibilities was developed by the British geneticist Reginald Crundall Punnett (1875–1967). The Punnett square is a grid configuration that depicts genotype and phenotype. In Punnett squares the genotypes of parents are represented with four letters. There are two alleles for each trait. Genotypes of haploid gametes are represented with two letters. Gametes will contain one allele for each trait in every possible combination, and all possible fertilizations are calculated. Punnett squares are used to compute the cross of a single gene or two genes and their alleles, but they become extremely complicated when used to predict the offspring of more than two alleles.

To construct a Punnett square, the alleles in the gametes of one parent are placed along the top, and the alleles in the gametes of the other parent are placed along the left side of the grid. The number of characteristics considered determines the size of the Punnett square. A monohybrid cross looks at one trait and has a two-by-two structure, with four

FIGURE 2.18

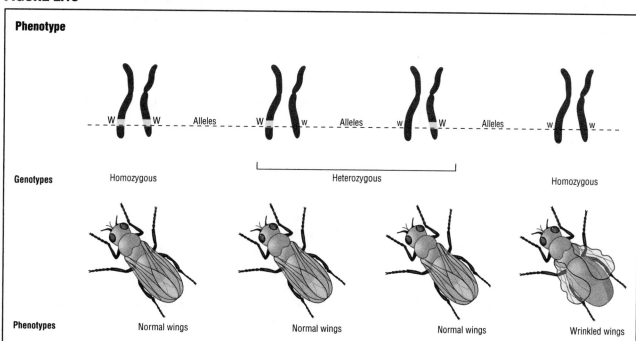

Phenotype

SOURCE: "Phenotype," in *Talking Glossary of Genetic Terms*, U.S. Department of Health and Human Services, National Institutes of Health, National Human Genome Research Institute, Division of Intramural Research, undated, http://www.genome.gov/Glossary/index.cfm?id=152 (accessed September 7, 2010)

FIGURE 2.19

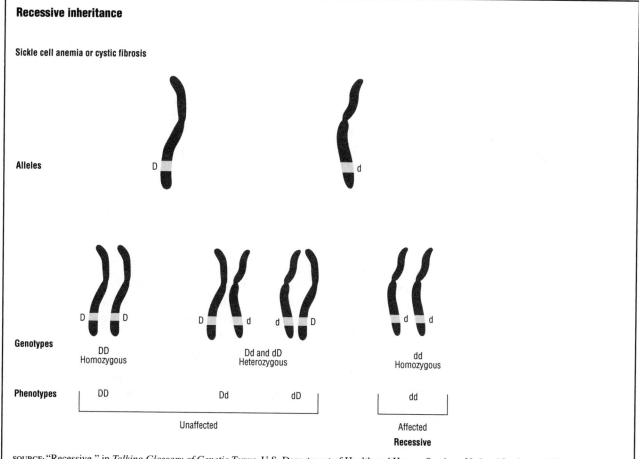

Recessive inheritance

SOURCE: "Recessive," in *Talking Glossary of Genetic Terms*, U.S. Department of Health and Human Services, National Institutes of Health, National Human Genome Research Institute, Division of Intramural Research, undated, http://www.genome.gov/Glossary/index.cfm?id=172 (accessed September 7, 2010)

FIGURE 2.20

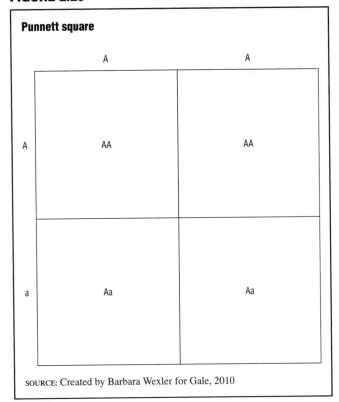

Punnett square

SOURCE: Created by Barbara Wexler for Gale, 2010

possible genotypes resulting from the cross. A dihybrid cross looks at two traits and has a four-by-four structure, providing 16 possible genotypes. Figure 2.20 is a simple Punnett square that shows one parent as being homozygous for the AA allele and another that is heterozygous for the Aa allele, with the four possible offspring: 50% AA and 50% Aa. This is an example of offspring that shows just one phenotype even though two genotypes are present. Figure 2.21 is a Punnett square that shows the predicted patterns of recessive inheritance. When both parents are carriers of the trait and dominant inheritance, the offspring is likely to arise from an affected parent and a normal parent.

Mitochondrial Inheritance

Mitochondria are the organelles involved in cellular metabolism and energy production and conversion. Mitochondria are inherited exclusively from the mother and contain their own DNA, known as mtDNA. Because some of the mtDNA multiplies during the organism's growth and development, it is more susceptible to mutation (changes in DNA sequence). The fact that mitochondria contain their own DNA has prompted scientists to speculate that they originally existed as independent one-celled organisms that over time developed interdependent relationships with more complex, eukaryotic cells.

Mitochondrial inheritance looks much like Mendelian inheritance (or genetic inheritance as described by Gregor Mendel's [1822–1884] laws), with two important exceptions. First, all maternal offspring are usually affected, whereas even in autosomal dominant disorders (those not related to the sex genes) only 50% of offspring are expected to be affected. (See Figure 2.22.) Second, mitochondrially inherited traits are never passed through the male parent. Males are as likely to be affected as females, but their offspring are not at risk. In other words, when there is a mutation in a mitochondrial gene, it is passed from a mother to all of her children; sons will not pass it on, but daughters will pass it on to all their children.

Mutations in mtDNA have been linked to the development of disease in humans. Leber's hereditary optic neuropathy, a painless loss of vision that afflicts people between the ages of 12 and 30 years, was the first human disease to be associated with a mutation in mtDNA. Many diseases that are linked to mtDNA affect the nervous system, heart or skeletal muscles, liver, or kidneys—sites of energy usage.

DETERMINING SEX

From the moment of fertilization, the new organism is assigned a sex, and its growth will proceed to develop either as a male or a female organism. The first clues that prompted scientists to consider that the determination of sex was influenced by genetics came from two key observations. The first was the fact that there is a general tendency toward a one-to-one ratio of males to females in all species. The second was the realization that the determination of sex followed the principles of Mendelian genetics—that gender was predictable as expected when individuals pure for a recessive trait were crossed with individuals that were hybrid.

The determination of sex occurs in all complex organisms, but the processes vary, even among animals. In humans 22 of the 23 pairs of chromosomes are as likely to be found in males as in females. These 22 chromosomes are known as the autosomes and the 23rd chromosome is called the sex chromosome. The sex chromosomes of females are identical and are called X chromosomes. (See Figure 2.23.) In males the sex chromosomes consist of an X chromosome and a smaller Y chromosome.

Chromosome Theory of Sex Determination

The genetic influence on gender is called the chromosome theory of sex determination, which states that:

- Sex is determined by the sex chromosome.

- In females the sex chromosomes are identical—both are X chromosomes.

- Because females have an XX genotype, all egg cells contain an X chromosome.

FIGURE 2.21

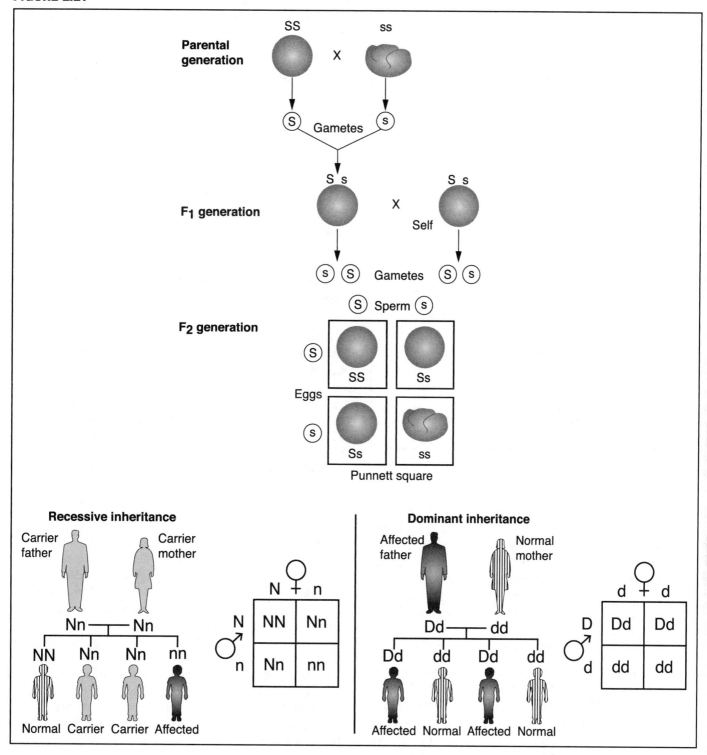

Punnett square with predicted patterns of recessive inheritance *Argosy Publishing, Cengage Gale.*

- In males the sex chromosomes are not identical; one is X and one is Y.

- Because males have an XY genotype, half of all sperm cells contain an X chromosome and half contain a Y chromosome.

- After fertilization, the egg may receive either an X or a Y chromosome from the sperm. Because all egg cells contain an X chromosome, the determination of gender is wholly dependent on the chromosomal composition of the sperm. Sperm carrying the Y chromosome are known as androsperm; those containing the X chromosome are called gynosperm. If the sperm carries the Y chromosome, the offspring will be male (XY); if it carries an X chromosome, the offspring will be female (XX).

FIGURE 2.22

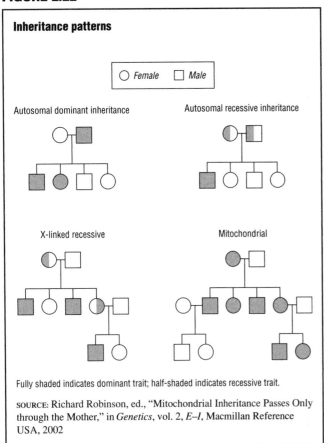

Inheritance patterns

○ Female □ Male

Autosomal dominant inheritance

Autosomal recessive inheritance

X-linked recessive

Mitochondrial

Fully shaded indicates dominant trait; half-shaded indicates recessive trait.

SOURCE: Richard Robinson, ed., "Mitochondrial Inheritance Passes Only through the Mother," in *Genetics*, vol. 2, *E–I*, Macmillan Reference USA, 2002

The determination of sex occurs at conception with the designation of chromosomal composition that is either XX or XY. However, a number of other genetic and environmental influences determine sex differentiation (the way in which the genetically predetermined sex becomes a reality). Differentiation translates the genetically coded message for sex into the physical traits, such as the hormones that influence the development of male and female genitalia, body functions, and behaviors that are associated with sex-linked and gender identity.

Interestingly, humans have an inherent tendency toward female development. Research conducted during the 1940s and 1950s confirmed that in many animals individuals with just a single X chromosome developed as females, although in many instances they did not develop completely and were sterile (unable to reproduce). The absence of the Y chromosome results in female development, whereas the presence of the Y chromosome sets in motion the series of events that result in male development. These findings led to the premise that female development is the default option in the process of sex determination.

Distribution of Males and Females in the Population

Because human males produce equal numbers of sperm bearing either the X or the Y chromosome, and fertilization is a random event, then it stands to reason that in each generation equal numbers of males and females should be born. An examination of birth statistics in the United States and in other countries where reliable statistics have been compiled over time show that every year there are more births of males than females. For example, according to Joyce A. Martin et al. of the Centers for Disease Control and Prevention, in "Births: Final Data for 2007" (*National Vital Statistics Reports*, vol. 58, no. 24, August 9, 2010), there were 2,208,071 live male births in 2007, compared with 2,108,162 live female births, a ratio of 1,047 males per 1,000 females for births to mothers of all races in the United States. Table 2.1 shows that births by sex varied by race, from a low of 1,033 males per 1,000 females among African-Americans to a high of 1,066 males per 1,000 females among Asians or Pacific Islanders. Martin et al. report that the annual distribution of births by sex has remained essentially unchanged over the past 60 years, with only small annual variations.

For years this difference was attributed to the idea that males were inherently stronger than females and better able to survive pregnancy and birth. This theory was dispelled when researchers found that nearly three times as many male fetuses spontaneously abort (dying before birth). In fact, male life expectancy is less than female life expectancy at every age, from conception to adulthood. (See Table 2.2.) The explanation for the higher proportion of male births appears to be that more male offspring are conceived—possibly as many as 125 to every 100—because the prenatal death rate for males is so high; at birth the gap closes to about 105 to every 100.

One explanation for the higher number of males conceived is that the smaller and stronger Y sperm are better able to swim quickly and successfully reach the egg cell. Along with androsperm size and mobility, environmental conditions influence gender determination and the chances of fetal survival. For example, the mother's age and general health are strongly linked to favorable outcomes of conception and pregnancy and are less strongly linked to, but are associated with, sex of her offspring. Younger mothers are more likely to conceive male offspring, by a ratio as high as 120 to 100, and unfavorable prenatal conditions such as poor health or maternal illness are more likely to compromise the survival of the male fetus than the female fetus.

Sex-Linked Characteristics

The two sex chromosomes also differ in terms of the genes they contain, which relate to many traits other than sex. The Y chromosome is quite small and carries few genes other than the one that determines male gender. One of the few confirmed traits linked to the Y chromosome is the hairy ear trait, a characteristic that is distinctive but largely unrelated to health. Because this trait is located exclusively on the Y chromosome, it only appears in males.

FIGURE 2.23

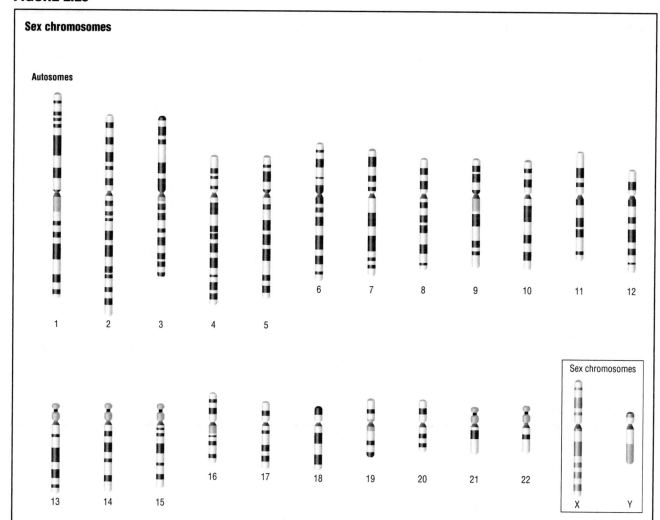

Sex chromosomes

Autosomes

SOURCE: "Sex Chromosomes," in *Talking Glossary of Genetic Terms*, U.S. Department of Health and Human Services, National Institutes of Health, National Human Genome Research Institute, Division of Intramural Research, undated, http://www.genome.gov/Glossary/index.cfm?id=181 (accessed September 7, 2010)

The X chromosome is larger and holds many genes that are as necessary for males as they are for females. The genes on the X chromosome are called X-linked, and characteristics or conditions arising from these genes are called X-linked traits or conditions. Most people, male and female, likely have several so-called defective genes with the potential to produce harmful characteristics or conditions, but these genes are usually recessive and are not expressed in the phenotype unless they are combined with a similar recessive gene on the corresponding chromosome. For this to occur, both parents would have to contribute the same defective gene. Species are also protected from the harmful effects of single defective genes by virtue of the fact that most traits are multigenic (controlled by more than one gene).

Even though X-linked dominant alleles affect males and females, males are more strongly affected because they inherit just one X chromosome and do not have a counterbalancing normal allele. Huntington's disease, an inherited neuropsychiatric disease that affects the body and mind, is an example of a disorder caused by an X-linked dominant allele. Males are affected more frequently and more severely than females by X-linked recessive alleles. The male receives his X chromosome from his mother. Because males have just one X chromosome, all the alleles it contains are expressed, including those that cause serious and sometimes lethal medical disorders. Examples of disorders caused by X-linked recessive alleles range from relatively harmless conditions such as red-green color blindness to the always-fatal Duchenne muscular dystrophy (DMD), one of a group of muscular dystrophies, which is characterized by the enlargement of selected muscles, particularly in the calves. This enlargement is termed pseudohypertrophy and occurs when fat and connective tissue replace muscle, giving the impression of enlarged muscles.

TABLE 2.1

Total births, by race of mother and selected demographic characteristics, 2007

[Birth rates are live births per 1,000 population. Fertility rates are computed by relating total births, regardless of age of mother, to women aged 15–44 years. Total fertility rates are sums of birth rates for 5-year age groups multiplied by 5. Populations estimated as of July 1. Mean age at first birth is the arithmetic average of the age of mothers at the time of the birth, computed directly from the frequency of first births by age of mother.]

Characteristic	All races	White	Black	American Indian or Alaskan Native	Asian or Pacific Islander
Number					
Births	4,316,233	3,336,626	675,676	49,443	254,488
Rate					
Birth rate	14.3	13.7	16.9	15.3	17.2
Fertility rate	69.5	68.8	72.7	64.9	71.3
Total fertility rate	2,122.0	2,111.5	2,168.0	1,866.5	2,038.5
Sex ratio*	1,047	1,049	1,033	1,038	1,066
Percent					
All births					
Births to mothers under 20 years	10.5	9.5	17.2	18.4	3.1
4th- and higher-order births	11.3	10.8	14.9	19.1	6.4
Births to unmarried mothers	39.7	34.8	71.2	65.3	16.6
Mothers born in the 50 states and DC	75.1	77.2	84.7	93.3	19.4
Mean					
Age of mother at first birth	25.0	25.2	22.7	21.8	28.6

*Male live births per 1,000 female live births.

Notes: Race and Hispanic origin are reported separately on birth certificates. Race categories are consistent with the 1977 Office of Management and Budget standards. Twenty-seven states reported multiple race data for 2007. The multiple-race data for these states were bridged to the single race categories of the 1977 Office of Management and Budget standards for comparability with other states. In this table all women (including Hispanic women) are classified only according to their race.

SOURCE: Joyce A. Martin et al., "Table 14. Total Number of Births, Rates (Birth, Fertility, and Total Fertility), and Selected Demographic Characteristics, by Race of Mother: United States, 2007," in "Births: Final Data for 2007," *National Vital Statistics Reports*, vol. 58, no. 24, August 2010, http://www.cdc.gov/nchs/data/nvsr/nvsr58/nvsr58_24.pdf (accessed September 7, 2010)

DMD is one of the most prevalent types of muscular dystrophy and involves rapid progression of muscle degeneration early in life. It is X-linked, affects mainly males, and is the most common neuromuscular disease of childhood, with an estimated overall prevalence (the number of cases of a disease that are present in a particular population at a given time) of 63 cases per 1 million births. The National Institute of Neurological Disorders and Stroke indicates in *Muscular Dystrophy: Hope through Research* (December 29, 2010, http://www.ninds.nih.gov/disorders/md/detail_md.htm#110183171) that one out of every 3,500 to 5,000 boys worldwide are diagnosed with it. Another X-linked recessive allele causes Tay-Sachs disease, a disease that is most common among people of Jewish descent and results in neurological disorders and death in childhood.

All the male offspring of females who carry X-linked recessive alleles will be affected by the recessive allele. Female children are not as likely to express harmful recessive X-linked traits because they have two X chromosomes. Fifty percent of the female offspring will receive the recessive allele from a mother who carries the allele and an unaffected father. (See Figure 2.22.)

COMMON MISCONCEPTIONS ABOUT INHERITANCE

There are many myths and misunderstandings about genetics and inheritance. For example, some people mistakenly believe that in any population dominant traits are inevitably more common than recessive traits. This is simply not true, as evidenced by the observation that, among humans, the allele that produces six fingers and six toes on each hand and foot, respectively, is dominant over the allele for five fingers and five toes, but the incidence of polydactyly (extra digits) is actually quite low.

Another lingering misconception is that sex-linked diseases occur only in males. This is untrue but it is easy to understand the source of the misunderstanding. For years it was thought that hemophilia (a disease that is characterized by uncontrolled bleeding) did not occur in females. The observation seemed reasonable because there were no reported cases of the disease among females. Even though it was true that there were no females with the disease, the reasoning was incorrect. For a female to suffer from hemophilia, she would require a defective recessive gene on both of her X chromosomes, meaning her mother was carrying the gene and her father was affected with the disease. Because

TABLE 2.2

Life expectancy at selected ages, by race and gender, selected years 1900–2006

[Data are based on death certificates]

Specified age and year	All races			White			Black or African American[a]		
	Both sexes	Male	Female	Both sexes	Male	Female	Both sexes	Male	Female
At birth				Remaining life expectancy in years					
1900[b, c]	47.3	46.3	48.3	47.6	46.6	48.7	33.0	32.5	33.5
1950[c]	68.2	65.6	71.1	69.1	66.5	72.2	60.8	59.1	62.9
1960[c]	69.7	66.6	73.1	70.6	67.4	74.1	63.6	61.1	66.3
1970	70.8	67.1	74.7	71.7	68.0	75.6	64.1	60.0	68.3
1980	73.7	70.0	77.4	74.4	70.7	78.1	68.1	63.8	72.5
1990	75.4	71.8	78.8	76.1	72.7	79.4	69.1	64.5	73.6
1995	75.8	72.5	78.9	76.5	73.4	79.6	69.6	65.2	73.9
1998	76.7	73.8	79.5	77.3	74.5	80.0	71.3	67.6	74.8
1999	76.7	73.9	79.4	77.3	74.6	79.9	71.4	67.8	74.7
2000	76.8	74.1	79.3	77.3	74.7	79.9	71.8	68.2	75.1
2001	76.9	74.2	79.4	77.4	74.8	79.9	72.0	68.4	75.2
2002	76.9	74.3	79.5	77.4	74.9	79.9	72.1	68.6	75.4
2003	77.1	74.5	79.6	77.6	75.0	80.0	72.3	68.8	75.6
2004	77.5	74.9	79.9	77.9	75.4	80.4	72.8	69.3	76.0
2005	77.4	74.9	79.9	77.9	75.4	80.4	72.8	69.3	76.1
2006	77.7	75.1	80.2	78.2	75.7	80.6	73.2	69.7	76.5
At 65 years									
1950[c]	13.9	12.8	15.0	—	12.8	15.1	13.9	12.9	14.9
1960[c]	14.3	12.8	15.8	14.4	12.9	15.9	13.9	12.7	15.1
1970	15.2	13.1	17.0	15.2	13.1	17.1	14.2	12.5	15.7
1980	16.4	14.1	18.3	16.5	14.2	18.4	15.1	13.0	16.8
1990	17.2	15.1	18.9	17.3	15.2	19.1	15.4	13.2	17.2
1995	17.4	15.6	18.9	17.6	15.7	19.1	15.6	13.6	17.1
1998	17.8	16.0	19.2	17.8	16.1	19.3	16.1	14.3	17.4
1999	17.7	16.1	19.1	17.8	16.1	19.2	16.0	14.3	17.3
2000	17.6	16.0	19.0	17.7	16.1	19.1	16.1	14.1	17.5
2001	17.7	16.2	19.0	17.8	16.3	19.1	16.2	14.2	17.6
2002	17.8	16.2	19.1	17.9	16.3	19.2	16.3	14.4	17.7
2003	17.9	16.4	19.2	18.0	16.5	19.3	16.4	14.5	17.9
2004	18.2	16.7	19.5	18.3	16.8	19.5	16.7	14.8	18.2
2005	18.2	16.8	19.5	18.3	16.9	19.5	16.8	14.9	18.2
2006	18.5	17.0	19.7	18.6	17.1	19.8	17.1	15.1	18.6
At 75 years									
1980	10.4	8.8	11.5	10.4	8.8	11.5	9.7	8.3	10.7
1990	10.9	9.4	12.0	11.0	9.4	12.0	10.2	8.6	11.2
1995	11.0	9.7	11.9	11.1	9.7	12.0	10.2	8.8	11.1
1998	11.3	10.0	12.2	11.3	10.0	12.2	10.5	9.2	11.3
1999	11.2	10.0	12.1	11.2	10.0	12.1	10.4	9.2	11.1
2000	11.0	9.8	11.8	11.0	9.8	11.9	10.4	9.0	11.3
2001	11.1	9.9	11.9	11.1	9.9	11.9	10.5	9.1	11.4
2002	11.0	9.9	11.9	11.1	9.9	11.9	10.5	9.2	11.4
2003	11.1	10.0	11.9	11.1	10.0	11.9	10.6	9.3	11.5
2004	11.4	10.3	12.2	11.4	10.3	12.2	10.8	9.5	11.7
2005	11.3	10.2	12.1	11.4	10.3	12.1	10.8	9.5	11.7
2006	11.6	10.5	12.3	11.5	10.5	12.3	11.1	9.8	12.0

—Data not available.

[a]Data shown for 1900–1960 are for the nonwhite population.

[b]Death registration area only. The death registration area increased from 10 states and the District of Columbia (D.C.) in 1900 to the coterminous United States in 1933.

[c]Includes deaths of persons who were not residents of the 50 states and D.C.

Notes: Populations for computing life expectancy for 1991–1999 are 1990-based postcensal estimates of U.S. resident population. In 1997 life table methodology was revised to construct complete life tables by single years of age that extend to age 100. Previously, abridged life tables were constructed for 5-year age groups ending with 85 years and over. Life table values for 2000 and later years were computed using a slight modification of the new life table method due to a change in the age detail of populations received from the U.S. Census Bureau. Values for data years 2000–2006 are based on a newly revised methodology that uses vital statistics death rates for ages under 66 and modeled probabilities of death for ages 66 to 100 based on blended vital statistics and Medicare probabilities of dying and may differ from figures previously published. The revised methodology is similar to that developed for the 1999–2001 decennial life tables. In 2003, seven states reported multiple-race data. In 2004, 15 states reported multiple-race data. In 2005, 21 states and D.C. reported multiple-race data. In 2006, 25 states and D.C. reported multiple-race data. The multiple-race data for these states were bridged to the single-race categories of the 1977 Office of Management and Budget Standards for comparability with other states. Some data have been revised and differ from previous editions of *Health, United States.* Data for additional years are available.

SOURCE: "Table 24. Life Expectancy at Birth, at 65 Years of Age, and at 75 Years of Age, by Race and Sex: United States, Selected Years 1900–2006," in *Health, United States, 2009. With Special Feature on Medical Technology,* U.S. Department of Health and Human Services, Centers for Disease Control and Prevention, National Center for Health Statistics, 2010, http://www.cdc.gov/nchs/data/hus/hus09.pdf#listtables (accessed September 7, 2010)

most people with hemophilia died young, few lived to produce offspring. In other words, female hemophiliacs were rare because the pairings that might produce them

were infrequent. During the 1950s the first cases of hemophilia in females were documented, and the theory was discarded.

Finally, the idea that humans are entirely unique in their genetic makeup is false. In fact, human beings share much of their genetic composition with other organisms in the natural world. Furthermore, most human genetic variation is relatively insignificant. Even variations that alter the sequence of amino acids in a protein often produce no discernable influence on the action of the protein. Differences in some portions of DNA with as yet unknown functions appear to have no impact at all.

CHAPTER 3
GENETICS AND EVOLUTION

The essence of Darwinism lies in a single phrase: natural selection is the major creative force of evolutionary change. No one denies that natural selection will play a negative role in eliminating the unfit. Darwinian theories require that it create the fit as well.

—Stephen Jay Gould, "The Return of Hopeful Monsters" (*Natural History*, vol. 86, no. 6, June–July 1977)

The term *evolution* has multiple meanings. It is most generally used to describe the theory that all organisms are linked via descent to a common ancestor. Evolution also refers to the gradual process during which change occurs. In biology it is the theory that groups of organisms, such as species, change or develop over long periods of time so that their descendants differ from their ancestors morphologically (in form, structure, and physiology) in terms of their life processes, activities, and functions. (Species are the smallest groups into which most living things that share common characteristics are divided. A key characteristic that defines a species is that its members can breed within the group but not outside it.)

It is important to understand that not all change is considered evolution; evolution encompasses only those changes that are inheritable and may be passed on to the next generation. For example, evolution does not explain why humans are taller and bigger today than they were a century ago. This phenotypic (observable) change is attributable to changes in the environment—that is, improvements in nutrition and medicine—and is not inherited. Similarly, it should also be noted that even though evolution leads to increasing complexity, it does not necessarily signify progress, because an adaptation, trait, or strategy that is successful at one time may be unsuccessful at another.

In genetic terms evolution can be defined as any change in the gene pool of a population over time or changes in the frequency of alleles in populations of organisms from generation to generation. Evolution requires genetic variation, and the incremental and often uneven changes described by the process of evolution arise in response to an organism's or species' genetic response to environmental influences.

Evidence of evolution has been derived from fossil records, genetics studies, and changes observed among organisms over time. The process produces the transformations that generate new species only able to survive if they can respond quickly and favorably enough to environmental changes. Population genetics is the discipline that considers variation and changing ratios of genetic types within populations to explain how populations evolve. Such changes within a population are called microevolution. In contrast, macroevolution describes larger-scale changes that produce entirely new species. Even though some researchers speculate that the two processes are different, many scientists believe macroevolutionary change is simply the final outcome of the collected effects of microevolution.

Molecular evolution is the term used to describe the period before cellular life developed on the earth. Scientists speculate that specific chemical reactions occurred that created information-containing molecules that contributed to the origin of life on the planet. Theories about molecular evolution presume that these early information-containing molecules were precursors to genetic structures capable of replication (duplication of deoxyribonucleic acid [DNA] by copying specific nucleic acid sequences) and mutation (change in DNA sequence).

NATURAL SELECTION

Natural selection is a mechanism of evolution. The principles of organic evolution by means of natural selection were described by the British naturalist Charles Darwin (1809–1882). Much of his early research focused on geology, and he developed theories about the origin of different land formations when he went on a five-year expedition around the world aboard the HMS *Beagle*.

During his travels he developed an interest in population diversity.

When Darwin identified 12 different species of finches on the Galápagos Islands off the coast of Ecuador, he speculated that the birds must have descended from a common ancestor even though they differed in terms of beak shape and overall size. The birds became known as "Darwin's finches" and are examples of a process called adaptive radiation, in which species from a common ancestor successfully adapt to their environment via natural selection. Darwin suspected that the finches had become geographically isolated from one another and after years of adapting to their distinctive environments had developed and gradually evolved into separate species that were incapable of interbreeding.

To explain this occurrence, Darwin relied on his own observations of the existence of variation in and between species, his knowledge of animal breeding, and the results of zoological research conducted by the French naturalist Jean Baptiste Lamarck (1744–1829). Lamarck suggested four laws to explain how animal life might change:

- The life force tends to increase the volume of the body and to enlarge its parts.

- New organs can be produced in a body to satisfy a new need.

- Organs develop in proportion to their use.

- Changes that occur in the organs of an animal are transmitted to that animal's progeny.

Darwin famously took issue with the fourth law, Lamarck's theory of acquired traits, particularly his suggestion that giraffes that make their necks longer by stretching to reach the uppermost leaves on tall trees would then pass on longer necks to their offspring. However, even though Darwin discredited the specifics of Lamarck's theories concerning evolution, he agreed with Lamarck's idea that species change over time, and he acknowledged Lamarck as an important forerunner and influence on his own work.

Darwin's ideas were also influenced by *An Essay on the Principle of Population, as It Affects the Future Improvement of Society* (1798), written by the British economist Thomas Robert Malthus (1766–1834). In this work Malthus put forth his hypothesis that unchecked population growth always exceeds the growth of the means of subsistence (the food supply needed to sustain it). In other words, if there were no outside factors stopping population growth, there would inevitably be more people than food. According to Malthus, actual population growth is kept in line with food supply growth by "positive checks," such as starvation and disease, which increase the death rate, and "preventive checks," such as postponement of marriage, which reduces the birthrate.

Malthus's hypothesis suggested that the human population always tended to rise above the food supply, but that historically overpopulation had been prevented by wars, famine, and epidemics of disease.

Darwin also knew that farmers had been able to modify species of domestic animals for hundreds of years. Cattle breeders produced breeds that yielded exceptional milk production by selectively breeding their best milk producers. Superior egg-laying hens had been bred using the same technique. Because it was possible for farmers to modify a species by artificially selecting those members permitted to reproduce, Darwin hypothesized that nature might have a comparable mechanism for determining which characteristics might be passed on to future generations.

He also realized that even though individual organisms in every species have the potential to produce many offspring, the natural population of any species remains relatively constant over time. Darwin concluded that the natural environment acts as a natural selector by determining over long periods of time which variations are best suited to survive and, by virtue of their survival, reproduce and pass on traits and adaptations that improve health and longevity.

Applying the principles of natural selection to the question of giraffes' neck lengths provides an explanation that is different from the one proposed by Lamarck. Short-necked giraffes were less able to obtain food, so they faced starvation. As such, the genes linked to the potential to develop long necks were more likely to be passed to the next generation than the genes for short necks. Over time, the process of natural selection resulted in a population of giraffes with long necks.

Laboratory research and observation also refuted the theory of inheritance of acquired characteristics. When white mice had their tails cut off and were permitted to reproduce, each new generation was born with tails. Children of parents who had suffered amputations or disfiguring accidents did not share their parents' disabilities. Darwin's belief in evolution by natural selection was based on four premises:

- Individuals within a species are variable.

- Some of these variations are passed on to offspring.

- In every generation more offspring are produced than can survive.

- The survival and reproduction of individuals are not random. The individuals who survive and reproduce or reproduce the most are those with the most favorable variations. They are naturally selected.

As support for the theory of acquired inheritance diminished, appreciation of the underlying assumptions for the role of natural selection in evolution grew.

Attacks on Darwin's Theories

Darwin's theories were met with criticism from scientists and members of the clergy. Even some scientists who subscribed to evolutionary theory took issue with the concept of natural selection. Lamarck's followers were among the most outspoken opponents of Darwin's theories. This was especially ironic because it was Lamarck's work that had inspired Darwin.

Other objections raised by scientists were related to how poorly inheritance was understood at that time. The notion of blending inheritance was popular. This is the idea that an organism blends together the traits it inherits from its parents. Those who endorsed it observed that, according to Darwin's assumptions, any new variation would mix with existing traits and would no longer exist after several generations. Even though Gregor Mendel (1822–1884) had published the paper "Versuche über Pflanzen-Hybriden" ("Experiments in Plant Hybridization") in 1865 that proposed particulate as opposed to blended inheritance, his theory was not widely accepted until 1900, when it was revisited and confirmed by scientists.

The other objection to Darwin's theories was the argument that variation within species was limited and that, once the existing variation was exhausted, natural selection would cease abruptly. In 1907 the American geneticist Thomas Hunt Morgan (1866–1945) and his colleagues effectively dispelled this objection. Their experiments with fruit flies demonstrated that new hereditary variation occurs in every generation and in every trait of an organism.

The clergy were even more vociferous adversaries. Darwin's major works, *On the Origin of Species by Means of Natural Selection, or the Preservation of Favoured Races in the Struggle for Life* (1859) and *The Descent of Man, and Selection in Relation to Sex* (1871), were published during a period of heightened religious fervor in England. Many religious leaders were aghast at Darwin's assertion that all life had not been created by God in one fell swoop. Moral outrage and opposition to Darwinian theory persisted into the 20th century. Even though society grew more tolerant, and many religions accepted and incorporated evolutionary theory into their beliefs, in the early 21st century the debate was revived, as many fundamentalist Christian denominations in the United States became more vocal about their creationist beliefs.

In some instances opponents protested the teaching of evolution in schools and continued to defend creationism. In July 1925 the science teacher John Thomas Scopes (1900–1970) was tried in a Tennessee court for teaching his high school class Darwin's theory of evolution. Scopes had violated the Butler Act, which prohibited teaching evolutionary theory in public schools in Tennessee. Dubbed the "Monkey Trial" because of the simplified interpretation of Darwin's idea that humans evolved from apes, the court-room drama pitted the defense attorney Clarence Seward Darrow (1857–1938) against the prosecutor William Jennings Bryan (1860–1925) in a debate that began over the teaching of evolution but became a conflict of deeply held intellectual, religious, and social values. In his acerbic account of the trial proceedings, the American writer H. L. Mencken (1880–1956) wrote in "'The Monkey Trial': A Reporter's Account" (January 30, 2007, http://www.law.umkc.edu/faculty/projects/ftrials/scopes/menk.htm):

> The Scopes trial, from the start, has been carried on in a manner exactly fitted to the anti-evolution law and the simian imbecility under it. There hasn't been the slightest pretense to decorum. The rustic judge, a candidate for re-election, has postured the yokels like a clown in a ten-cent side show, and almost every word he has uttered has been an undisguised appeal to their prejudices and superstitions.... Darrow has lost this case. It was lost long before he came to Dayton. But it seems to me that he has nevertheless performed a great public service by fighting it to a finish and in a perfectly serious way. Let no one mistake it for comedy, farcical though it may be in all its details. It serves notice on the country that Neanderthal man is organizing in these forlorn backwaters of the land, led by a fanatic, rid of sense and devoid of conscience. Tennessee, challenging him too timorously and too late, now sees its courts converted into camp meetings and its Bill of Rights made a mock of by its sworn officers of the law. There are other States that had better look to their arsenals before the Hun is at their gates.

At the end of deliberations, Darrow requested a guilty verdict so that the case could be heard before the Tennessee Supreme Court on appeal. The jury complied, and the presiding judge fined Scopes $100. A year later, the Tennessee Supreme Court overturned the verdict on a technicality. Wanting to close the case once and for all, the court dismissed it altogether.

Misuse of Darwin's Theories

After Darwin's theories became well known, some people made use of his terminology and concepts to argue that certain groups of human beings were naturally superior to others. The term *social Darwinism* is used to refer to these ideas, but it is important to note that Darwin himself did not believe in social Darwinism.

One example of social Darwinist thinking would be arguing that the rich and successful members of society are fitter or superior or are in some way more highly evolved than the poor. Social Darwinism has also been used to justify racism and colonialism, as in the 19th and early 20th centuries, when many white Europeans and Americans asserted that they were naturally superior to Africans and Asians and used this claim to justify taking control of their land and resources. Some argued further that it was not just a right but an obligation—the "white man's burden"—for Europeans and Americans to rule over and "civilize" people in less industrialized parts of the world.

None of these social Darwinist theories are scientific in nature, and all are false. Modern genetics shows that there is no group of human beings that is more evolved or otherwise better than the rest of humanity. Despite having been discredited by scientific research, social Darwinism continues to be used in attempts to justify various prejudices and inequalities.

THE MODERN
EVOLUTION-CREATIONISM DEBATE

Arguments over the accuracy and importance of Darwin's theories have continued into the 21st century. Creationists believe the biblical account of the earth's creation as it appears in the book of Genesis. Some acknowledge microevolution (changes in a species over time in response to natural selection), but they generally do not believe in speciation (that one species can beget or become another over time). There are various gradations of creationist beliefs, but all reject evolution and its argument that the interaction of natural selection and environmental factors explain the diversity of life on the earth.

At the core of the conflict is the observation that evolution threatens the view that human beings have a special place in the universe. Many creationists find it disturbing to contemplate the idea that human existence is a random occurrence, or that universal order is a chance occurrence rather than a response to a divine decree or plan.

In "Was Darwin Wrong?" (*National Geographic*, November 2004), David Quammen reports that despite seemingly overwhelming evidence, many Americans believe evolution is simply an unproven speculation rather than an explanatory statement that fits the evidence. Even though observation and experiment support evolutionary theory, its apparent contradiction with many religious tenets renders it unacceptable to those Americans who choose to believe that a higher power, and not evolution, produced human life on the earth.

Nearly Half of Americans Believe Humans Did Not Evolve

More than a century and a half after Darwin's publication of *On the Origin of Species by Means of Natural Selection*, his theory remains highly controversial. Scientists assert that evolution is well established by scientific evidence, but Gallup Organization surveys repeatedly reveal that a substantial portion of Americans do not believe that the theory of evolution best explains the origins of human life.

For example, in *Four in 10 Americans Believe in Strict Creationism* (December 17, 2010, http://www.gallup.com/poll/145286/Four-Americans-Believe-Strict-Creationism.aspx), Frank Newport of the Gallup Organization reports that in 2010, 40% of the U.S. population believed in creationism. (See Figure 3.1.) Newport explains that a belief in creationism is affected by

FIGURE 3.1

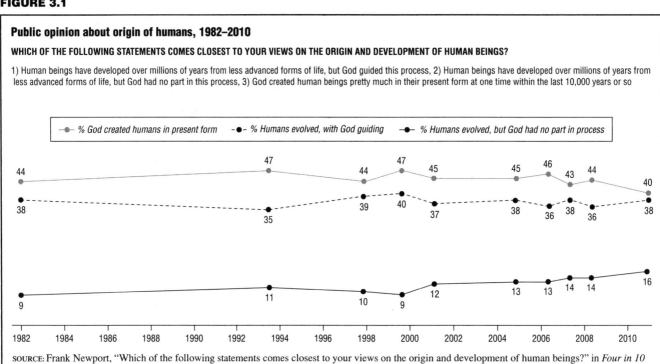

Public opinion about origin of humans, 1982–2010

WHICH OF THE FOLLOWING STATEMENTS COMES CLOSEST TO YOUR VIEWS ON THE ORIGIN AND DEVELOPMENT OF HUMAN BEINGS?

1) Human beings have developed over millions of years from less advanced forms of life, but God guided this process, 2) Human beings have developed over millions of years from less advanced forms of life, but God had no part in this process, 3) God created human beings pretty much in their present form at one time within the last 10,000 years or so

SOURCE: Frank Newport, "Which of the following statements comes closest to your views on the origin and development of human beings?" in *Four in 10 Americans Believe in Strict Creationism*, The Gallup Organization, December 2010, http://www.gallup.com/poll/145286/Four-Americans-Believe-Strict-Creationism.aspx (accessed January 13, 2011). Copyright © 2010 by The Gallup Organization. Reproduced by permission of The Gallup Organization.

TABLE 3.1

TABLE 3.2

Public opinion about belief in evolution, by educational attainment, 2010

WHICH OF THE FOLLOWING STATEMENTS COMES CLOSEST TO YOUR VIEWS ON THE ORIGIN AND DEVELOPMENT OF HUMAN BEINGS?

	Humans evolved, God guided process	Humans evolved, God had no part in process	God created humans in present form within last 10,000 years
	%	%	%
Postgraduate	49	25	22
College graduate	38	21	37
Some college	36	16	44
High school or less	34	9	47

SOURCE: Frank Newport, "Which of the following statement comes closest to your views on the origin and development of human beings?" in *Four in 10 Americans Believe in Strict Creationism*, The Gallup Organization, December 2010, http://www.gallup.com/poll/145286/Four-Americans-Believe-Strict-Creationism.aspx (accessed January 13, 2011). Copyright © 2010 by The Gallup Organization. Reproduced by permission of The Gallup Organization.

Public opinion about origin of humans, by church attendance, 2010

WHICH OF THE FOLLOWING STATEMENTS COMES CLOSEST TO YOUR VIEWS ON THE ORIGIN AND DEVELOPMENT OF HUMAN BEINGS?

	Humans evolved, God guided process	Humans evolved, God had no part in process	God created humans in present form within last 10,000 years
	%	%	%
Attend church weekly	31	2	60
Attend church almost every week/monthly	47	9	41
Attend church seldom/never	39	31	24

SOURCE: Frank Newport, "Which of the following statements comes closest to your views on the origin and development of human beings?" in *Four in 10 Americans Believe in Strict Creationism*, The Gallup Organization, December 2010, http://www.gallup.com/poll/145286/Four-Americans-Believe-Strict-Creationism.aspx (accessed January 13, 2011). Copyright © 2010 by The Gallup Organization. Reproduced by permission of The Gallup Organization.

educational attainment, religiosity as measured by church attendance, and political affiliation.

Table 3.1 reveals that a belief in evolution as a process in which God played no role increased in direct response to educational attainment, rising from just 9% of those with a high school education or less to 16% of people who attended college and to 21% of people with a college degree. Interestingly, more people believe that human evolution is a process guided by God. Nearly one-third (34%) of respondents with a high school education or less, 36% with some college, 38% with college degrees, and 49% with postgraduate degrees said that humans evolved under God's guidance.

As might be anticipated, Newport indicates that belief in the theory of evolution as a process in which God had no part was inversely related to church attendance. Just 2% of those who attend church weekly said they believe in this definition of evolution, compared with 31% of people who said they seldom or never attend church. (See Table 3.2.)

Views about the origins of human life also varied along political party lines, with more Republicans (52%) than Democrats (34%) or Independents (34%) believing that God created humans in their present form within the last 10,000 years. (See Table 3.3.) Just 8% of Republicans said humans evolved and God had no part in the process, compared with 20% of Democrats and 21% of Independents.

Interestingly, when the Gallup Organization asked survey respondents to identify the theory with which they associate Darwin, in 2009 more than one-third (34%)

TABLE 3.3

Public opinion about belief in evolution, by political party affiliation, 2010

WHICH OF THE FOLLOWING STATEMENTS COMES CLOSEST TO YOUR VIEWS ON THE ORIGIN AND DEVELOPMENT OF HUMAN BEINGS?

	Humans evolved, God guided process	Humans evolved, God had no part in process	God created humans in present form within last 10,000 years
	%	%	%
Republican	36	8	52
Independent	39	21	34
Democrat	40	20	34

SOURCE: Frank Newport, "Which of the following statements comes closest to your views on the origin and development of human beings?" in *Four in 10 Americans Believe in Strict Creationism*, The Gallup Organization, December 2010, http://www.gallup.com/poll/145286/Four-Americans-Believe-Strict-Creationism.aspx (accessed January 13, 2011). Copyright © 2010 by The Gallup Organization. Reproduced by permission of The Gallup Organization.

were unsure or did not know and an additional 10% gave an incorrect response. (See Figure 3.2.) Even though correctly identifying the theory of evolution as Darwin's was related to the likelihood of believing in the theory, 29% of people who attributed the theory of evolution to Darwin reported that they did not believe in evolution. (See Figure 3.3.)

According to Jon D. Miller, Eugenie C. Scott, and Shinji Okamoto, in "Science Communication: Public Acceptance of Evolution" (*Science*, vol. 313, no. 5788, August 11, 2006), between 1985 and 2002 people in the United States and in 32 European countries were asked whether they considered the statement, "Human beings, as we know them, developed from earlier species of animals," to be true or false. Of the nations surveyed, the United States had the second-highest percentage of adults who said the statement was false and the second-lowest percentage of adults who said the statement was true. The researchers conclude that "the acceptance of evolution is lower in the United States than in Japan or Europe, largely because of widespread fundamentalism and the politicization of science in the United States."

In "God or Darwin? The World in Evolution Beliefs" (*Guardian* [London, England], July 1, 2009), Simon Rogers reports on a 2009 poll that was conducted by the British Council and Ipsos MORI, a market research company. The poll, which surveyed over 10,000 adults in 10 countries, found that the percentages of people who had heard of Darwin ranged from highs of 93% in Russia, 91% in both Great Britain and Mexico, 90% in China, 86% in Argentina, and 84% in the United States to lows of 27% in South Africa, 38% in Egypt, and 62% in India. More than half of the survey respondents in China (55%), Mexico (52%), and Great Britain (51%) agreed that scientific evidence for evolution exists, compared with just 39% of the survey respondents in Spain, 33% in the United States, and 8% in both Egypt and South Africa. Interestingly, 85% of the survey respondents in India, 65% in Mexico, 62% in Argentina, and 54% in Great Britain, Russia, and South Africa said they thought it is possible to believe in God and evolution simultaneously. Just 53% of the survey respondents in the United States, 46% in Spain, 45% in Egypt, and 39% in China said it is possible to believe in God and evolution simultaneously.

Some States Move to Teach Creationism in Public Schools

Because less than half of Americans say they believe in evolution, it is not surprising that the decision about which theory—evolution or creationism—should be taught in public schools has been hotly contested. The American Institute of Biological Sciences, a nonprofit organization that is dedicated to advancing biological research and education, reports in "AIBS State News on Teaching Evolu-

FIGURE 3.2

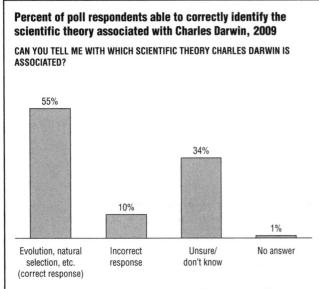

Percent of poll respondents able to correctly identify the scientific theory associated with Charles Darwin, 2009

CAN YOU TELL ME WITH WHICH SCIENTIFIC THEORY CHARLES DARWIN IS ASSOCIATED?

SOURCE: Frank Newport, "Can you tell me with which scientific theory Charles Darwin is associated?" in *On Darwin's Birthday, Only 4 in 10 Believe in Evolution*, The Gallup Organization, February 2009, http://www.gallup.com/poll/114544/Darwin-Birthday-Believe-Evolution.aspx (accessed September 8, 2010). Copyright © 2010 by The Gallup Organization. Reproduced by permission of The Gallup Organization.

FIGURE 3.3

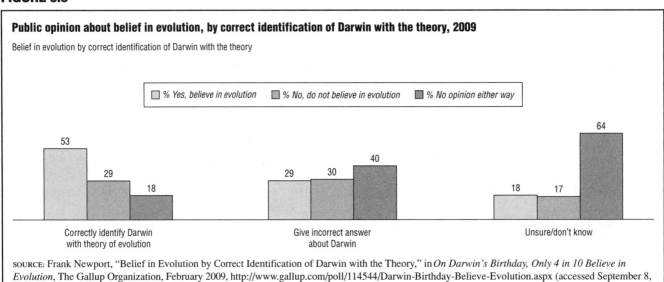

Public opinion about belief in evolution, by correct identification of Darwin with the theory, 2009

Belief in evolution by correct identification of Darwin with the theory

SOURCE: Frank Newport, "Belief in Evolution by Correct Identification of Darwin with the Theory," in *On Darwin's Birthday, Only 4 in 10 Believe in Evolution*, The Gallup Organization, February 2009, http://www.gallup.com/poll/114544/Darwin-Birthday-Believe-Evolution.aspx (accessed September 8, 2010). Copyright © 2010 by The Gallup Organization. Reproduced by permission of The Gallup Organization.

tion" (December 2010, http://www.aibs.org/public-policy/evolution_state_news.html) that during 2009 and 2010 there were active school board controversies over this issue in at least five states: California, Louisiana, Ohio, South Carolina, and Texas. The issue remained unresolved in other states, as debates favoring and opposing antievolution and anti-creationism legislation continued.

Some opponents of evolutionary theory propose an alternate theory known as intelligent design (ID), an explanation that credits intelligence, rather than an undirected process such as natural selection, as the source of life on the earth. ID proponents disagree with a basic tenet of evolutionary theory: Darwin's claim that the complex design of biological systems resulted by chance. They contend that direction from an intelligent designer (a supernatural being) is necessary to explain adequately the origins and complexity of life on the earth, particularly human life.

Throughout the United States school boards have considered whether they want to teach students evolution, creationism, or alternatives such as ID. In 2005, 80 years after the Scopes trial, the school board in Dover, Pennsylvania, became the first in the nation to require that students be taught an alternative explanation to evolution. By 2006 this mandate had been overturned, but its brief success emboldened other states to follow suit. For example, in 2009 Louisiana amended the Louisiana Science Education Act to permit school districts and teachers to present information about creationism and ID in science classes. As of January 2011, no state school board had mandated the teaching of ID exclusively; however, many school boards had moved to include it in the science curricula.

Even some staunch advocates of evolutionary theory do not necessarily wish to exclude the teaching of creationism or ID—they simply prefer that these alternative explanations be taught in the context of religion classes rather than in science classes. There are even those who advocate both theories. The Vatican has stated that it does not consider evolution to be in conflict with the Christian faith. Francis S. Collins (1950–), a committed Christian who served as the director of the Human Genome Institute at the National Institutes of Health until August 2008, has repeatedly expressed his view that God created the universe and chose the remarkable mechanism of evolution to create plants, animals, and humans.

In *Science, Evolution, and Creationism* (2008, http://www.nap.edu/catalog.php?record_id=11876#toc), the National Academy of Sciences and the Institute of Medicine of the National Academies, Science, Evolution, and Creationism offer guidance for school boards, educators, and the courts. The authors not only present the overwhelming evidence that supports biological evolution and assess the alternative views offered by advocates of creationism but also suggest that science and religion should

be viewed as different ways of understanding the world rather than as constructs that are in conflict with one another. They also assert that the evidence for evolution should not be perceived as opposing, but as completely compatible with religious beliefs. They do, however, contend that nonscientific explanations should not be presented in science classrooms and advise against presenting creationist ideas in science class to prevent students from becoming confused about the distinctions between science and religion.

Randy Moore and Sehoya Cotner of the University of Minnesota, Minneapolis, indicate in "The Creationist down the Hall: Does It Matter When Teachers Teach Creationism?" (*BioScience*, vol. 59, no. 5, May 2009) that they surveyed over 1,000 students entering introductory biology classes at the Minneapolis campus of the University of Minnesota. They wanted to answer the following questions:

- Are incoming biology majors more or less likely to accept evolution or creationism than are nonmajors?

- Are incoming biology majors more likely to have been taught evolution or creationism in their high school biology courses than were nonmajors?

- How are biology majors' and nonmajors' views of evolution and creationism associated, if at all, with the treatment of these topics in their high school biology courses?

Moore and Cotner find that nonmajors were twice as likely as biology majors to accept the claim that the theory of evolution is speculative and not scientific. The researchers also find that independent of a student's major, those whose high school biology class included instruction on creationism (with or without evolution) were more likely to accept creationism-based responses than were students who were only taught about evolution in high school biology. Similarly, students taught evolution in high school biology and not creationism were more likely to accept evolution-based statements than were students who did not learn about evolution in high school biology. Moore and Cotner also observe that "creationism remains popular among high school biology teachers," with 29% of biology majors reporting that their high school biology classes included both evolution and creationism.

VARIATION AND ADAPTATION

Effective adaptations and variations are perpetuated in a species and tend to be incorporated into the normal or predominant phenotype for most individuals in the species. Variation persists, but it ranges around an evolutionarily determined norm. This is called adaptive radiation. For example, over time Darwin's finches developed beaks best suited to their functions. The finches that eat grubs have long, thin beaks to enter holes in the ground and pull out the grubs. Finches that eat buds and fruit have clawlike beaks to grind their food, giving them a survival advantage in environs where buds are the only available food source. In another example of adaptive radiation, present-day giraffes have necks of varying lengths, but most tend to be long.

Ancient humans underwent many evolutionary changes. An example of adaptive radiation in humans is the development and refinement of the upper limbs to perform fine motor skills necessary to make and use complex tools. Another adaptation is the quantity of the pigment melanin present in the skin. People from areas near the equator have more melanin, an adaptation that darkens their skin and protects them from the sun. Anatomical structure also appears to have responded to the environment. Those who thrive in colder climates produce offspring that are shorter and broader than those who live in warmer climates. This adaptive stature, with its relatively low surface area, enables those people to conserve rather than lose body heat. For example, native Alaskans, who are generally of short stature, are well suited to their cold climate.

Natural selection does not produce uniformity or perfection. Instead, it generates variability that persists when it helps a species to adapt to, and thrive in, its environment. It acts on outward appearance (phenotypes), not on internal coding (genotypes), and it is not a process that always discards individual genes in favor of others that might produce traits better suited for survival. For traits attributable to multiple genes, many different combinations of gene pairs may produce the same or comparable phenotypes. Multiple phenotypes may be neutral or even beneficial in terms of survival in a given environment, and there is no reason for such variation to be eliminated by natural selection.

Furthermore, natural selection does not completely or rapidly eliminate genes that produce traits unsuited for adaptation or survival. Even though some individuals with harmful traits die young or do not reproduce, some do reproduce and pass their genes and traits to the next generation. Culling out these genes may require several generations. In other instances, seemingly harmful genes may be retained in the gene pool because there may be circumstances or environments in which their presence would improve survival. For example, the recessive allele that causes sickle-cell diseases (sickle-cell anemia and sickle B-thalassemia, in which the red blood cells contain abnormal hemoglobin) may have had a role in survival in some parts of the world. People who are pure recessive for this trait become ill and die prematurely, but those who are hybrid for the trait may have retained a survival advantage in areas where malaria is present because people who have the sickle-cell trait or anemia are immune to the effects of malaria. This phenomenon is called balanced polymorphism and is an example of the seemingly counterintuitive actions of natural selection. Even though the sickle-cell trait is not beneficial on its own, historically it was advantageous in areas where malaria was a greater threat to survival than sickle-cell diseases.

By definition, natural selection is an unending, continuous process. The popular understanding of natural selection as "survival of the fittest" is somewhat misleading because organisms and species with phenotypes most suited to survive in their environments are not necessarily the "fittest." Some examples of natural selection include the evolution of bacteria that are antibiotic-resistant and insects that resist extermination with pesticides.

Increasing and Decreasing Genetic Variation

A gene pool encompasses the alleles for all the genes in a population. Gene pools in natural populations contain considerable variation, and for evolution to proceed there must be mechanisms to create and increase genetic variation. Mutation (a change in a gene) serves to create or increase genetic variation. Recombination, a process that creates new alleles and new combinations of alleles, also increases genetic variation. Another way for new alleles to enter a gene pool is by migrating from another population. In closely related species new organisms can enter a population, mate within it, and produce fertile hybrids. This action is known as gene flow.

The tendency toward increased genetic variation within a population is balanced by other mechanisms that act to decrease it. For example, some variations have the effect of limiting an organism's ability to reproduce. Any such variations, as well as those incidentally paired with them, will be nonadaptive because they will have a lower probability of being passed to the next generation. Under such circumstances, the process of natural selection serves to reduce genetic variation. In fact, differences in reproductive capability are often called natural selection. Whenever natural selection weeds out nonadaptive alleles by limiting an organism's reproductive capability, it is depleting genetic variation within the population.

Genetic Drift

Other evolutionary mechanisms contribute to genetic variation. Genetic drift is random change in the genetic composition of a population. It may occur when two groups of a species are separated and as a result cannot reproduce with each other. The gene pool of these groups will naturally differ over time. When the two groups are reunited and reproduce, gene migration occurs as their genetic differences combine, serving to increase genetic diversity. If they remain separate, their genetic differences may become so great that they develop into two separate species.

In small populations genetic drift can cause relatively rapid change because each individual's alleles constitute a large proportion of the gene pool, and when an individual does not reproduce, the results are felt more acutely in smaller, rather than larger, populations. When small populations are affected by genetic drift, they may suffer a loss of valuable diversity. This is why scientists and others involved in conservation, such as zoo curators of endangered species, make every effort to ensure that populations are large enough to withstand the effects of genetic drift.

MUTATION

Variation is the essence of life, and mutations are the source of all genetic variation. A staggering number and variety of alterations, rearrangements, and duplications of genetic material have occurred since the first living cells, in which there is an incredible range of life forms, from amoebas and fruit flies to whales and humans. These dramatically different life forms were all produced using genetic material that was present and reproduced from the first living cells. They resulted from the process of mutation, an alteration in genetic material.

Perhaps because of their negative depiction in science fiction and horror films, mutations are widely considered to be harmful, dramatic, and deleterious (harmful in subtle or unexpected ways). Most mutations are not harmful, and the same limited number of mutations has probably been recurring in each species for millions of years. Many helpful mutations have already been incorporated into the normal genotype through natural selection. Harmful mutations do occur, and while they tend to be eliminated through natural selection, they do recur randomly. It is a mistake, however, to consider mutation as sudden or exclusively harmful. Mutation cannot generate sudden, drastic changes—it requires many generations to generalize throughout a species population. Furthermore, if the changes caused by mutation did not favor survival, natural selection would work against their generalization throughout the species population. In the absence of mutation, there would have been no development of life and no evolution would have occurred.

The effects of mutation vary greatly. Even though mutations are changes in genetic material, they do not necessarily affect an individual organism's phenotype. When phenotype is affected, it is because the code for protein synthesis has been changed. Whether mutation will affect phenotype and the extent to which it will be influenced depends on how protein manufacture is affected, when and where the mutation occurs, and the complexity of the genetic controls governing the selected trait.

When a trait is governed by the interaction of many genes, with each exerting about the same influence, the effects of a mutation might be negligible. In traits controlled by a single pair of genes, mutation is likely to exert a much greater influence. For traits governed by the interaction between a major controlling gene pair and several other less influential gene pairs, the location of the mutation will determine its impact. Other considerations such as whether the mutated gene is dominant or recessive also determine whether it will directly act on an individual's phenotype—a mutated recessive gene might not appear in the phenotype for several generations.

Mutation can occur in any cell in an organism's body, but only germinal mutations (those that affect the cells that give rise to sperm or eggs) are passed to the next generation. When mutation occurs in somatic cells in the body such as muscle, liver, or brain cells, only cells that derive from mitotic division of the affected cell will contain the mutation. Even though mutations in somatic cells can cause disease, most do not have a significant impact, because if the mutated gene is not involved in the specialized function of the affected cell, these "silent mutations" will not be detected. Furthermore, mutations that appear only in somatic cells disappear when the organism dies; they are not passed on to subsequent generations and do not enter the gene pool that is the source of genetic variation for the species.

Another reason that many mutations are not expressed in the phenotype is that they affect only one copy of the gene, leaving diploid organisms with an intact copy of the gene. These types of mutations have a recessive inheritance pattern and do not affect phenotype unless an individual inherits two copies of the mutation. There is also the question of the probability of a mutation in a gamete affecting offspring. The mutation may occur in a single gamete and so may only be passed on if that particular gamete is involved in conception. For example, human semen contains more than 50 million sperm per ejaculate, so it is unlikely that a mutation carried by a single sperm will be passed on. When mutation occurs during embryonic (before birth) development and all gametes are affected, there is a greater chance that it might influence the phenotype of future generations. Alternatively, if, like many mutations, the gene defect occurs with advancing age, and the affected individuals are beyond their reproductive years, then it will have no impact on future generations.

How Different Types of Genetic Mutations Occur

Mutation is a normal and fairly frequent occurrence, and the opportunity for a mutation to take place exists every time a cell replicates. In general, the cells that divide many times throughout the course of an organism's life (e.g., skin cells, bone marrow cells, and the cells that line the intestines) are at greater risk for mutation than those that divide less frequently (e.g., adult brain and muscle cells).

DNA nearly always reproduces itself accurately, but even a minor alteration produces a mutation that may alter a protein, prevent its production, or have no effect at all. There are five broad classes, and within them many varieties, of mutations, and each is named for the error or action that causes it:

- Point mutations are substitutions, deletions, or insertions in the sequence of DNA bases in a gene. The most common point mutation in mammals is called a base substitution and occurs when an A-T pair replaces a G-C pair. Base substitutions are further classified as either transitions or transversions. Transitions occur when one pyrimidine (C or T) is substituted for the other and one purine (A or G) is substituted on the other strand of DNA. Transversions occur when a purine replaces a pyrimidine. Sickle-cell anemia results from a transversion in which T replaces A in the gene for a component of hemoglobin.

- Structural chromosomal aberrations occur when the DNA in chromosomes is broken. The broken ends may remain loose or join those occurring at another break to form new combinations of genes. When movement of a chromosome section from one chromosome to another takes place, it is called translocation. Translocation between human chromosomes 8 and 21 has been implicated in the development of a specific type of leukemia (cancer of the white blood cells). It has also been shown to cause infertility (inability to sexually reproduce) by hindering the distribution of chromosomes during meiosis.

- Numerical chromosomal aberrations are changes in the number of chromosomes. In a duplication mutation, genes are copied, so the new chromosome contains all of its original genes plus the duplicated one. Polyploidy is a numerical chromosomal aberration in which the entire genome has been duplicated and an individual who is normally diploid (having two of each chromosome) becomes tetraploid (containing four of each chromosome). Polyploidy is responsible for the creation of thousands of new species because it increases genetic diversity and produces species that are bigger, stronger, and more able to resist disease.

- Aneuploidy refers to occasions when just one or a few chromosomes are involved and generally describes the loss of a chromosome. Examples of aneuploidy are Down syndrome (multiple, characteristic physical and cognitive disabilities), in which there is an extra chromosome 21 (usually caused by an error in cell division called nondisjunction), and Turner's syndrome, in which there is only one X chromosome. Common characteristics of Turner's syndrome include short stature and lack of ovarian development as well as increased risk of cardiovascular problems, kidney and thyroid problems, skeletal disorders such as scoliosis (curvature of the spine) or dislocated hips, and hearing and ear disturbances.

- Transposon-induced mutations involve sections of DNA that copy and insert themselves into new locations on the genome. Transposons usually disrupt and inactivate gene function. In humans selected types of hemophilia have been linked to transposon-induced mutations.

Frequency and Causes of Mutation

In the absence of external environmental influences, mutations occur rarely and are seldom expressed because many forms of mutation are expressed by a recessive allele. The most common naturally occurring mutations arise simply as accidents. Susceptibility to mutation varies during the life cycle of an organism. For example, among humans mutation of egg cells increases with advancing age—the older the mother the more likely she is to carry gametes with mutations. Susceptibility also varies among members of a species such as humans based on their geographic location and ethnic origin.

The mutation rate is the frequency of new mutations per generation in an organism or a species. Mutation rates vary widely from one gene to another within an organism and between organisms. The mutation rate for bacteria is 1 per 100 million genes per generation. Despite this relatively low rate, the enormous number of bacteria (there are more than 20 billion produced in the human intestines each day) translates into millions of new mutations to the bacteria population every day. The human mutation rate is estimated at 1 per 10,000 genes per generation, and human mutation rates are comparable throughout the world. The only exceptions are populations that have been exposed to factors known as mutagens that cause a change in DNA structure and as a result increase the mutation rate. With approximately 25,000 genes, a typical human contains three mutations. Considering the fact that human genes mutate approximately once every 30,000 to 50,000 times they are duplicated, and in view of the complexity of the gene replication process, it is surprising that so few "mistakes" are made.

Gene size, gene-base composition, and the organism's capacity to repair DNA damage are closely linked to how many mutations occur and remain in the genome. Larger genes are more susceptible than smaller ones because there are more opportunities and potential sites for mutation. Organisms better able to repair and restore DNA sequences, such as many kinds of yeast and bacteria, will be less likely to have high mutation rates.

The overwhelming majority of human mutations arise in the father, as opposed to the mother. When compared with egg cells, there are more opportunities for mutation during the many cell divisions needed to produce sperm. Sperm are produced late in the life of males, whereas females produce eggs earlier (during embryonic development) and are thought to be born with their full complement of eggs. Thus, the frequency of mutations that are passed to the next generation increases with parental age.

The mutation rate is partly under genetic control and is strongly influenced by exposure to environmental mutagens. Some mutagens act directly to alter DNA, and others act indirectly, by triggering chemical reactions in the cell that result in the breakage of a gene or a group of genes. Radiation or ultraviolet light (from natural sources such as the sun or radioactive material in the earth or from manmade sources such as X-rays) can significantly accelerate the rate of mutations.

The link between radiation and mutation was first identified during the 1920s by Hermann Joseph Muller (1890–1967), who was a student of Thomas Hunt Morgan and worked in the famous "Fly Room." Muller discovered that he could increase the mutation rate of fruit flies more than a hundredfold by exposing reproductive cells to high doses of radiation. His pioneering work prompted medical and dental professionals to exercise caution and minimize patients' exposure to radiation. Because mutagens in the reproductive cells are likely to affect heredity, special precautions such as donning a lead apron to block exposure are used when dental X-rays or other diagnostic imaging studies are performed. Similarly, special care is taken to prevent radiation exposure to the embryo or fetus because a mutation during this period of development when cells are rapidly proliferating might be incorporated in many cells and could result in birth defects.

MODERN SYNTHESIS OF EVOLUTIONARY GENETICS

Twenty-first-century theories of evolutionary genetics are indebted to Darwin for his groundbreaking descriptions of organisms, individuals, and speciation. Modern theory differs considerably in that it addresses evolutionary mechanisms at the level of populations, genes, and phenotypes and incorporates understanding of actions, such as genetic drift, that Darwin had not considered.

The modern synthesis of evolutionary theory differs from Darwinism by identifying mechanisms of evolution that act in concert with natural selection, such as random genetic drift. It asserts that characteristics are inherited as discrete entities called genes and that variation within a population results from the presence of multiple alleles of a gene. The most controversial tenet of modern evolutionary theory is its contention that speciation is usually the result of small, gradual, and incremental genetic changes. In other words, macroevolution is simply the cumulative effect of microevolution. Some evolutionary biologists have instead embraced the theory of punctuated equilibrium set forth by the American paleontologists Niles Eldredge (1943–) and Stephen Jay Gould (1941–2002) in 1972.

Punctuated equilibrium suggests that long periods of stasis (stability) are followed by rapid speciation. It posits that new species arose rapidly during a period of a few thousand years and then remained essentially unchanged for millions of years before the next period of adaptation. Punctuated equilibrium also proposes that change occurred in a small portion of the population, rather than uniformly throughout the population. Even though it is different from Darwin's theory of speciation, it is not inconsistent with natural selection; it simply presumes different mechanisms and timetables for the development of new species.

CHAPTER 4
GENETICS AND THE ENVIRONMENT

Nature prevails enormously over nurture when the differences in nurture do not exceed what is commonly to be found among persons of the same rank in society and in the same country.

—Francis Galton, "The History of Twins, as a Criterion of the Relative Powers of Nature and Nurture" (*Journal of the Anthropological Institute*, vol. 5, 1876)

Even though genetics clearly controls many of an organism's traits, it is simplistic and incorrect to assume that organisms, including humans, are completely defined by their genes. There are some phenotypes (observable traits) that are exclusively controlled by either genetics or environment, but most are influenced by a complex interaction of the two. Modern genetics study defines environment as every influence other than genetic—such as air, water, diet, radiation, and exposure to infection—and it subscribes to the overarching assumption that every trait of every organism is the product of some set of interactions between genes and the environment. Epigenetic changes, which are inherited factors that alter the physical properties of deoxyribonucleic acid (DNA) but do not change the DNA sequence of nucleotides, also influence how genes are expressed. (Figure 1.8 in Chapter 1 presents some epigenetic mechanisms.)

NATURE VERSUS NURTURE

The debate over the relative contributions of genetics and environment—nature versus nurture—remains unresolved in many fields of study, from education to animal behavior to human disease. Historically, scientists assumed opposing viewpoints, choosing to favor either nature or nurture, rather then exploring the ways in which both play critical and complementary roles. Nature proponents argued that human characteristics are uniquely and primarily conditioned by genetics, whereas nurture proponents asserted that environmental influences and experiences determined differences between individuals and populations.

Genes versus Environment

When researchers analyze the origins of disease, the terms used to describe causation are *genetic* versus *environmental*, but the issues are the same as those in the nature-versus-nurture debate. Conditions considered to be primarily genetic are ones in which the presence or absence of genetic mutations determines whether an individual or population will develop a disease, independent of environmental exposures or circumstances. A disease considered to be primarily environmental is one in which people of virtually any genetic background can develop the disease when they are exposed to the specific environmental factors that cause it.

Even the conditions and diseases once believed to be at either end of the continuum (caused by either purely genetic or purely environmental factors) may not be exclusively attributable to one or the other. For example, an automobile accident that results in an injury might be deemed entirely environmentally caused, but many geneticists would contend that risk-taking behaviors such as the propensity to exceed the speed limit are probably genetically mediated. Furthermore, the course and duration of rehabilitation and recovery from an injury or illness is also likely genetically influenced.

At the other end of the continuum are diseases believed to be predominantly genetic in origin, such as sickle-cell anemia. Even though this disease does not have an environmental cause, there are environmental triggers that may determine when and how seriously the disease will strike. For example, sickle-cell attacks are more likely when the body has an insufficient supply of oxygen, so people with the sickle-cell trait who live at high altitudes or those who engage in intense aerobic exercise may be at increase risk of attacks. There are many more conditions for which the risk of developing the disease is strongly influenced by both genetic and environmental factors. Multiple genes and environmental factors may be involved in causing a given condition and its expression.

Asthma: A Disease with Genetic and Environmental Causes

Asthma (a disorder of the lungs and airways that causes wheezing and other breathing problems) is a good example of how genes and the environment can interact to cause diseases. The scientific evidence that asthma had a genetic basis came from studies that were conducted between the 1970s and the 1990s, which found patterns of inheritance in families and genetic factors that were involved in the severity and triggers for asthma attacks. For example, in "Family Concordance of IgE, Atopy, and Disease" (*Journal of Allergy and Clinical Immunology*, vol. 73, no. 2, February 1984), Michael D. Lebowitz, Robert Barbee, and Belton Burrows looked at 350 families and found that in families where neither parent had asthma, 6% of the children had asthma; when one parent had the condition, 20% of the children were affected; and when both parents had the condition, 60% of the children were affected.

In "Genetic Factors in the Presence, Severity, and Triggers of Asthma" (*Archives of Disease in Childhood*, vol. 73, no. 2, August 1995), a study of pairs of twins in which at least one twin had asthma, Edward P. Sarafino and Jarrett Goldfedder of Trenton State College discovered that genetics and environment both made strong contributions to the development of the illness. If asthma was directed solely by genes, then 100% of the monozygotic (identical) twins, who are exactly the same genetically, would be expected to have asthma (the concordance rate is the rate of agreement, when both members of the pair of twins have the same trait). Instead, the researchers found that 59% of twins both had asthma. If asthma was entirely environmentally caused, then genes should make no difference at all—the concordance rate would be the same for monozygotic and dizygotic (fraternal, meaning nonidentical) twins. Sarafino and Goldfedder noted that the concordance rate of 59% was more than twice as high in monozygotic twins as in dizygotic twins (24%). This research offers clear confirmation that asthma has a significant genetic component.

Asthma is an example of a condition that is polygenic (controlled by more than one gene). Familial studies demonstrate that asthma does not conform to simple Mendelian patterns of inheritance and that multiple independent segregating genes are required for phenotypic expression (polygenic inheritance). Nevertheless, it has long been known that the complex phenotypes of asthma have a significant genetic contribution. Researchers speculate that several genes combine to increase susceptibility to asthma, and that there are also genes that lower susceptibility to developing the condition. Genes also affect asthma severity and the way people with asthma respond to various medications.

Environmental triggers for asthma include allergens such as air pollution, tobacco smoke, dust, and animal dander. Other environmental factors linked to its development are a diet high in salt, a history of lung infections, and the lack of siblings living at home. Researchers theorize that younger children in families with older siblings are less likely to develop asthma because their early exposures to foreign substances (such as dirt and germs) brought into the home by older siblings heighten their immune system. The enhanced immune response is believed to have a protective effect against asthma and other illnesses.

Michael E. Wechsler and Elliot Israel acknowledge in "The Genetics of Asthma" (*Seminars in Respiratory and Critical Care Medicine*, vol. 23, no. 4, August 2002) some of the difficulties that researchers face when they study a complex disease such as asthma. Wechsler and Israel describe the following challenges faced by geneticists seeking to pinpoint the genetics of asthma:

- Population studies suggest that asthma is a polygenic disease, with many diverse locations of possible asthma genes already identified.

- The definition of asthma can vary from one health care practitioner to another. Some make the diagnosis based on changes in airway reactivity, others on levels of airway function, and still others on clinical symptoms. As a result, phenotyping methods must be examined and carefully compared because the variety of clinical symptoms a patient with asthma may present, such as coughing, having shortness of breath, wheezing, and experiencing a tightness in the chest, are also common to several other conditions such as bronchitis or heart failure, and confusion may result in misdiagnosis.

- Even though a clinical history of asthma symptoms or phenotypes often suggests a diagnosis of asthma, there is no definitive, specific definition that classifies an individual as having or not having asthma. As a result, some people may be incorrectly labeled or identified, potentially yielding false data or nonreplicable results.

- Epidemiological (the study of the spread of disease in a population) studies conducted during the late 1990s suggest that asthma may have many different phenotypic expressions at different ages as assessed by their risk factors and prognosis. For example, children under the age of six years who at various times experience wheezing are labeled as asthmatic. However, most of these children do not continue to have asthma symptoms as they age—they seem to outgrow the condition.

- Asthma differs in terms of severity (mild, moderate, or severe and intermittent or persistent), suggesting different genetic or environmental influences and triggers. There are also several different subgroups of asthma patients, including aspirin-sensitive asthmatics and exercise-induced asthmatics. Each of these variations may have a different biological mechanism that accounts for each individual's phenotype.

Wechsler and Israel conclude that even though a small percentage of cases of asthma may result from a single gene defect or a single environmental factor, asthma is a complex genetic disease that cannot be explained by single-gene models. In most instances it appears to result from the interaction of multiple genetic and environmental factors. Similar to other complex diseases such as diabetes and hypertension (high blood pressure), the complexity of asthma genetics may be characterized by the contribution of different genes and different environments in different populations. Population studies of asthma face challenges comparable to genetics studies of other common complex traits. They are complicated by genetic factors such as incomplete penetrance (transmission of disease genes without the appearance of the disease), genetic heterogeneity (mutations in any one of several genes that may result in similar phenotypes), epistasis (when the effects of multiple genes have a greater effect on phenotype than individual effects of single genes), polygenic inheritance (mutations in multiple genes simultaneously producing the affected phenotype), and gene-environment interactions.

Donata Vercelli of the University of Arizona, Tucson, reports in "Gene-Environment Interactions: The Road Less Traveled by in Asthma Genetics" (*Journal of Allergy and Clinical Immunology*, vol. 123, no. 1, January 2009) that the genetics of asthma made great strides in 2007 with the publication of the results of the first genome-wide association study (GWAS) of childhood asthma. GWASs look at sets of single nucleotide polymorphisms (SNPs) throughout the genome to identify the most frequently occurring genetic variants that are associated with a disease and the heritable traits that are associated with a risk of developing the disease. Figure 4.1 shows how a GWAS compares DNA from people with a disease and those without by looking for SNP markers of genetic variation. The GWAS for childhood asthma found that variants of chromosome 17q21 were associated with an increased risk of asthma. Vercelli also observes that it is now well understood that the environment strongly influences the effect of genetic variants and suggests that there are gene-environment interactions that have not been identified.

FIGURE 4.1

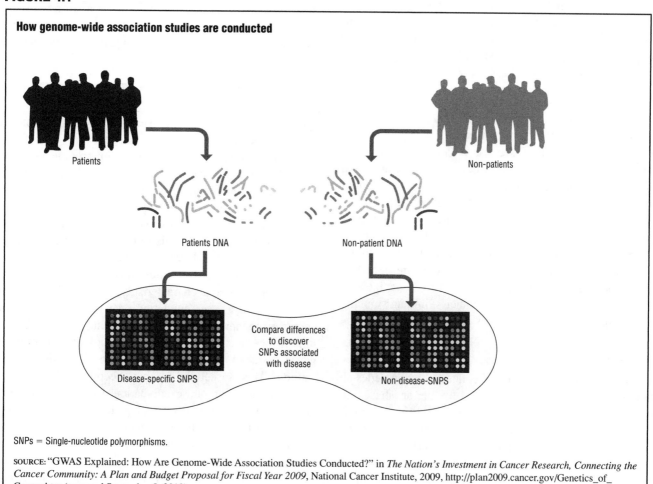

How genome-wide association studies are conducted

Patients

Non-patients

Patients DNA

Non-patient DNA

Disease-specific SNPS

Compare differences to discover SNPs associated with disease

Non-disease-SNPS

SNPs = Single-nucleotide polymorphisms.

SOURCE: "GWAS Explained: How Are Genome-Wide Association Studies Conducted?" in *The Nation's Investment in Cancer Research, Connecting the Cancer Community: A Plan and Budget Proposal for Fiscal Year 2009*, National Cancer Institute, 2009, http://plan2009.cancer.gov/Genetics_of_Cancer.htm (accessed September 8, 2010)

In "Genes in Asthma: New Genes and New Ways" (*Current Opinion in Allergy and Clinical Immunology*, vol. 8, no. 5, October 2008), Miriam F. Moffatt of Imperial College notes that the identification of two new asthma susceptibility genes, ORMDL3 and CHI3L1, in 2008 enhanced understanding of how genetic and environmental factors interact to produce asthma. Genes (ADRB2, IL13, and IL4) located on chromosome 5q and a gene (ORMDL3) on chromosome 17 appear to influence the development of childhood asthma.

Shuk-Mei Ho of the University of Cincinnati College of Medicine observes in "Environmental Epigenetics of Asthma: An Update" (*Journal of Allergy and Clinical Immunology*, vol. 126, no. 3, September 2010) that asthma is influenced by a complex interplay of genetic and environmental factors now known to be mediated by epigenetics. Ho states that asthma risk is influenced by genetics: having one parent with asthma doubles a child's risk of asthma and having two affected parents increases the risk fourfold. Furthermore, multiple studies of the genetic association of asthma have identified 43 asthma genes. Nonetheless, Ho opines that genetics alone is insufficient to fully explain the dramatic increase in the prevalence of asthma during the past three decades and the marked increase in the frequency of occupational asthma. Ho posits that these increases are attributable to gene-environment interactions that are mediated by epigenetic mechanisms and that epigenetics regulate the immune responses that are associated with asthma.

For example, living in a developed country is a strong risk factor for asthma. Ho explains that this increased risk may be in part due to outdoor air pollutants that have been shown to trigger asthma such as environmental tobacco smoke, diesel exhaust particles, pollens, molds, and viral and bacterial pathogens (disease-causing agents); to indoor pollutants; or to other environmental exposures such as diet. According to Ho, researchers are expanding their understanding of the epigenetic actions of some of these environmental risk factors. For example, DNA methylation (the process in which methyl groups are added to nucleotides in genomic DNA) may be the epigenetic mechanism that explains how a mother's lifelong smoking may influence her unborn child's asthma risk.

Genetic Susceptibility

Genetic susceptibility is the concept that the genes an individual inherits affect how likely he or she is to develop a particular condition or disease. When an individual is genetically susceptible to a particular disease, his or her risk of developing the disease is higher. Genetic susceptibility interacts with environmental factors to produce disease, but genes and environment do not necessarily make equal contributions to causation. Figure 4.2 shows how genes and environment interact to

FIGURE 4.2

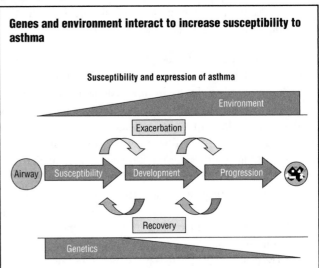

Genes and environment interact to increase susceptibility to asthma

SOURCE: George D. Leikauf, "Figure 1. Susceptibility and Expression of Asthma," in "Hazardous Air Pollutants and Asthma," *Environmental Health Perspectives Supplements*, vol. 110, no. S4, August 2002, http://ehp.niehs.nih.gov/members/2002/suppl-4/505-526leikauf/leikauf-full.html (accessed September 8, 2010)

increase the susceptibility, development, and progression of asthma. Genes can cause a slight or a strong susceptibility. When the genetic contribution is weak, the environmental influence must be strong to produce disease, and vice versa.

In most instances a susceptibility gene strongly influences the risk of developing a disease only in response to a specific environmental exposure. If the environmental exposure occurs infrequently, the gene will be of low penetrance, and it may seem that the environmental exposure is the primary cause of the disease, even though the gene is required for developing the disease. For this reason, even when environmental agents are suspected to be a major cause of a particular disease, there is still the possibility that genetic factors also play a major part, particularly genetic mutations with low penetrance. Similarly, a critical mix of nature and nurture is likely to determine individual traits and characteristics. Genetic factors may be considered as the foundation on which environmental agents exert their influence. For example, individuals who inherit a certain array of multiple alleles of susceptibility genes have an increased risk from birth of developing asthma. This susceptibility may become evident when initial sensitization and exacerbation occur in early childhood and when immunity develops to allergens. Whether asthma diminishes in frequency and severity or progresses depends largely on each individual's environmental exposures. Based on this premise, it is now widely accepted that even though certain environmental factors alone and certain genetic factors alone may explain the origins of some traits and diseases, most of the time the interaction of both genetic and environmental factors will be required for their expression.

Since the 1990s researchers have identified more and more genes that influence an individual's susceptibility to disease. Scientists have already linked specific DNA variations with increased risk of common diseases and conditions, including cancer, diabetes, hypertension, and Alzheimer's disease (a progressive neurological disease that causes impaired thinking, memory, and behavior). The questions that persist in the genes-versus-environment debate no longer focus on whether a particular trait or disease is caused exclusively by a specific gene. Instead, researchers continue to explore the extent to which genes and environment influence the development of specific traits, especially conditions that are linked to health and susceptibility to disease.

Examples of the kinds of questions about susceptibility that geneticists and other medical researchers seek to answer include:

- Why do some smokers live long, healthy lives, whereas others develop lung cancer and die early?

- Why do some people who are repeatedly exposed to the human immunodeficiency virus (HIV), the virus that causes acquired immunodeficiency syndrome (AIDS), resist contracting the virus?

- Why do common allergens (substances that cause allergies) cause moderate discomfort for some people and life-threatening asthma for others?

TWIN AND FAMILIAL STUDIES

The British anthropologist Francis Galton (1822–1911), a cousin of Charles Darwin (1809–1882), conducted some of the first reported twin studies. Advancing his cousin's theory of evolution, he performed twin studies in 1876 to investigate the extent to which the similarity of twins changes over the course of development. He is considered the originator of the field of medical genetics because of his considerable contributions to the nature-versus-nurture debate and the research methods he developed for evaluating heritability. Genetic determination is the combination of genes that creates a trait or characteristic, and heritability is what causes differences in those characteristics.

Twin studies, meaning comparisons of monozygotic twins to dizygotic twins, are performed to estimate the relative contributions of genes and environment—that is, the extent to which genetic-versus-environmental influences operate on specific traits. Twin studies also help determine the proportion of the variability in a trait that might be due to genetic factors. The studies aim to identify the causes of familial resemblance by comparing the concordance rates of monozygotic and dizygotic twins. Monozygotic twins share the same genetic material (they have 100% of their genes in common) and dizygotic twins share only half their genetic material (they have 50% of their genes in common). Most twin studies report their results in terms of pairwise concordance rates, which measure the number of pairs, or probandwise concordance rates, which count the number of individuals.

Monozygotic twins serve as excellent subjects for controlled experiments because they share prenatal environments, and those reared together also share common family, social, and cultural environments. Furthermore, twin studies can point to hereditary effects and estimate heritability, a term that describes the magnitude of the genetic effect. The limitations of such studies include the potential to confuse or mistakenly attribute to genetics the divergent impact of environmental influences. In addition, in some studies it has been difficult to control for other potential causes or sources of variation.

Some of the most conclusive twin study research has analyzed monozygotic and dizygotic twins who were raised apart. Researchers have sought to establish whether characteristics such as personality traits, aptitudes, and occupational preferences are the products of nature or nurture. Similar characteristics among monozygotic twins reared apart might indicate that their genes played a major role in developing that trait. Different characteristics might indicate the opposite—that environmental influences assume a much stronger role. By comparing monozygotic and dizygotic twins, investigators can test their hypotheses and confirm the findings of earlier research. For example, if monozygotic twins raised in different homes have many similarities, but dizygotic twins raised apart have little in common, researchers may conclude that genes are more important than environment in determining specific characteristics, traits, susceptibilities, and diseases.

In "Environmental and Heritable Factors in the Causation of Cancer—Analyses of Cohorts of Twins from Sweden, Denmark, and Finland" (*New England Journal of Medicine*, vol. 343, no. 2, July 13, 2000), Paul Lichtenstein et al. indicate that they examined nearly 45,000 pairs of twins to determine if the likelihood of developing certain kinds of cancers is more closely linked to environmental exposures than genetics. The results of this large-scale study reveal that environmental factors are linked to twice as many cancers as genetic factors. In fact, the risk of developing only three types of cancer (albeit some of the most common cancers)—breast, colorectal, and prostate cancers—show a significant genetic correlation.

Prostate cancer is found to have the strongest genetic link, with 42% of risk explained by genetic factors and 58% by environmental factors. The other cancers with a demonstrable genetic link, breast and colorectal cancers, are found to have less than a 35% link to genetics.

Lichtenstein et al. conclude that inherited genetic factors make a minor contribution to susceptibility to most types of cancer.

In "Breast Tissue Composition and Susceptibility to Breast Cancer" (*Journal of the National Cancer Institute*, vol. 102, no. 16, August 18, 2010), Norman F. Boyd et al. explain that breast density is consistently associated with breast cancer risk, more strongly than most other risk factors for the disease, and assert that high breast density may account for a substantial fraction of breast cancer. Boyd et al. observe that twin studies demonstrate that breast density is a highly heritable trait and posit that the risk of breast cancer may increase when women with extensive breast density are exposed to mitogens (substances that induce mitosis and cell transformation) and mutagens (substances that cause changes in DNA such as radioactive compounds, X-rays, ultraviolet radiation, and certain chemicals).

Twin and familial studies have also identified genetic factors that influence susceptibility to mental illnesses such as bipolar disorder (a serious mood disorder in which people alternate between periods of mania and depression). In "Genetics of Bipolar Disorder: Successful Start to a Long Journey" (*Trends in Genetics*, vol. 25, no. 2, February 2009), Nick Craddock and Pamela Sklar report that GWAS has helped identify chromosome regions and specific genes that are strongly associated with the susceptibility to bipolar disorder.

Epigenetic Changes Pass from One Generation to the Next

There is also emerging evidence that exposure to traumatic experiences (especially those that increase the risk of developing emotional and behavioral disorders) early in life may be transmitted from one generation to the next. Tamara B. Franklin et al. discuss in "Epigenetic Transmission of the Impact of Early Stress across Generations" (*Biological Psychiatry*, vol. 68, no. 5, September 1, 2010) how they exposed mice to the stress of separation from their mothers during the first 14 days of their lives. As adults, the mice not only displayed symptoms of anxiety such as irritability and jumpiness but also had altered gene function when compared with mice that had not experienced stress early in life. Furthermore, these mice passed on these genetic changes to their offspring, and even though the offspring were reared normally and did not experience the stress that their parents had experienced, they displayed the same symptoms of anxiety. Franklin et al. conclude that "these findings highlight the negative impact of early stress on behavioral responses across generations and on the regulation of DNA methylation in the germline."

Do Genes Govern Sexual Orientation?

The role of genetics in establishing sexual orientation (the degree of sexual attraction to men or women) and its link to homosexuality have been hotly debated in the relevant scientific literature and the media. Studies of monozygotic twins reveal that sexual orientation, like the overwhelming majority of human traits and characteristics, is not exclusively governed by genetics, but is more likely the result of a gene-environment interaction. For example, if homosexuality was exclusively controlled by genes, then either both members of a set of monozygotic twins would be homosexual or neither would be. Multiple studies show that if one twin is homosexual his or her sibling is also homosexual less than 40% of the time.

J. Michael Bailey, Michael P. Dunne, and Nicholas G. Martin systematically evaluated gender identity and sexual orientation of twins and reported their findings in "Genetic and Environmental Influences on Sexual Orientation and Its Correlates in an Australian Twin Sample" (*Journal of Personality and Social Psychology*, vol. 78, no. 3, March 2000). The researchers observe that both male and female homosexuality appears to run in families and that studies of unseparated twins suggest that this is primarily because of genetic rather than familial environmental influences. They also observe that previous research suffers from limitations such as recruiting subjects via publications aimed at homosexuals or by word of mouth—strategies likely to bias the samples and results.

To overcome these limitations, Bailey, Dunne, and Martin assessed twins from the Australian Twin Registry rather than recruiting twins especially for the purpose of their research. Using probandwise concordance (an estimate of the probability that a twin is nonheterosexual given that his or her co-twin is nonheterosexual), they find lower rates of twin concordance for nonheterosexual orientation than in previous studies. The most striking difference is between the researchers' probandwise concordance rates and those of past twin studies of sexual orientation. Previously, the lowest concordances for single-sex monozygotic twins were 47% for women and 48% for men. This study documents concordances of just 20% for women and 24% for men, which are significantly lower than the rates reported for the two largest previous twin studies of sexual orientation. Bailey, Dunne, and Martin conclude that sexual orientation is familial; however, their study does not provide statistically significant support for the importance of genetic factors for this trait. They caution that this does not mean that their results entirely exclude heritability. In fact, they consider their findings consistent with moderate heritability for male and female sexual orientation, even though their male monozygotic concordance suggests that any major gene for homosexuality has either low penetrance or low frequency.

Bailey, Dunne, and Martin attribute their markedly different results to the observation that in previous studies twins deciding whether to participate in research that was clearly designed to study homosexuality probably considered the sexual orientation of their co-twin before agreeing to participate. In contrast, the more general focus of the Bailey, Dunne, and Martin study and its anonymous response format made such considerations less likely.

Even though it remains unclear in 2011 whether concordance is closer to 30% or 50%, all researchers concur that it is not 100%. This finding suggests that the influence of genes on sexual orientation is indirect and influenced by environment. Neil Whitehead and Briar Whitehead claim in *My Genes Made Me Do It* (1999) that "genes make proteins, not preferences." Similarly, they contend that "genes create a tendency, rather than a tyranny" and conclude that all the monozygotic twin studies reveal that in terms of determining sexual orientation, neither genetic nor family-related factors are overwhelming. Furthermore, Whitehead and Whitehead believe that all influences—genetic and environmental—are subject to change and that it is possible to "foster or foil genetic or family influences."

In "Genetic and Environmental Effects on Same-Sex Sexual Behavior: A Population Study of Twins in Sweden" (*Archives of Sexual Behavior*, vol. 39, no. 1, February 2010), Niklas Långström et al. report the results of a twin study of same-sex sexual behavior in men and women. Because it was a true population-based study as opposed to a small sample, it overcame some of the biases present in previous research such as self-selection. According to the researchers, this study revealed heritability estimates of 39% for any lifetime same-sex partner in men. Furthermore, the genetic influences on any lifetime same-sex partner and total number of same-sex partners were weaker in women than in men. The results of this study support Bailey, Dunne, and Martin's findings, including the premise that same-sex behavior arises not only from heritable but also from specific environmental sources. Långström et al. conclude that their results are "consistent with moderate, primarily genetic, familial effects, and moderate to large effects of the nonshared environment (social and biological) on same-sex sexual behavior."

Even though there is evidence that homosexuality is heritable, many questions about the genetic contribution to same-sex preferences persist. The article "Potential Evolutionary Role for Same-Sex Attraction" (ScienceDaily.com, February 4, 2010) poses some of these unanswered questions, such as: "Because homosexual men are much less likely to produce offspring than heterosexual men, shouldn't the genes for this trait have been extinguished long ago?" Paul Vasey and Doug P. VanderLaan endeavor to answer this question based on their observations of Samoan male homosexuals. In "An Adaptive Cognitive Dissociation between Willingness to Help Kin and Nonkin in Samoan Fa'afafine" (*Psychological Science*, vol. 21, no. 2, February 2010), the researchers observe that Samoan male homosexuals (who are known locally as fa'afafine)

display more caring toward their nieces and nephews than do Samoan women and heterosexual men. The researchers suggest that Samoan male homosexuals have an evolutionary advantage because they serve as valuable "helpers-in-the-nest" by caring for nieces and nephews. Vasey and VanderLaan explain that this may explain the enduring survival of their genes but it does not compensate for the fact that they do not reproduce.

IS THERE A GAY GENE? Brian S. Mustanski et al. examine in "A Genomewide Scan of Male Sexual Orientation" (*Human Genetics*, vol. 116, no. 4, March 2005) the entire human genetic makeup (genetic information on all chromosomes) in an effort to identify possible genetic determinants of male sexual orientation. The researchers looked at the genetic makeup of 456 men from 146 families with two or more gay brothers and found the same genetic patterns among the gay men on three chromosomes: 7, 8, and 10. Sixty percent of the gay men in the study shared these common genetic patterns, which was slightly more than the 50% expected by chance alone. Patterns involving chromosomes 7 and 8 were associated with sexual orientation regardless of whether the man received them from his father or his mother; however, the areas on chromosome 10 were only associated with male sexual orientation if they were inherited from the mother. The identification of these regions has spurred further research to identify the individual genes that are linked to sexual orientation.

An animal study conducted by geneticists in Korea points to a gene that governs sexual preference in mice. In "Male-Like Sexual Behavior of Female Mouse Lacking Fucose Mutarotase" (*BMC Genetics*, vol. 11, no. 62, July 2010), Dongkyu Park et al. report that when they removed a single gene—the fucose mutarotase (FucM) gene—from female mice, they changed the sexual orientation of the mice. Female mice with the FucM gene were attracted to male mice, and those without the gene displayed male-like sexual behavior, preferring other female mice. Park et al. speculate that because the FucM gene governs levels of alpha-fetoprotein (AFP) in the blood, female mice without the gene had lower levels of AFP, which in turn reduced levels of estradiol (a female sex hormone).

GENETIC AND ENVIRONMENTAL INFLUENCES ON INTELLIGENCE

The role of genetics in determining a person's intelligence is a controversial subject. Few would deny that genes play some role, but many are uncomfortable with the idea that genes determine intelligence. For if intelligence is a genetic trait, the implication is that some people are born to be smart, others are not, and education and upbringing cannot change it. The results of many studies and contentious debate in the scientific community have produced little consensus about the relationship between genetics and intelligence. At least part of the problem stems from the fact that the term *intelligence* is defined differently by different people.

This issue has been argued since the 1870s—when Galton proposed his controversial and arguably racist notions about the heritability of intelligence—however, the debate was reignited during the 1990s, when Richard J. Herrnstein (1930–1994) and Charles Murray (1943–) published *The Bell Curve: Intelligence and Class Structure in American Life* (1994). Herrnstein and Murray expressed their beliefs that between 40% and 80% of intelligence is determined by genetics and that it is intelligence levels, not environmental circumstances, poverty, or lack of education, that are at the root of many of the world's social problems. Critics argued that Herrnstein and Murray not only manipulated and misinterpreted data to support their contention that intelligence levels differ among ethnic groups but also reintroduced outdated and harmful racial stereotypes. Many observers did agree with Herrnstein and Murray's premises that intellect is spread unevenly among individuals and population subgroups; that innate intelligence is distributed through the entire population on a "bell curve," with most people near the average and fewer at the high and low ends; and even that the distribution varies by race and ethnicity. However, few have been willing to accept the idea that intelligence is entirely genetically encoded, permanently fixed, and unresponsive to environmental influences.

One traditional measure of intelligence is a standardized intelligence quotient (IQ) test, which measures an individual's ability to reason and solve problems. Nearly all studies that focus on the link between intelligence and genetics rely on results obtained from IQ tests, which generally provide an overall score along with measures of verbal ability and performance ability. Even though there are several versions of IQ tests available, test takers generally perform comparably on all of them, presumably indicating that they all measure similar aspects of cognitive ability. Critics of IQ tests as measures of intelligence contend that they do not measure all abilities in the complex realm of intelligence. They observe that the tests measure one small segment of the diverse abilities that make up intelligence, evaluate only analytic abilities, fail to assess creative or practical abilities, and measure only a small sample of the skills that define the domain of intelligent human behavior.

Despite concern about whether IQ tests fully measure intelligence, nearly all scientific study of the contributions of genetics and environment to intelligence has focused on measuring IQ and examining individuals who differ in their familial relationships. For example, if genetics plays the predominant role in IQ, then monozygotic twins should have IQs that compare more closely than the IQs of dizygotic twins, and siblings' IQs should correlate more

highly than those of cousins. In "Genetics of Childhood Disorders, II: Genetics and Intelligence" (*Journal of the American Academy of Child and Adolescent Psychiatry*, vol. 38, no. 5, May 1999), Alan S. Kaufman of Yale University observes that scientists who argue in favor of genetic determination of IQ cite these data to support their assertion that:

- Monozygotic twins' IQs are more similar than those of dizygotic twins.

- IQs of siblings correlate more highly than IQs of half-siblings, which, in turn, correlate higher than IQs of cousins.

- IQ correlations between a biological parent and child living together are higher than those between an adoptive parent and child living together.

Kaufman asserts that the following results support scientists who believe that environment more strongly determines IQ:

- IQs of dizygotic twins correlate more highly than IQs of siblings of different ages despite the same degree of genetic similarity.

- Unrelated siblings reared together, such as biological and adopted children, have IQs that are more similar than biological siblings reared apart.

- Correlations between IQs of an adoptive parent and a child living together are similar to correlations of a biological parent and a child living apart.

- Siblings reared together have IQs that are more similar than siblings reared apart, and the same finding holds for parents and children, when they live together or apart.

The preponderance of evidence from twin, familial, and adoption studies supports increasing heritability of intelligence over time, ranging from 20% in infancy to 60% in adulthood, along with environmental factors estimated to contribute about 30%. Kaufman suggests that heredity is important in determining a person's IQ, but that environment is also crucial. Based on twin studies, the heritability percentage for IQ is approximately 50, which is comparable to the heritability value for body weight. Kaufman believes the genetic contribution to weight and intelligence are comparable. He observes that many overweight people have a genetic predisposition for a large frame and a metabolism that promotes weight gain, whereas naturally thin people have the opposite genetic predisposition. Nonetheless, for most people environmental factors such as diet and exercise have a substantial impact on weight. Similarly, genetics and environment interact to determine IQ, and people with genetic, familial relationships such as parents and siblings frequently share common environments.

In "Virtual Twins: New Findings on Within-Family Environmental Influences on Intelligence" (*Journal of Educational Psychology*, vol. 92, no. 3, September 2000), Nancy L. Segal of California State University, Fullerton, examines virtual twins (genetically unrelated siblings [typically adopted] of the same age who are reared together from early infancy) to assess environmental influences on intelligence. The results of this study reveal that the IQ correlation of virtual twins fell considerably below correlations reported for monozygotic twins, dizygotic twins, and full siblings. Segal interprets these results as a demonstration of the modest effects of environment on intellectual development and as supporting a predominantly genetic role in determining intelligence.

Eric Turkheimer et al. consider the relationship between genetics and socioeconomic status (an environmental variable) and intelligence in "Socioeconomic Status Modifies Heritability of IQ in Young Children" (*Psychological Science*, vol. 14, no. 6, November 1, 2003). The researchers looked at twins raised in families living near or below the poverty level and twins reared in more affluent homes and analyzed genetic influences along with shared and nonshared environmental influences. Turkheimer et al. find that nearly all the variation in IQ test scores observed among children raised in wealthier families could be attributed to genetics. In contrast, twins raised in poorer environments appeared to suffer the effects of their environment and failed to realize their genetic potential. Turkheimer et al. opine that the children raised in more prosperous circumstances were probably more likely to have received the intellectual stimulation that is necessary for cognitive development.

The precise contributions of genetics and environmental influences remain unresolved. Ian J. Deary, Wendy Johnson, and Loma M. Houlihan of the University of Edinburgh review current understanding of the relationship between genetics, environment, and variations in human intelligence in "Genetic Foundations of Human Intelligence" (*Human Genetics*, vol. 126, no. 1, March 2009). Deary, Johnson, and Houlihan state that about half of the variation in a wide range of cognitive abilities is attributable to a general factor g, which is highly heritable. However, they note that researchers have not yet pinpointed the specific individual genes and their variants that contribute to differences in intelligence.

In "Human Intelligence and Polymorphisms in the DNA Methyltransferase Genes Involved in Epigenetic Marking" (*PLoS One*, vol. 5, no. 6, June 25, 2010), Paul Haggerty et al. posit that because known genetic influences do not adequately explain the high degree of heritability of intelligence and because epigenetic mechanisms have been associated with some types of mental impairment, these same mechanisms may be involved in intelligence. To test this hypothesis, the researchers looked at SNPs in four DNA methyltransferases involved in epigenetic markers and then applied these SNPs to a test group, which consisted of 1,542 people born in Scotland in 1936. All the study subjects had taken the same IQ test at the age of 11 years and were retested at the age of 70 years. The researchers find that childhood and adult intelligence were associated with one of the SNPs. Haggerty et al. conclude that "the potential involvement of epigenetics, and imprinting in particular, raises the intriguing possibility that even the heritable component of intelligence could be modifiable by factors such as diet during early development."

Scientists Identify the Gene Involved in Human Brain Development

The genetic underpinnings of human intelligence were further supported by the discovery in 2006 of a key gene (HAR1F) that helped the human brain evolve from its primate ancestors. Kerri Smith reports in "Honing in on the Genes for Humanity" (*Nature*, vol. 442, no. 7104, August 17, 2006) that after examining 49 areas of genetic code that have changed the most between the human and chimpanzee genomes, James Sikela et al. pinpointed an area of the human genome that appears to have evolved 70 times faster than the balance of human genetic code. Sikela et al. observed 18 differences in the HAR1F gene that took place in humans but not in chimps and may have prompted the rapid growth of the human brain's cerebral cortex (the part of the brain where thought processes occur). It may also explain why human brains are three times the size of chimp brains.

Even though the specific genes for human intelligence have not been identified, there is evidence that certain characteristics of the human brain that are associated with intelligence are heritable. In "Brain Plasticity and Intellectual Ability Are Influenced by Shared Genes" (*Journal of Neuroscience*, vol. 30, no. 16, April 21, 2010), Rachel G. H. Brans et al. report the results of a twin study that looked at brain size using magnetic resonance imaging (MRI is a radiological technique that provides detailed images of organs and tissues). The researchers find that thinning of the frontal cortex (the portion of the brain that is located behind the forehead where thinking and planning are believed to occur) and thickening of the medial cortex of the brain (where introspection and self-referential mental activity are thought to occur) with advancing age is heritable and associated with cognitive ability and the capacity to think, perceive, and perform mental processes. Brans et al. observe that people with higher intelligence, as measured by IQ, show less cortical thinning and more cortical thickening as they age than people with average or below average intelligence. Because these changes appear to be heritable, Brans et al. posit that genes govern these brain changes and their relationship to intelligence.

HOW DO GENES INFLUENCE BEHAVIOR AND ATTITUDES?

The question of interest is no longer whether human social behavior is genetically determined; it is to what extent.

—Edward O. Wilson, *On Human Nature* (1978)

Most people accept the premise that genes at least in part influence personality and behavior. Families have long decried a characteristic "bad temper" or "wild streak" appearing in a new generation, or boasted about inherited musical or artistic talents. It seems intuitively correct to assume that some of our behaviors and attitudes, both the desirable and less desirable ones, are in part genetically mediated. In "The DNA Age: That Wild Streak? Maybe It Runs in the Family" (*New York Times*, June 15, 2006), Amy Harmon asserts that an increasing understanding of genetics has renewed consideration about the degree to which individuals can actually control how they behave.

Studies of families and twins strongly suggest genetic influences on the development and expression of specific behaviors, but there is no conclusive research demonstrating that genes determine behaviors. Michael Rutter of Kings College in London, England, observes in "The Interplay of Nature, Nurture, and Developmental Influences: The Challenge ahead for Mental Health" (*Archives of General Psychiatry*, vol. 59, no. 11, November 2002) that a range of mental health disorders from autism and schizophrenia to attention deficit hyperactivity disorder involve at least indirect genetic effects, with heritability ranging from 20% to 50%. He further asserts that genetically influenced behaviors also bring about gene-environment correlations.

Rutter explains the mechanism of genetic influence on behavior: genes affect proteins, and through the effects of these proteins on the functioning of the brain there are resultant effects on behavior. Rutter views environmental influences as comparable to genetic influences in that they are strong and pervasive but do not determine behaviors. Furthermore, studies of environmental effects show that there are individual differences in response. Some individuals are severely affected and others experience few repercussions from environmental factors. This has given rise to the idea of varying degrees of resiliency—that people vary in their relative resistance to the harmful effects of psychosocial adversity—as well as to the premise that genetics may offer protective effects from certain environmental influences.

Peter D. Kramer opines in "Scarred DNA and How It Might Heal" (*Psychology Today*, May 12, 2008) that epigenetics may help explain why some people are resilient in the face of a challenge or threat and why others suffer lifelong problems. He cites a study in which a group of genetically identical small mice were exposed to and bullied by larger, aggressive mice. Several of the small mice frequently responded by displaying behaviors that are associated with "social defeat," such as anxiety and depression. However, some of the small mice responded differently to the bullying by remaining confident and seemingly unaffected. Even though the mice had the exact same genes, their DNA segments revealed differing levels of activity, specifically differences in the methylation of a part of a gene that regulates production of brain-derived neurotrophic factor (a protein that stimulates growth and adaptability of some nerve cells), which means that the mice were actually epigenetically distinct from one another. Conceivably, the epigenetic changes associated with defeated outlooks or resiliency would be heritable and evident in their offspring.

Traditional psychological theory holds that attitudes are learned and most strongly influenced by environment. In "The Heritability of Attitudes: A Study of Twins" (*Journal of Personality and Social Psychology*, vol. 80, no. 6, June 2001), James M. Olson et al. examine whether there is a genetic basis for attitudes by reviewing earlier studies and conducting original research on monozygotic and dizygotic twins. The researchers argue that the premise that attitudes are learned is not incompatible with the idea that biological and genetic factors also influence attitudes. Olson et al. hypothesize that genes probably influence predispositions or natural inclinations, which then shape environmental experiences in ways that increase the likelihood of the individual developing specific traits and attitudes. For example, children who are small for their age might be teased or taunted by other children more than their larger peers. As a result, these children might develop anxieties about social interaction, with consequences for their personalities, such as shyness, low self-esteem, or discomfort with large groups.

Amy Abrahamson, Laura Baker, and Avshalom Caspi examine genetic influences on attitudes of adolescents and report their findings in "Rebellious Teens? Genetic and Environmental Influences on the Social Attitudes of Adolescents" (*Journal of Personality and Social Psychology*, vol. 83, no. 6, December 2002). The purpose of their study was to investigate sources of familial influence on adolescent social attitudes in an effort to understand whether and how families exert an influence on the attitudes of adolescents. They wanted to pinpoint the age when genetic influences actually emerge and to determine the extent to which parents and siblings shape teens' views about controversial issues. Abrahamson, Baker, and Caspi explored genetic and environmental influences in social attitudes in 654 adopted and nonadopted children and their biological and adoptive relatives in the Colorado Adoption Project. Conservatism and religious attitudes were measured in the children annually from ages 12 to 15 and in the parents during the visit with 12-year-olds.

The researchers find that both conservatism and religious attitudes are strongly influenced by shared-family environmental factors throughout adolescence. Familial resemblance for conservative attitudes arises from both genetic and common environmental factors, and familial influence on religious attitudes is almost entirely in response to shared-family environmental factors. These findings are different from previous findings in twin studies, which suggests that genetic influence on social attitudes does not emerge until adulthood. In contrast, Abrahamson, Baker, and Caspi detect significant genetic influence in conservatism as early as age 12, but find no evidence of genetic influence on religious attitudes during adolescence. The researchers conclude that genetic factors exert an influence on social attitudes much earlier than previously indicated. For example, significant genetic influences on variations in conservatism are identified as early as age 12. Abrahamson, Baker, and Caspi provide further evidence that shared environmental factors contribute significantly to individual differences in social attitudes during adolescence.

Lindon J. Eaves and Peter K. Hatemi of Virginia Commonwealth University confirm in "Transmission of Attitudes toward Abortion and Gay Rights: Effects of Genes, Social Learning, and Mate Selection" (*Behavior Genetics*, vol. 38, no. 3, May 2008) the role of genetics in shaping social attitudes. The researchers consider the biological and social transmission attitudes toward abortion and gay rights. Eaves and Hatemi find that genetics appeared to be largely responsible for the transmission of these attitudes and that the nongenetic impact of parents' attitudes played a relatively minor role.

Behaviors related to risk taking also appear to have a strong genetic contribution. In "Monoamine Oxidase A Gene (MAOA) Associated with Attitude towards Longshot Risks" (*PLoS One*, vol. 4, no. 12, December 31, 2009), Songfa Zhong et al. look at the heritability of risk-taking behavior. The researchers observe that even though twin studies have confirmed the heritability of economic risk taking, there has not been a direct association of this behavior with a specific gene. Zhong et al. examined 350 subjects and their attitudes toward risk taking and the genotype of a specific polymorphism, monoamine oxidase A (MAOA is an enzyme involved in the degradation of dopamine and norepinephrine, which are neurotransmitters that exert a strong influence on behavior).

The researchers offered the subjects the option of choosing an insurance program, which was a relatively low economic risk, or a lottery purchase, which was considered riskier—a long-shot preference. Zhong et al. find that more subjects with the MAOA high-activity allele preferred the long-shot lottery purchase than did the subjects with the MAOA low-activity allele and conclude that the high activity of the MAOA polymorphism is significantly associated with economic risk taking and preference for the long-shot and higher-risk option.

ENVIRONMENTAL GENOME PROJECT

In many instances it is difficult to understand how genes manage to assert themselves over the countless complications that are imposed by the environment. The Environmental Genome Project (EGP) was launched by the National Institute of Environmental Health Sciences in 1998. The EGP aims to improve understanding of human genetic susceptibility to environmental exposures, including gaining an understanding of how individuals differ in their susceptibility to environmental agents and how these susceptibilities change over time.

Ongoing EGP research activities focus on biostatistics and bioinformatics; DNA sequencing; ethical, legal, and social implications; population-based epidemiology; and technology development. For example, the EGP reports in "Welcome to the NIEHS SNPs Program" (June 18, 2009, http://egp.gs.washington.edu/welcome.html) on a University of Washington project that is systematically identifying and genotyping SNPs in environmental response genes. The project aims to find common sequence variation in human genes that are involved in DNA repair and cell cycle pathways. When the project is completed, it will have produced maps of human genes that may be used to assess the risk of human disease that is associated with specific environmental exposures.

CHAPTER 5
GENETIC DISORDERS

We could wish that ... life-histories were found in every family, showing the health and diseases of its different members. We might thus in time find evidence of pathological connections and morbid liabilities not now suspected.

—Sir William Gull, 1896

It has long been known that heredity affects health. Genetics, the study of single genes and their effects on the body and mind, explains how and why certain traits such as hair color and blood types run in families. Genomics, a discipline that is only about two decades old, is the study of more than single genes; it considers the functions and interactions of all the genes in the genome. In terms of health and disease, genomics has a broader and more promising range than genetics. The science of genomics relies on knowledge of and access to the entire genome and applies to common conditions, such as breast and colorectal cancer, Parkinson's disease, and Alzheimer's disease. It also plays a role in infectious diseases once believed to be entirely environmentally caused such as the human immunodeficiency virus (HIV) and tuberculosis. Like most diseases, these frequently occurring disorders result from the interactions of multiple genes and environmental factors. Genetic variations in these disorders may have a protective or a causative role in the expression of diseases.

It is commonly accepted that diseases fall into one of three broad categories: those that are primarily genetic in origin, those that are largely attributable to environmental causes, and those in which genetics and environmental factors make comparable, though not necessarily equal, contributions. As understanding in genomics and epigenetics (heritable changes that influence gene expression and are caused by mechanisms other than by changes in DNA sequence) advance and scientists identify genes and heritable changes in gene function that are linked to a growing number of diseases, the distinctions between these classes of disorders are diminishing. This chapter considers some of the disorders believed to be predominantly genetic in origin and some that are the result of genes acted on by environmental factors.

There are two types of genes: dominant and recessive. When a dominant gene is passed on to offspring, the feature or trait it determines will appear regardless of the characteristics of the corresponding gene on the chromosome inherited from the other parent. If the gene is recessive, the feature it determines will not show up in the offspring unless both the parents' chromosomes contain the recessive gene for that characteristic. Similarly, among diseases and conditions primarily attributable to a gene or genes, there are autosomal dominant disorders and autosomal recessive disorders (an autosomal chromosome is a chromosome that is not involved in the determination of sex).

Another way to characterize genetic disorders is by their pattern of inheritance, as single gene, multifactorial, chromosomal, or mitochondrial. Single-gene disorders (also called Mendelian or monogenic) are caused by mutations in the deoxyribonucleic acid (DNA) sequence of one gene. Because genes code for proteins, when a gene is mutated so that its protein product can no longer carry out its normal function, it may produce a disorder. According to Human Genome Project Information, in "Genetic Disease Information" (July 21, 2008, http://www.ornl.gov/sci/techresources/Human_Genome/medicine/assist.shtml), there are more than 6,000 known single-gene disorders that occur in about 1 out of every 200 births. Examples are cystic fibrosis, sickle-cell anemia, Huntington's disease, and hereditary hemochromatosis (a disorder in which the body absorbs too much iron from food; rather than the excess iron being excreted, it is stored throughout the body, and the iron deposits damage the pancreas, liver, skin, and other tissues). Figure 5.1 shows the cystic fibrosis gene and its location on chromosome 7; Figure 5.2 shows the sickle-cell anemia gene found on chromosome 11; and Figure 5.3 shows the

FIGURE 5.1

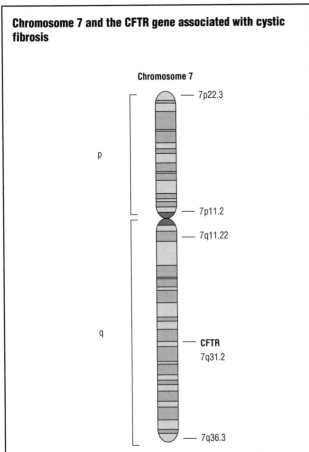

Chromosome 7 and the CFTR gene associated with cystic fibrosis

SOURCE: "CFTR: The Gene Associated with Cystic Fibrosis," in *Gene Gateway—Exploring Genes and Genetic Disorders*, U.S. Department of Energy Office of Science, Office of Biological and Environmental Research, Human Genome Project, 2003, http://www.ornl.gov/TechResources/Human_Genome/posters/chromosome/cftr.html (accessed September 24, 2010)

FIGURE 5.2

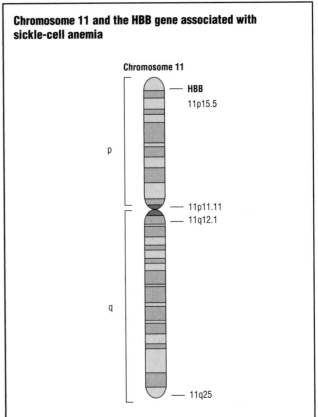

Chromosome 11 and the HBB gene associated with sickle-cell anemia

SOURCE: "HBB: The Gene Associated with Sickle Cell Anemia," in *Gene Gateway—Exploring Genes and Genetic Disorders*, U.S. Department of Energy Office of Science, Office of Biological and Environmental Research, Human Genome Project, 2003, http://www.ornl.gov/TechResources/Human_Genome/posters/chromosome/hbb.html (accessed September 24, 2010)

hereditary hemochromatosis gene located on chromosome 6. Single-gene disorders are the result of either autosomal dominant, autosomal recessive, or X-linked inheritance.

Multifactorial or polygenic disorders result from a complex combination of environmental factors and mutations in multiple genes. For example, different genes that influence breast cancer susceptibility have been found on seven different chromosomes, rendering it more difficult to analyze than single-gene or chromosomal disorders. Some of the most common chronic diseases are multifactorial in origin. Examples include heart disease, Alzheimer's disease, arthritis, diabetes, and cancer.

Chromosomal disorders are produced by abnormalities in chromosome structure, missing or extra copies of chromosomes, or errors such as translocations (movement of a chromosome section from one chromosome to another). Down syndrome or trisomy 21 is a chromosomal disorder

that results when an individual has an extra copy, or a total of three copies, of chromosome 21. Mitochondrial disorders result from mutations in the nonchromosomal DNA of mitochondria, which are organelles involved in cellular respiration. When compared with the three other patterns of inheritance, mitochondrial disorders occur infrequently.

There are significant differences between the 19th-century germ theory of disease and the 21st-century genomic theory of disease. By the middle of the 20th century it became possible to improve the quality of life and to save the lives of people with some genetic diseases. Effective treatment included changes in diet to prevent or manage conditions such as phenylketonuria (PKU) and glucose galactose malabsorption (GGM). PKU is an inherited error of metabolism caused by a deficiency in the enzyme phenylalanine hydroxylase. (See Figure 5.4.) It may result in mental retardation, organ damage, and unusual posture. Dietary changes are also used to treat GGM, a rare metabolic disorder caused by a lack of the enzyme that converts galactose into glucose. For people with severe cases of GGM, it is vital to avoid lactose (milk sugar), sucrose (table sugar), glucose, and galactose. Other

FIGURE 5.3

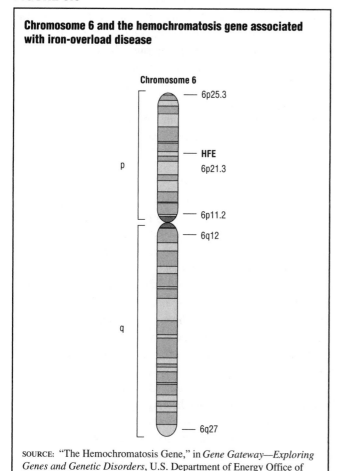

Chromosome 6 and the hemochromatosis gene associated with iron-overload disease

Chromosome 6

- 6p25.3

p

- HFE
  6p21.3

- 6p11.2

- 6q12

q

- 6q27

SOURCE: "The Hemochromatosis Gene," in *Gene Gateway—Exploring Genes and Genetic Disorders*, U.S. Department of Energy Office of Science, Office of Biological and Environmental Research, Human Genome Project, http://www.ornl.gov/TechResources/Human_Genome/posters/chromosome/hfe.html (accessed September 24, 2010)

FIGURE 5.4

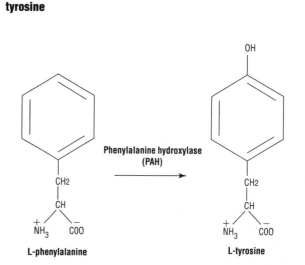

A normal level of the enzyme phenylalanine hydroxylase (PAH) converts the amino acid phenylalanine to the amino acid tyrosine

Phenylalanine hydroxylase (PAH)

L-phenylalanine

L-tyrosine

SOURCE: Adapted from "The Enzyme Phenylalanine Hydroxylase Converts the Amino Acid Phenylalanine to Tyrosine," in "Nutritional and Metabolic Diseases," in *Genes and Disease*, National Institutes of Health, U.S. National Library of Medicine, National Center for Biotechnology Information, 1998–2010, http://www.ncbi.nlm.nih.gov/books/bv.fcgi?rid=gnd.section.213&ref=toc (accessed September 24, 2010)

therapeutic measures may involve surgery to correct deformities and avoidance of environmental triggers, as in some types of asthma.

At the start of the second decade of the 21st century the possibility of preventing and changing genetic legacies appears within reach of modern medical science. Genomic medicine predicts the risk of disease in the individual, whether highly probable, as in the case of some of the well-established single-gene disorders, or in terms of an increased susceptibility likely to be influenced by environmental factors. The promise of genomic medicine is to make preventive medicine more powerful and treatment more specific to the individual, enabling investigation and treatments that are custom-tailored to an individual's genetic susceptibilities, or to the characteristics of the specific disease or disorder.

In "Genomic Medicine—An Updated Primer" (*New England Journal of Medicine*, vol. 362, no. 21, May 27, 2010), W. Gregory Feero, Alan E. Guttmacher, and Francis S. Collins assert that "the wealth of scientific discovery generated over the past 10 years is unparalleled in the history of biomedicine. Moreover, the rate of discovery is accelerating. At the outset of the past decade, the identification of gene mutations causing single-gene disorders was largely a cottage industry, conducted by individual laboratories studying extended families. By the end of the decade, genomewide association studies targeting gene associations for complex conditions involved collaborations of research groups spanning the globe."

COMMON GENETICALLY INHERITED DISEASES

Even though many diseases, disorders, and conditions are termed *genetic*, classifying a disease as genetic simply means there is an identified genetic component to either its origin or its expression. Many medical geneticists contend that most diseases cannot be classified as strictly genetic or environmental. The phenotype of genetic diseases can sometimes be modified, even to the point of nonexpression, by controlling environmental factors. Similarly, environmental (infectious) diseases may not be expressed because of some genetic predisposition to immunity. Each disease, in each individual, exists along a continuum between a genetic disease and an environmental disease.

Many diseases are believed to have strong genetic contributions, including:

- Heart disease—coronary atherosclerosis, hypertension (high blood pressure), and hyperlipidemia (elevated blood levels of cholesterol and other lipids)

- Diabetes—abnormally high blood glucose resulting from the body's inability to use blood glucose for energy

- Cancer—retinoblastomas, colon, stomach, ovarian, uterine, lung, bladder, breast, skin (melanoma), pancreatic, and prostate

- Neurological disorders—Alzheimer's disease, amyotrophic lateral sclerosis (a degenerative neurological condition commonly known as Lou Gehrig's disease), Gaucher's disease (a rare disorder of lipid metabolism), Huntington's disease, multiple sclerosis (a chronic progressive disorder in which the immune system destroys the myelin sheath that protects nerve fibers), narcolepsy (a disorder characterized by sudden, uncontrollable episodes of deep sleep), neurofibromatosis (a disease characterized by tumors of the fibrous covering of peripheral nerves and other developmental problems), Parkinson's disease (a disorder that affects nerve cells in the part of the brain that controls muscle movement), Tay-Sachs disease, and Tourette's syndrome (a neurological disorder characterized by repetitive, involuntary movements and vocalizations; a syndrome is a set of symptoms or conditions that taken together suggest the presence of a specific disease or an increased risk of developing the disease)

- Mental illnesses, mental retardation, and behavioral conditions—alcoholism, anxiety disorders, attention deficit hyperactivity disorder, eating disorders, Lesch-Nyhan syndrome (a disorder caused by a deficiency of the enzyme hypoxanthine-guanine phosphoribosyltransferase, which causes a buildup of uric acid in all body fluids and produces symptoms such as severe gout, poor muscle control, and moderate retardation), and schizophrenia (a mental disorder characterized by delusions, hallucinations, incoherence, and physical agitation)

- Other genetic disorders—cleft lip and cleft palate, clubfoot, cystic fibrosis, Duchenne muscular dystrophy, GGM, hemophilia, Hurler's syndrome (a disease of metabolism in which a person cannot break down long chains of sugar molecules called glycosaminoglycans), Marfan's syndrome (a disorder of connective tissue that affects the skeletal system, cardiovascular system, eyes, and skin), PKU, sickle-cell disease, and thalassemia (a disorder characterized by the abnormal production of hemoglobin that can lead to growth failure, bone deformities, and enlarged liver and spleen)

- Other medical conditions—including alpha-1-antitrypsin deficiency (a lack of a liver protein that blocks the destructive effects of certain enzymes that may lead to emphysema and liver disease), arthritis, asthma, baldness, congenital adrenal hyperplasia (a disorder of the adrenal gland that causes male characteristics to appear early or inappropriately), migraine headaches, obesity, periodontal disease, porphyria (a disorder in which a crucial part of hemoglobin, called heme, does not develop properly, which can lead to abdominal pain, light sensitivity, rashes, and nerve and muscle problems), and some speech disorders

ALZHEIMER'S DISEASE

Alzheimer's disease (AD) is a progressive, degenerative disease that affects the brain and results in severely impaired memory, thinking, and behavior. The National Center for Biotechnology Information reports in *Genes and Disease: Blood and Lymph Diseases* (2008, http://www.ncbi.nlm.nih.gov/books/bookres.fcgi/gnd/gnd.pdf) that AD is the fourth-leading cause of death in adults and that the incidence of the disease rises with age. In *2010 Alzheimer's Disease Facts and Figures* (March 2010, http://www.alz.org/documents_custom/report_alzfactsfigures2010.pdf), the Alzheimer's Association estimates that in 2010 there were about 5.3 million Americans with AD. In 2007 it was the sixth-leading cause of death for all ages (not just adults, as estimated by the previously mentioned National Center for Biotechnology Information) in the United States. (See Figure 5.5.) AD also contributes to many more deaths that are attributed to other causes, such as heart and respiratory failure.

AD has become a disease of particular concern in the United States because the nation's older adult population is growing rapidly. The U.S. Census Bureau projects in *The Next Four Decades—the Older Population in the United States: 2010 to 2050* (May 2010, http://www.census.gov/prod/2010pubs/p25-1138.pdf) that by 2050 the United States will have approximately 88.5 million people over the age of 65 years. The Alzheimer's Association estimates in *2010 Alzheimer's Disease Facts and Figures* that by 2050 the number of cases of AD in the United States will increase to as many as 16 million. Prevalence (the number of people with a disease at a given time) is partially determined by the length of time people with AD survive. Even though the average survival is eight years after diagnosis, some AD patients have lived longer than 20 years with the disease. Therefore, improvements in AD care, as well as increased length of life of the older adult population in general, will increase the numbers of AD patients.

Genetic Causes of AD

AD is not a normal consequence of growing older, and scientists continue to seek its cause. Researchers have found some promising genetic clues to the disease. Mutations in more than 10 genes, located on chromosomes 1, 6,

FIGURE 5.5

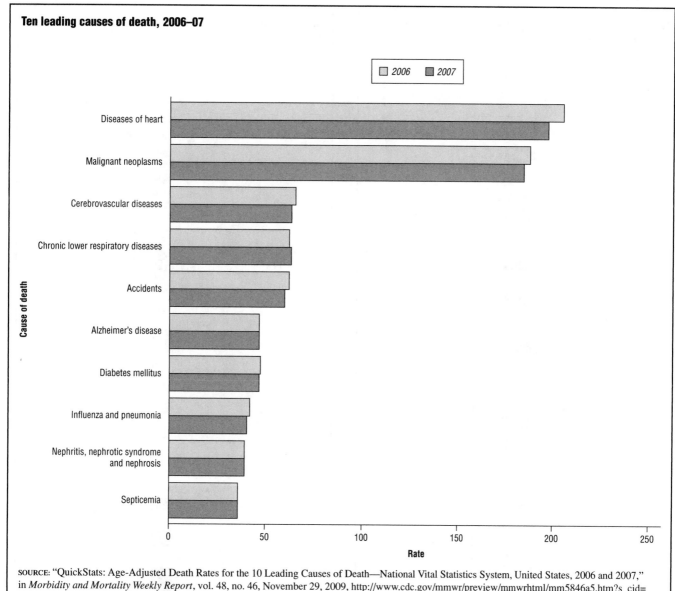

Ten leading causes of death, 2006–07

Legend: ☐ 2006 ■ 2007

(Y-axis — Cause of death; X-axis — Rate)

- Diseases of heart
- Malignant neoplasms
- Cerebrovascular diseases
- Chronic lower respiratory diseases
- Accidents
- Alzheimer's disease
- Diabetes mellitus
- Influenza and pneumonia
- Nephritis, nephrotic syndrome and nephrosis
- Septicemia

SOURCE: "QuickStats: Age-Adjusted Death Rates for the 10 Leading Causes of Death—National Vital Statistics System, United States, 2006 and 2007," in *Morbidity and Mortality Weekly Report*, vol. 48, no. 46, November 29, 2009, http://www.cdc.gov/mmwr/preview/mmwrhtml/mm5846a5.htm?s_cid= mm5846a5_e (accessed September 24, 2010)

14, 19, and 21, are thought to be involved in the disease, and the best described are PS1 (or AD3) on chromosome 14 and PS2 (or AD4) on chromosome 1. (See Figure 5.6 and Figure 5.7.)

The formation of lesions made of fragmented brain cells surrounded by amyloid-family proteins is characteristic of the disease. Interestingly, these lesions and their associated proteins are closely related to similar structures found in Down syndrome. Tangles of filaments largely made up of a protein associated with the cytoskeleton have also been observed in samples taken from AD brain tissue.

The first genetic breakthrough was reported by Alison Goate et al. in "Segregation of a Missense Mutation in the Amyloid Precursor Protein Gene with Familial Alzheimer's Disease" (*Nature*, vol. 349, no. 6311, February 21, 1991).

The researchers stated they had discovered that a mutation in a single gene could cause this progressive neurological illness. Scientists found the defect in the gene that directs cells to produce a substance called amyloid protein. They also discovered that low levels of the brain chemical acetylcholine contribute to the formation of hard deposits of amyloid protein that accumulate in the brain tissue of AD patients. In unaffected people the protein fragments are broken down and excreted by the body. Amyloid protein is found in cells throughout the body. Researchers do not know why it becomes a deadly substance in the brain cells of some people and not others.

In 1995 three more genes linked to AD were identified. One gene appears to be related to the most devastating form of AD, which can strike people in their 30s. When defective, the gene may prevent brain cells from correctly processing a

FIGURE 5.6

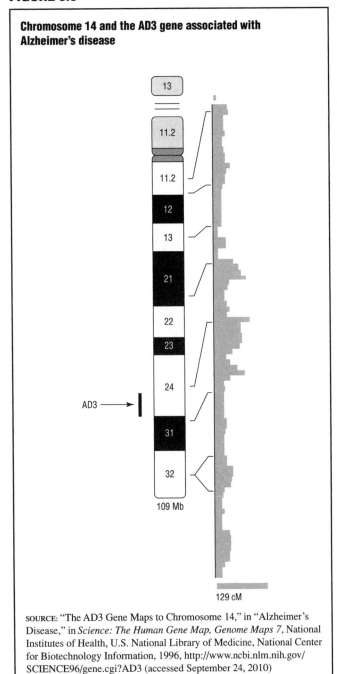

Chromosome 14 and the AD3 gene associated with Alzheimer's disease

SOURCE: "The AD3 Gene Maps to Chromosome 14," in "Alzheimer's Disease," in *Science: The Human Gene Map, Genome Maps 7*, National Institutes of Health, U.S. National Library of Medicine, National Center for Biotechnology Information, 1996, http://www.ncbi.nlm.nih.gov/SCIENCE96/gene.cgi?AD3 (accessed September 24, 2010)

FIGURE 5.7

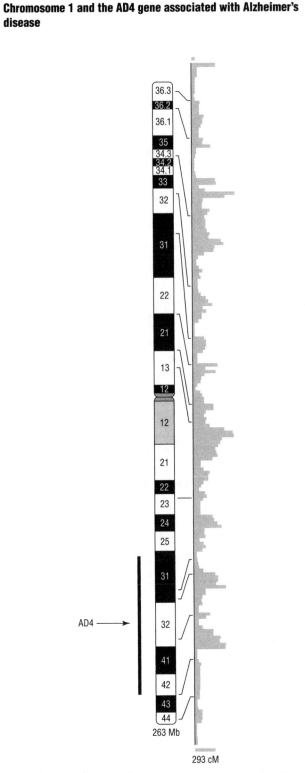

Chromosome 1 and the AD4 gene associated with Alzheimer's disease

SOURCE: "The AD4 Gene Maps to Chromosome 1," in "Alzheimer's Disease," in *Science: The Human Gene Map, Genome Maps 7*, National Institutes of Health, U.S. National Library of Medicine, National Center for Biotechnology Information, 1996, http://www.ncbi.nlm.nih.gov/SCIENCE96/gene.cgi?AD4 (accessed September 24, 2010)

substance called beta amyloid precursor protein. The second gene is linked to another early-onset form of AD that strikes before the age of 65 years. This gene also appears to be involved in producing beta amyloid.

The third gene, known as apolipoprotein E (apoE), is located on the long arm of chromosome 19 and plays several roles. It regulates lipid metabolism and helps redistribute cholesterol. In the brain apoE participates in repairing nerve tissue that has been injured. There are three forms (alleles) of the gene: apoE-2, apoE-3, and apoE-4. ApoE-2 is a variant that may confer protection against developing AD or may delay its onset. The apoE-3 variant does not appear to influence the risk of developing AD. People with one copy of the apoE-4 variant have an increased risk of developing AD. People with two copies of the variant, one from each parent, have an even greater risk. The apoE-4 gene variant may also determine when a person develops AD and is associated with earlier memory loss and other symptoms.

Other candidate genes have been linked to the risk of developing AD. Denise Harold et al. identify in "Genome-wide Association Study Identifies Variants at CLU and PICALM Associated with Alzheimer's Disease, and Shows Evidence for Additional Susceptibility Genes" (*Nature Genetics*, vol. 41, no. 10, October 2009) two more genes, CLU and PICALM, that have a statistically significant association with AD. The PICALM gene in particular may be involved in the breakdown of the synapses, the places where nerve impulses are conveyed from one nerve cell to another. In "Genetic Variation and Neuroimaging Measures in Alzheimer Disease" (*Archives of Neurology*, vol. 67, no. 6, June 2010), Alessandro Biffi et al. identify two additional gene variants, BIN1 and CNTN5, that are linked to the risk for AD.

Adam C. Naj et al. report in "Dementia Revealed: Novel Chromosome 6 Locus for Late-Onset Alzheimer Disease Provides Genetic Evidence for Folate-Pathway Abnormalities" (*PLoS Genetics*, vol. 6, no. 9, September 23, 2010) that people with a variant of the gene MTHFD1L are nearly twice as likely to develop AD as those without the gene variant. MTHFD1L on chromosome 6 plays a role in regulating levels of homocysteine, a naturally occurring amino acid that is found in blood plasma. High levels of homocysteine increase the risk of developing late-onset AD as well as heart disease, stroke, and osteoporosis.

In "Meta-analysis Confirms CR1, CLU, and PICALM as Alzheimer Disease Risk Loci and Reveals Interactions with APOE Genotypes" (*Archives of Neurology*, vol. 67, no. 12, December 2010), Gyungah Jun et al. performed a meta-analysis that considered 7,070 cases of AD in an effort to better understand the relationship between genetic variations of CR1, CLU, and PICALM and AD risk and whether the risk associated with these genes is influenced by apoE. The researchers find that CR1, CLU, and PICALM increase AD risk in populations with European ancestry and that PICALM and apoE interact synergistically—that is, their combined effect is greater than the sum of the two effects.

Jean-Charles Lambert et al. report in "The CALHM1 P86L Polymorphism Is a Genetic Modifier of Age at Onset in Alzheimer's Disease: A Meta-analysis Study" (*Journal of Alzheimer's Disease*, August 30, 2010) that a specific variant, P86L, of the calcium homeostasis modulator 1 gene (CALHM1) is associated with a risk of developing AD. The researchers performed a meta-analysis of 7,873 AD cases and 13,274 controls of Caucasian origin from Belgium, Finland, France, Italy, Spain, Sweden, the United Kingdom, and the United States. Lambert et al. find that the CALHM1 variant does not determine whether an individual will develop AD but may influence when the disease begins through an interaction with apoE-4.

Testing for AD

A complete physical, psychiatric, and neurological evaluation can usually produce a diagnosis of AD that is about 90% accurate. For many years the only sure way to diagnose the disease was to examine brain tissue under a microscope, which was not possible while the AD victim was still alive. An autopsy of someone who has died of AD reveals a characteristic pattern that is the hallmark of the disease: tangles of fibers (neurofibrillary tangles) and clusters of degenerated nerve endings (neuritic plaques) in areas of the brain that are crucial for memory and intellect. Also, the cortex of the brain is shrunken.

Diagnostic tests for AD include analysis of blood and spinal fluid as well as the use of magnetic resonance imaging (a radiological technique that provides detailed images of organs and tissues) to measure the volume of brain tissue in areas of the brain that are used for memory, organizational ability, and planning to accurately identify people with AD and predict who will develop AD in the future.

The presence of specific proteins called biomarkers in spinal fluid and blood can assist in diagnosing people with AD and may also be able to predict the disease up to a decade before symptoms appear. In "Diagnosis-Independent Alzheimer Disease Biomarker Signature in Cognitively Normal Elderly People" (*Archives of Neurology*, vol. 67, no. 8, August 2010), Geert De Meyer et al. examine the levels of three biomarkers—cerebrospinal fluid–derived beta-amyloid protein 1-42, total cerebrospinal fluid tau protein, and phosphorylated tau181P protein—in the cerebrospinal fluid of three groups of older adults: people with AD, people with mild memory impairment, and cognitively normal older adults. De Meyer et al. find that by measuring the levels of biomarkers they could accurately identify 90% of the AD patients. The levels of biomarkers associated with AD were also found in 72% of the study subjects with mild memory loss and in 36% of those who were cognitively normal. The results suggest that these three biomarkers not only accurately diagnose AD but also may accurately predict its onset.

In "A Serum Protein-Based Algorithm for the Detection of Alzheimer Disease" (*Archives of Neurology*, vol. 67, no. 9, September 2010), Sid E. O'Bryant et al. report successful diagnosis of AD using a blood serum biomarker. A blood serum biomarker is preferable to a spinal fluid biomarker because it is easier to obtain a blood sample than a spinal fluid sample, which requires lumbar puncture (also known as a spinal tap), a procedure in which a needle is inserted into the spinal canal. O'Bryant et al. developed a biomarker risk score that considers several protein-based biomarkers, including the clotting protein fibrinogen; interleukin-10, a protein related to immune function; and C-reactive protein, which measures inflammation. The biomarker risk score accurately identified 80% of the subjects with AD and correctly classified 91% of the subjects without AD. When the biomarker risk score was combined with other information, such as the subject's age, sex, and educational attainment and the presence of the apoE variant that is associated with an increased risk of developing AD, the researchers were able to accurately identify 94% of the subjects with AD.

Some genetic testing for AD is available, although it is largely used for research purposes. According to the Alzheimer's Association, in "Genetic Testing" (November 2008, http://www.alz.org/national/documents/topic sheet_genetictesting.pdf), even though testing remains controversial, some people may still choose to test for apoE-4 to assess their risk. The National Institute on Aging explains in "Alzheimer's Disease Genetics Fact Sheet" (September 19, 2009, http://www.nia.nih.gov/ Alzheimers/Publications/geneticsfs.htm) that apoE testing is still largely reserved for research purposes rather than for assessing an individual's risk of developing AD.

An ideal test would be simple and accurate and would detect the most commonly occurring forms of AD (rather than the true familial forms of the disease, which account for less than 5% of cases) early enough for the use of experimental medications to slow the progression of the disease, as well as identify those at risk of developing AD. However, the availability of tests raises ethical and practical questions: Do patients really want to know their risks of developing AD? Should consent for genetic testing be obtained from all family members because the results of one person's test may have implications for other family members?

Treatments for AD

There is still no cure or prevention for AD, and treatment focuses on managing symptoms. Medication can lessen some of the symptoms, such as agitation, anxiety, unpredictable behavior, and depression. Physical exercise and good nutrition are important, as is a calm and highly structured environment. The object is to help the AD patient maintain as much comfort, normalcy, and dignity for as long as possible.

In 2011 just five prescription drugs were available to treat people who suffer from AD. Four of these— Razadyne, Exelon, Aricept, and Cognex—are cholinesterase inhibitors and are prescribed for the treatment of mild to moderate AD. These drugs produce some delay in the deterioration of memory and other cognitive skills in some patients; however, they offer mild benefits at best and may lose their effectiveness over time.

The fifth approved medication, Namenda, is prescribed for the treatment of moderate to severe AD. It acts to delay the progression of some symptoms of moderate to severe AD and may allow patients to maintain certain daily functions a little longer. It is thought to work by regulating glutamate, a chemical in the brain that, in excessive amounts, may lead to brain cell death.

According to Andrew Pollack, in "Hopes for Alzheimer's Drug Are Dashed" (*New York Times*, March 3, 2010), Dimebon, an antihistamine that has been used in Russia for 25 years and a once promising experimental drug for AD, was given a U.S. late-stage clinical trial and was found to have failed to slow or control the cognitive decline or behavioral problems associated with AD. Even though the drug reportedly showed some success in earlier trials in Russia, the U.S. clinical trial found that it was no more effective than a placebo (a substance with no pharmacological activity that is given as a control in clinical trials).

In "Alzheimer's Disease: Clinical Trials and Drug Development" (*Lancet Neurology*, vol. 9, no. 7, July 2010), Francesca Mangialasche et al. observe that even though drug treatment for AD has had modest success in managing symptoms, there have been no breakthroughs in disease-modifying drugs—drugs that would significantly slow, halt, or reverse the disease process. Mangialasche et al. opine that "a single cure for Alzheimer's disease is unlikely to be found" because AD does not result from a single defective gene or disease process. The future of successful drug development may be in compounds that target different AD-related mechanisms simultaneously.

CANCER

Cancer is a large group of diseases that are characterized by uncontrolled cell division and the growth and spread of abnormal cells. These cells may grow into masses of tissue called tumors. Tumors composed of cells that are not cancerous are called benign tumors. Tumors consisting of cancer cells are called malignant tumors. The dangerous aspect of cancer is that cancer cells invade and destroy normal tissue.

The mechanisms of action that disrupt the cell cycle are impairment of a DNA repair pathway, transformation of a normal gene into an oncogene (a hyperactive gene that stimulates cell growth), and the malfunction of a tumor-suppressor gene (a gene that inhibits cell division). Figure 5.8 shows the multiple systems that interact to control the cell cycle.

The spread of cancer cells occurs either by local growth of the tumor or by some of the cells becoming detached and traveling through the blood and lymphatic

FIGURE 5.8

The cell cycle and cancer

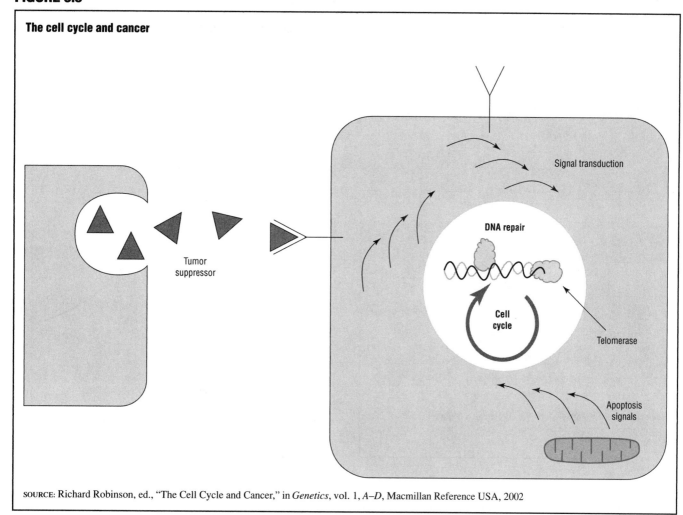

SOURCE: Richard Robinson, ed., "The Cell Cycle and Cancer," in *Genetics*, vol. 1, *A–D*, Macmillan Reference USA, 2002

system to seed additional tumors in other parts of the body. Metastasis (the spread of cancer cells) may be confined to a local region of the body, but if left untreated (and often despite treatment), the cancer cells can spread throughout the entire body, eventually causing death. It is perhaps the rapid, invasive, and destructive nature of cancer that makes it, arguably, the most feared of all diseases, even though it is second to heart disease as the leading cause of death in the United States. (Figure 5.5 uses the term *malignant neoplasms* to describe cancer.)

Cancer can be caused by both external environmental influences (chemicals, radiation, and viruses) and internal factors (hormones, immune conditions, and inherited mutations). These factors may act together or in sequence to begin or promote cancer. There is consensus in the scientific community that several cancer-promoting influences accrue and interact before an individual develops a malignant growth. With only a few exceptions, no single factor or risk alone is sufficient to cause cancer. As with other disorders that arise in response to multiple factors, susceptibility to certain cancers is often attributed to a

mutated gene. Figure 5.9 shows how the inheritance of a mutated gene increases susceptibility for retinoblastoma (cancer of the eye). According to the National Library of Medicine's Genetics Home Reference, in "Retinoblastoma" (January 9, 2011, http://ghr.nlm.nih.gov/condition/retinoblastoma), retinoblastoma affects about 250 to 300 children in the United States each year.

Genetic Research

Scientists and physicians have known for some time that predisposition to some forms of breast cancer are inherited and have been searching for the gene or genes responsible so that they can test patients and provide more careful monitoring for those at risk. In 1994 researchers identified the BRCA1 gene, and in late 1995 they also isolated the BRCA2 gene. The Genetics Home Reference notes in "Breast Cancer" (January 9, 2011, http://ghr.nlm.nih.gov/condition/breast-cancer) that variations of the BRCA1, BRCA2, CDH1, PTEN, STK11, and TP53 genes increase the risk of developing breast cancer and that the AR, ATM, BARD1, BRIP1, CHEK2, DIRAS3, ERBB2, NBN, PALB2, RAD50, and

FIGURE 5.9

Inheritance of a mutated retinoblastoma gene

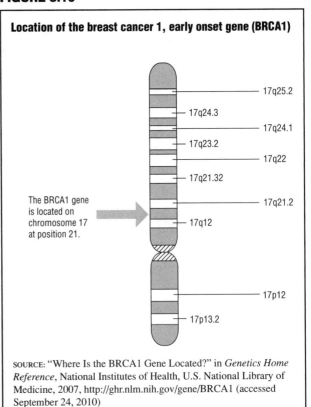

One hit — Cell remains normal

Two hits — Cancer

SOURCE: Richard Robinson, ed., "Inheritance of a Mutated Retinoblastoma Gene," in *Genetics*, vol. 1, *A–D*, Macmillan Reference USA, 2002

FIGURE 5.10

Location of the breast cancer 1, early onset gene (BRCA1)

The BRCA1 gene is located on chromosome 17 at position 21.

17q25.2
17q24.3
17q24.1
17q23.2
17q22
17q21.32
17q21.2
17q12
17p12
17p13.2

SOURCE: "Where Is the BRCA1 Gene Located?" in *Genetics Home Reference*, National Institutes of Health, U.S. National Library of Medicine, 2007, http://ghr.nlm.nih.gov/gene/BRCA1 (accessed September 24, 2010)

RAD51 genes are associated with breast cancer. Some of the genes that have been isolated from breast tumors, such as ERBB2, DIRAS3, and TP53, are somatic mutations, which means they are not inherited.

When a woman with a family history of breast cancer inherits a defective form of either BRCA1 or BRCA2, she has an estimated 80% to 90% chance of developing breast cancer. Figure 5.10 shows the location of the BRCA1 gene on the long arm of chromosome 17 at position 21. Figure 5.11 shows the location of the BRCA2 gene on the long arm of chromosome 13 at position 12.3. These two genes are also linked to ovarian, prostate, and colon cancer and the BRCA2 gene likely plays some role in breast cancer in men. Scientists suspect that these genes and other genes may also participate in some way in the development of breast cancer in women with no family history of the disease. Antonis C. Antoniou et al. report in "A Locus on 19p13 Modifies Risk of Breast Cancer in BRCA1 Mutation Carriers and Is Associated with Hormone Receptor-Negative Breast Cancer in the General Population" (*Nature Genetics*, vol. 42, no. 10, October 2010) that some people with mutations of BRCA1 also have other genetic variants that serve to further increase their risk. The researchers identify five single nucleotide polymorphisms (SNPs) that are

FIGURE 5.11

Location of the breast cancer 2, early onset gene (BRCA2)

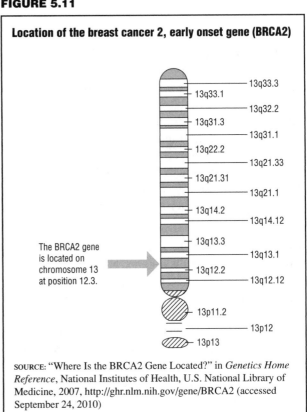

The BRCA2 gene is located on chromosome 13 at position 12.3.

13q33.3
13q33.1
13q32.2
13q31.3
13q31.1
13q22.2
13q21.33
13q21.31
13q21.1
13q14.2
13q14.12
13q13.3
13q13.1
13q12.2
13q12.12
13p11.2
13p12
13p13

SOURCE: "Where Is the BRCA2 Gene Located?" in *Genetics Home Reference*, National Institutes of Health, U.S. National Library of Medicine, 2007, http://ghr.nlm.nih.gov/gene/BRCA2 (accessed September 24, 2010)

FIGURE 5.12

Location of the tumor-producing gene ERBB2

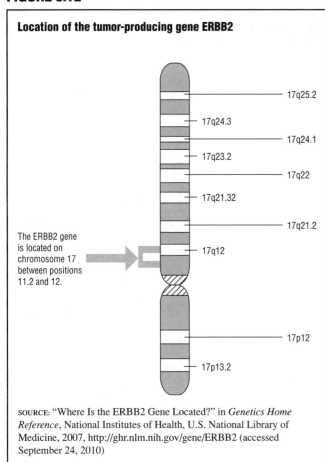

The ERBB2 gene is located on chromosome 17 between positions 11.2 and 12.

17q25.2
17q24.3
17q24.1
17q23.2
17q22
17q21.32
17q21.2
17q12
17p12
17p13.2

SOURCE: "Where Is the ERBB2 Gene Located?" in *Genetics Home Reference*, National Institutes of Health, U.S. National Library of Medicine, 2007, http://ghr.nlm.nih.gov/gene/ERBB2 (accessed September 24, 2010)

associated with triple-negative breast cancer (an aggressive form of breast cancer that does not express the genes for estrogen or progesterone receptors and so does not respond to hormonal therapy) in a region of chromosome 19p13.

According to François Bertucci et al., in "Gene Expression Profiling of Inflammatory Breast Cancer" (*Cancer*, vol. 116, no. 11, June 1, 2010), overexpression of the ERBB2 gene is associated with inflammatory breast cancer, an aggressive form of cancer that can cause death more quickly than other types of breast cancers because the cancer spreads very quickly. The ERBB2 gene produces a protein on the surface of cells that serves as a receiving point for growth-stimulating hormones. Figure 5.12 shows the location of the ERBB2 gene on the long arm of chromosome 17 between positions 11.2 and 12.

In "The BARD1 Cys557Ser Variant and Breast Cancer Risk in Iceland" (*PLoS Medicine*, vol. 3, no. 7, July 2006), Simon N. Stacey et al. report their discovery of another mutation that when present with BRCA1 or BRCA2 significantly increases the risk of developing breast cancer. Stacey et al. studied 1,090 Icelandic women who had breast cancer and compared them to 703 similar women without the dis-

ease. They found a specific mutation, BARD1, in 2.8% of women with beast cancer but only in 1.6% of women without cancer. They also found that women with both the BARD1 and BRCA2 mutations were twice as likely to develop breast cancer as women without these mutations. Irmgard Irminger-Finger of the University Hospitals Geneva explains in "BARD1, a Possible Biomarker for Breast and Ovarian Cancer" (*Gynecologic Oncology*, vol. 117, no. 2, May 2010) that a blood test based on BARD1, a protein that interacts with the breast cancer gene BRCA1, may be a biomarker for breast cancer that can be used to screen women and detect the disease early, when it is most amenable to treatment. Figure 5.13 shows the location of BARD1 on the long arm of chromosome 2 between positions 34 and 35.

Walter Bodmer and Ian Tomlinson point out in "Rare Genetic Variants and the Risk of Cancer" (*Current*

FIGURE 5.13

Location of the BARD1 gene that is associated with breast cancer

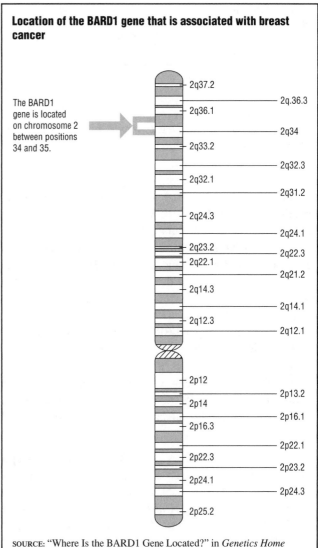

The BARD1 gene is located on chromosome 2 between positions 34 and 35.

2q37.2
2q.36.3
2q36.1
2q34
2q33.2
2q32.3
2q32.1
2q31.2
2q24.3
2q24.1
2q23.2
2q22.3
2q22.1
2q21.2
2q14.3
2q14.1
2q12.3
2q12.1
2p12
2p13.2
2p14
2p16.1
2p16.3
2p22.1
2p22.3
2p23.2
2p24.1
2p24.3
2p25.2

SOURCE: "Where Is the BARD1 Gene Located?" in *Genetics Home Reference*, National Institutes of Health, U.S. National Library of Medicine, 2007, http://ghr.nlm.nih.gov/gene/BARD1 (accessed September 24, 2010)

FIGURE 5.14

> **Location of the BRIP1 gene that increases the risk of developing breast cancer**
>
> The BRIP1 gene is located on chromosome 17 between positions 22 and 24.
>
> SOURCE: "Where Is the BRIP1 Gene Located?" in *Genetics Home Reference*, National Institutes of Health, U.S. National Library of Medicine, 2007, http://ghr.nlm.nih.gov/gene/BRIP1 (accessed September 24, 2010)

FIGURE 5.15

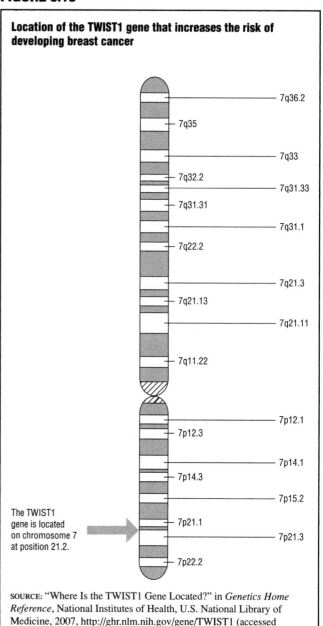

> **Location of the TWIST1 gene that increases the risk of developing breast cancer**
>
> The TWIST1 gene is located on chromosome 7 at position 21.2.
>
> SOURCE: "Where Is the TWIST1 Gene Located?" in *Genetics Home Reference*, National Institutes of Health, U.S. National Library of Medicine, 2007, http://ghr.nlm.nih.gov/gene/TWIST1 (accessed September 24, 2010)

Opinion in Genetics and Development, vol. 20, no. 3, June 2010) that rare, low-penetrance genetic variants, such as CHEK2, PALB2, and BRIP1, can double the risk of breast cancer in carriers. However, other gene mutations raise a woman's breast cancer risk even further. For example, mutations in BRCA1, BRCA2, and TP53 increase the carrier's risk of breast cancer 10- to 20-fold by the age of 60 years. Like some of the other known breast cancer genes such as BRCA1 and BRCA2, BRIP1 is a DNA repair gene, so women with a faulty version of this gene cannot repair damaged DNA correctly. According to the Genetics Home Reference, in "BRIP1" (January 9, 2011, http://ghr.nlm.nih.gov/gene/BRIP1), individuals with faulty DNA repair genes have an increased risk of cancer because their healthy cells are more likely to accumulate genetic damage that can trigger the cell to replicate uncontrollably. Figure 5.14 shows the location of BRIP1 on the long arm of chromosome 17 between positions 22 and 24.

In "Network Modeling Links Breast Cancer Susceptibility and Centrosome Dysfunction" (*Nature Genetics*, vol. 39, no. 11, October 2007), Miguel Angel Pujana et al. identify another gene, HMMR, that if mutated, as is the case in about 10% of the U.S. population, may interact with BRCA1 to increase the risk of developing breast cancer. Pelle Sahlin et al. report in "Women with Saethre-Chotzen Syndrome Are at Increased Risk of Breast Cancer" (*Genes Chromosomes Cancer*, vol. 46,

no. 7, July 2007) that mutations in the gene TWIST1 are found to increase the risk of breast cancer 20-fold. Figure 5.15 shows the location of TWIST1 on the short arm of chromosome 7 at position 21.2.

Ying Ni et al. explain in "Germline Mutations and Variants in the Succinate Dehydrogenase Genes in Cowden and Cowden-like Syndromes" (*American Journal of Human Genetics*, vol. 83, no. 2, August 2008) that another gene mutation, SDH, is found to increase breast cancer risk as well as the risk of developing thyroid and kidney cancers.

Clare Turnbull et al. discuss in "Genome-wide Association Study Identifies Five New Breast Cancer

Susceptibility Loci" (*Nature Genetics*, vol. 42, no. 6, May 2010) their process of identifying additional regions on chromosomes 9, 10, and 11 that are associated with increased breast cancer risk. The researchers genotyped 582,886 SNPs from 3,659 cases of breast cancer with a family history of the disease and 4,897 controls—women without breast cancer or a family history of breast cancer. Besides finding five areas that are associated with heightened risk, Turnbull et al. determine that when women have more than one gene variant or site associated with susceptibility, their risk increases. This finding is consistent with the premise of polygenic susceptibility for breast cancer.

CYSTIC FIBROSIS

Cystic fibrosis (CF) is the most common inherited fatal disease of children and young adults in the United States. According to the Genetics Home Reference, in "Cystic Fibrosis" (January 9, 2011, http://ghr.nlm.nih.gov/condition/cystic-fibrosis), CF occurs in about 1 out of 2,500 to 3,500 whites, 1 out of 17,000 African-Americans, and 1 out of 31,000 Asian-Americans. In "Learning about Cystic Fibrosis" (July 2, 2010, http://www.genome.gov/10001213), the National Human Genome Research Institute (NHGRI) notes that approximately 30,000 young people had the disease in 2010. More than 10 million Americans, almost all of whom are white, are symptomless carriers of the CF gene. Like sickle-cell disease, it is a recessive genetic disorder—to inherit this disease, a child must receive the CF gene from both parents.

In 1989 the CF gene was identified, and in 1991 it was cloned and sequenced. It is located on the long arm of chromosome 7 at position 31.2. (See Figure 5.16.) The gene was called cystic fibrosis transmembrane conductance regulator (CFTR) because it was discovered to encode a membrane protein that controls the transit of chloride ions across the plasma membrane of cells. Nearly 1,000 mutations of the large gene have been identified. Even though most are extremely rare, several account for more than two-thirds of all mutations. Figure 5.17 shows three types of defects in the CFTR gene that can cause CF. The most frequently occurring mutation causes faulty processing of the protein such that the protein is degraded before it reaches the cell membrane.

The mutated versions of the gene found in people with CF were seen to cause relatively modest impairment of chloride transport in cells. However, this seemingly minor defect can result in a multisystem disease that affects organs and tissues throughout the body, provoking abnormal, thick secretions from glands and epithelial cells. Ultimately, these secretions fill the lungs and cause affected children to die of respiratory failure.

FIGURE 5.16

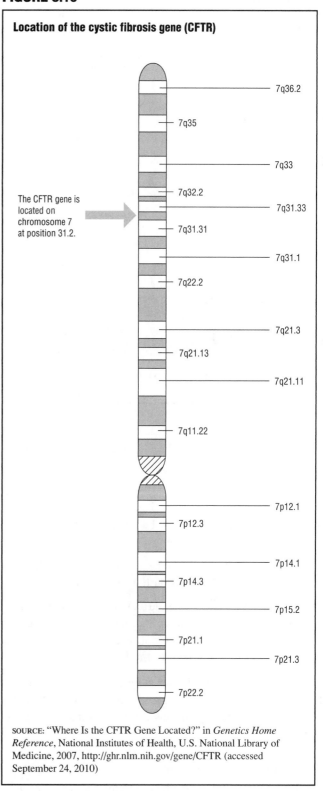

Location of the cystic fibrosis gene (CFTR)

The CFTR gene is located on chromosome 7 at position 31.2.

7q36.2
7q35
7q33
7q32.2
7q31.33
7q31.31
7q31.1
7q22.2
7q21.3
7q21.13
7q21.11
7q11.22
7p12.1
7p12.3
7p14.1
7p14.3
7p15.2
7p21.1
7p21.3
7p22.2

SOURCE: "Where Is the CFTR Gene Located?" in *Genetics Home Reference*, National Institutes of Health, U.S. National Library of Medicine, 2007, http://ghr.nlm.nih.gov/gene/CFTR (accessed September 24, 2010)

The progression from the defective gene and protein it encodes to life-threatening illness follows this complex path:

1. The defect in chloride passage across the cell membrane indirectly produces an accumulation of thick mucus secretions in the lungs.

FIGURE 5.17

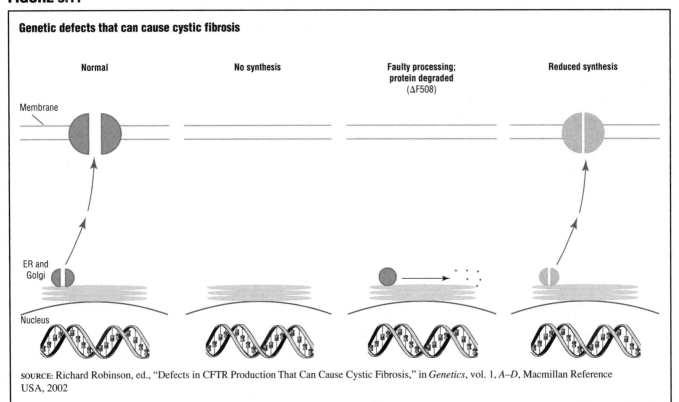

Genetic defects that can cause cystic fibrosis

SOURCE: Richard Robinson, ed., "Defects in CFTR Production That Can Cause Cystic Fibrosis," in *Genetics*, vol. 1, *A–D*, Macmillan Reference USA, 2002

2. The bacteria *Pseudomonas aeruginosa* grows in the mucus.

3. In a campaign to combat the bacterial invasion, the body's immune system is activated but is unable to access the bacteria because the thick mucus protects it.

4. The immune reaction persists and becomes chronic, resulting in inflammation that harms the lungs.

5. Ultimately, it is the affected individual's own immune response, rather than the defective CF gene, the protein, or the bacterial infection, that produces the often fatal damage to the lungs.

A simple sweat test is currently the standard diagnostic test for CF. The test measures the amount of salt in the sweat; abnormally high levels are the hallmark of cystic fibrosis.

CF Gene-Screening Falters

In August 1989 researchers isolated the specific gene that causes CF. The mutation of this gene accounts for about 70% of the cases of the disease. In 1990 scientists successfully corrected the biochemical defect by inserting a healthy gene into diseased cells grown in the laboratory—this was a major step toward developing new therapies for the disease. In 1992 they injected healthy genes into laboratory rats with a deactivated common cold virus as the vector, or delivery agent. The rats began to manufacture the missing protein, which regulates the chloride and sodium in the tissues, preventing the deadly buildup of mucus. Scientists were hopeful that within a few years CF would be eliminated as a fatal disease, giving many children the chance for a healthy and normal life.

However, optimism faded in 1993, when the medical community discovered that the CF gene was more complicated than expected. At the same time, they discovered that many people who have inherited mutated genes from both parents do not have cystic fibrosis. Alan W. Cuthbert of the University of Cambridge notes in "New Horizons in the Treatment of Cystic Fibrosis" (*British Journal of Pharmacology*, November 25, 2010) that by 2010 scientists found that the gene can be mutated at more than 1,700 points. With so many possible mutations, the potential combinations in a person who inherits one gene from each parent are immeasurable.

The combinations of different mutations create different effects. Some may result in crippling and fatal CF, whereas others may cause less serious disorders, such as infertility, asthma, or chronic bronchitis. To further complicate the picture, other genes can alter the way different mutations of the CF gene affect the body.

Researchers have found that CF mutations are much more common than previously thought. For example, in the landmark study "Abstracts from the Nineteenth Annual Education Conference of the National Society of Genetic Counselors (Savannah, Georgia, November 2000)" (*Journal of Genetic Counseling*, vol. 9, no. 6, December 2000), Stephanie A. Cohen and Lyn Hammond cite a study of 5,000 healthy women receiving prenatal care at Kaiser Permanente in California who were tested for the CF gene, which is thought to be present in less than 1% of the population. Of those screened, 11% had the mutation. This finding may indicate that many more common diseases, such as asthma, may be caused by mutations of the CF gene. Other scientists speculate that the frequency of CF carriers among people of European descent may have, at some point in time, conferred immunity to some other disorder, in much the same way that the sickle-cell carriers were protected from malaria.

Because only a few dozen of the mutations occur with a frequency that is greater than 1 out of 1,000 in the general population and approximately 20 mutations are identified in more than 80% of carriers (e.g., just seven mutations account for nearly all CF cases among Ashkenazi Jews), the American College of Obstetricians and Gynecologists recommends in *Cystic Fibrosis Testing: The Decision Is Yours* (2001) routine screening for the 29 most common mutations. Nearly three-quarters of CF cases among whites are caused by the DF508 mutation and the W1282X mutation is most common among Ashkenazi Jews.

The Cystic Fibrosis Foundation (CFF) supports clinical research studies in human gene therapy. In "Drug Development Pipeline" (January 10, 2011, http://www.cff.org/research/DrugDevelopmentPipeline/), the CFF notes that one form of gene therapy uses a compacted DNA technology. This gene transfer system compacts single copies of the healthy CFTR gene so that they are small enough to pass through a cell membrane into the nucleus. The goal is for the DNA to produce the CFTR protein that is needed to correct the basic defect in CF cells. Paula Anderson of the University of Arkansas for Medical Sciences describes in "Emerging Therapies in Cystic Fibrosis" (*Therapeutic Advances in Respiratory Disease*, vol. 4, no. 3, June 7, 2010) the development of drugs that target specific defects in CFTR transcription, processing, or functioning. Anderson also explains that clinical trials are under way to establish the safety and efficacy of a type of gene therapy that uses plasmid-DNA and nonviral vectors, because using viruses to deliver the therapy have proven ineffective and inefficient.

DIABETES

Diabetes is a disease that affects the body's use of food, causing blood glucose (sugar levels in the blood) to become too high. Normally, the body converts sugars, fats, starches, and proteins into a form of sugar called glucose. The blood then carries glucose to all the cells throughout the body. In the cells, with the help of the hormone insulin, which facilitates the entry of glucose into the cells, the glucose is either converted into energy for use immediately or stored for the future. Beta cells of the pancreas, a small organ located behind the stomach, manufacture the insulin. The process of turning food into energy via glucose is important because the body depends on glucose for its energy source.

In a person with diabetes, food is converted to glucose, but there is a problem with insulin. In one type of diabetes the pancreas does not manufacture enough insulin, and in another type the body has insulin but cannot use the insulin effectively (this latter condition is called insulin resistance). When insulin is either absent or ineffective, glucose cannot get into the cells to be converted into energy. Instead, the unused glucose accumulates in the blood. If a person's blood-glucose level rises high enough, the excess glucose is excreted from the body via urine, causing frequent urination. This, in turn, leads to an increased feeling of thirst as the body tries to compensate for the fluid lost through urination.

Types of Diabetes

There are two distinct types of diabetes. Type 1 diabetes (formerly called juvenile diabetes) occurs most often in children and young adults. The pancreas stops manufacturing insulin, so the hormone must be injected daily. Type 2 diabetes is most often seen in adults. In this type the pancreas produces insulin, but it is not used effectively, and the body resists its effects. Table 5.1 compares the phenotype (presentation) and genotype of type 1 and type 2 diabetes.

In "Long-Term Trends in Diabetes" (July 2009, http://www.cdc.gov/diabetes/statistics/slides/long_term_trends.pdf), the Centers for Disease Control and Prevention (CDC) states that in 2008, 18.8 million Americans had been diagnosed with diabetes. Figure 5.18 shows that the number and percent of Americans diagnosed with diabetes has risen sharply since the mid-1990s. In 2007 diabetes was the seventh-leading cause of death. (See Figure 5.5.) The individuals at risk for type 2 diabetes are usually overweight, over 40 years old, and have a family history of diabetes. According to the National Diabetes Education Program, in "The Facts about Diabetes: America's Seventh Leading Cause of Death" (July 2008, http://ndep.nih.gov/media/FS_GenSnapshot.pdf), type 2 represents 90% to 95% of diabetes patients, whereas type 1 accounts for 5% to 10% of diabetes cases.

Causes of Diabetes

The causes of both type 1 and type 2 diabetes are unknown, but a family history of the disease increases the risk for both types, leading researchers to believe there is a genetic component. Some scientists believe a flaw in the body's immune system may also be a factor in type 1 diabetes.

TABLE 5.1

Comparison of type 1 and type 2 diabetes

	Type 1 diabetes	Type 2 diabetes
Phenotype (observable characteristics)	Onset primarily in childhood and adolescence Often thin or normal weight Prone to ketoacidosis Insulin administration required for survival Pancreas is damaged by an autoimmune attack Absolute insulin deficiency Treatment: insulin injections	Onset predominantly after 40 years of age* Often obese No ketoacidosis Insulin administration not required for survival Pancreas is not damaged by an autoimmune attack Relative insulin deficiency and/or insulin resistance Treatment: (1) healthy diet and increased exercise; (2) hypoglycemic tablets; (3) insulin injections
Genotype (genetic makeup)	Increased prevalence in relatives Identical twin studies: <50% concordance HLA association: yes	Increased prevalence in relatives Identical twin studies: usually above 70% concordance HLA association: no

*Type 2 diabetes is increasingly diagnosed in younger patients.
Note: HLA is human leukocyte antigen.

SOURCE: Adapted from Laura Dean and Johanna McEntyre, "Table 1. Comparison of Type 1 and Type 2 Diabetes," in *The Genetic Landscape of Diabetes*, U.S. Department of Health and Human Services, National Institutes of Health, U.S. National Library of Medicine, National Center for Biotechnology Information, 2004, http://www.ncbi.nlm.nih.gov/bookshelf/br.fcgi?book=diabetes&part=A3 (accessed September 27, 2010)

FIGURE 5.18

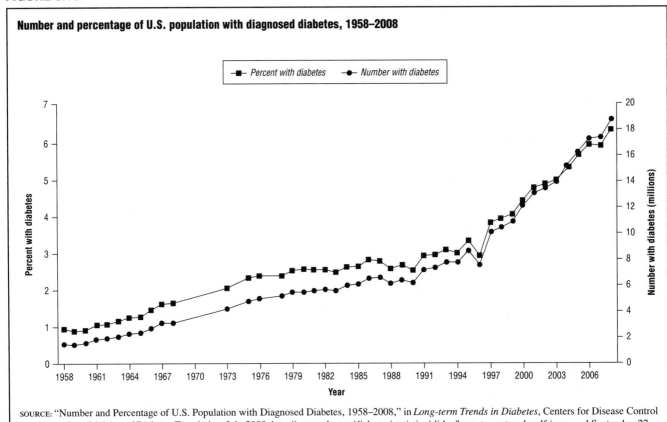

Number and percentage of U.S. population with diagnosed diabetes, 1958–2008

SOURCE: "Number and Percentage of U.S. Population with Diagnosed Diabetes, 1958–2008," in *Long-term Trends in Diabetes*, Centers for Disease Control and Prevention, Division of Diabetes Translation, July 2009, http://www.cdc.gov/diabetes/statistics/slides/long_term_trends.pdf (accessed September 27, 2010)

Mutations in many genes contribute to the origin and onset of type 1 diabetes. For example, an insulin-dependent diabetes mellitus (IDDM1) site on chromosome 6 may harbor at least one susceptibility gene for type 1 diabetes. The role of this mutation in increasing susceptibility is not yet known; however, because chromosome 6 also contains genes for antigens (the molecules that normally tell the immune system not to attack itself), there may be some interaction between immunity and diabetes. In type 1 diabetes the body's immune system mounts an immunological assault on its own insulin and the pancreatic cells that manufacture it. Some 10 sites in the human genome, including a gene at the locus IDDM2 on chromosome 11 and the gene for glucokinase (an enzyme that is crucial for

FIGURE 5.19

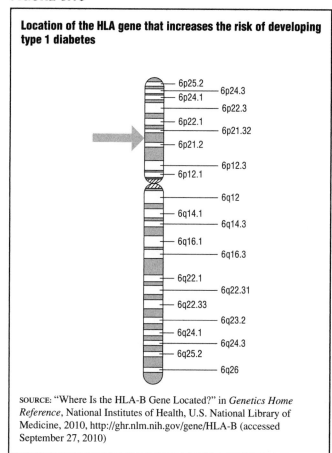

Location of the HLA gene that increases the risk of developing type 1 diabetes

SOURCE: "Where Is the HLA-B Gene Located?" in *Genetics Home Reference*, National Institutes of Health, U.S. National Library of Medicine, 2010, http://ghr.nlm.nih.gov/gene/HLA-B (accessed September 27, 2010)

glucose metabolism) on chromosome 7, appear to increase susceptibility to type 1 diabetes.

In "Genetics of Type 1 Diabetes: What's Next?" (*Diabetes*, vol. 59, no. 7, July 29, 2010), Flemming Pociot et al. report that genome-wide scans looking for links to type 1 diabetes have confirmed the linkage of type 1 diabetes to the human leukocyte antigen (HLA) gene on chromosome 6. Figure 5.19 shows the HLA-B gene on the short arm of chromosome 6 at position 21.3. Examples of other variants that are associated with type 1 diabetes include the insulin gene on chromosome 11p15.5, the lymphoid-specific protein tyrosine phosphatase gene on chromosome 1p13, and IL2RA on chromosome 10p15.1, which is linked to the development of other immune system disorders. There are also more than 50 other regions that affect the risk of developing type 1 diabetes.

In type 2 diabetes heredity also plays a role, but because the pancreas continues to produce insulin, the disease is considered a problem of insulin resistance, in which the body is not using the hormone efficiently. In people prone to type 2 diabetes, being overweight can set off the disease because excess fat prevents insulin from working correctly. Maintaining a healthy weight and keeping physically fit can usually prevent noninsulin-dependent diabetes. To date, insulin-dependent diabetes (type 1) cannot be prevented.

Karani S. Vimaleswaran and Ruth J. F. Loos of the Institute of Metabolic Science in Cambridge, England, note in "Progress in the Genetics of Common Obesity and Type 2 Diabetes" (*Expert Reviews in Molecular Medicine*, vol. 12, no. 7, February 26, 2010) that approximately 20 sites of genetic variants, which were identified using genome-wide association studies, are linked to susceptibility or an increased risk of developing type 2 diabetes. The contribution that each variant alone makes to increasing risk varies, but most are assumed to only slightly increase the risk of developing the disease. However, in combination they likely exert a greater effect.

Complications of Diabetes

Because diabetes deprives body cells of the glucose needed to function properly, complications can develop that threaten the lives of diabetics. Complications of diabetes include higher risk and rates of heart disease; circulatory problems, especially in the legs, often severe enough to require surgery or even amputation; diabetic retinopathy, a condition that can cause blindness; kidney disease that may require dialysis; dental problems; impaired healing and increased risk of infection; and problems during pregnancy. People who pay close attention to the roles of diet, exercise, weight management, and pharmacological control (proper use of insulin and other medications) to manage their disease suffer the fewest complications.

HUNTINGTON'S DISEASE

Huntington's disease (HD) is an inherited, progressive brain disorder. It causes the degeneration of cells in the basal ganglia, a pair of nerve clusters deep in the brain that affect both the body and the mind. HD is caused by a single dominant gene that affects men and women of all races and ethnic groups. Figure 5.20 shows the inheritance pattern of HD.

Gene Responsible for HD Found

The gene mutation that produces HD was mapped to chromosome 4 in 1983 and cloned in 1993. (See Figure 5.21.) In the HD gene, which is officially called the huntingtin gene, the mutation involves a triplet of nucleotides: cytosine (C), adenine (A), and guanine (G), known as CAG. The mutation is an expansion of a nucleotide triplet repeat in the DNA that codes for the protein huntingtin. In unaffected people the gene has 30 or fewer of these triplets, but HD patients have 40 or more. These increased multiples either destroy the gene's ability to make the necessary protein or cause it to produce a misshapen and malfunctioning protein. Either way, the defect results in the death of brain cells.

The number of repeated triplets is inversely related to the age when the individual first experiences symptoms—the more repeated triplets, the younger the age of onset

FIGURE 5.20

The inheritance pattern of Huntington's disease (HD)

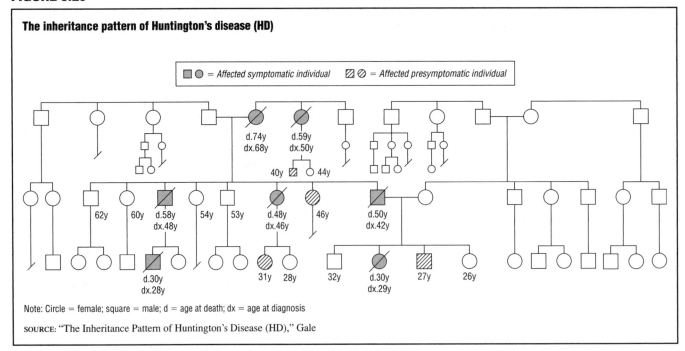

Note: Circle = female; square = male; d = age at death; dx = age at diagnosis

SOURCE: "The Inheritance Pattern of Huntington's Disease (HD)," Gale

of the disease. Like myotonic dystrophy, in which the symptoms of the disease often increase in severity from one generation to the next, the unstable triplet repeat sequence can lengthen from one generation to the next, with a resultant decrease in the age when symptoms first appear.

HD does not usually strike until mid-adulthood, between the ages of 30 and 50, although there is a juvenile form that can affect children and adolescents. Early symptoms, such as forgetfulness, a lack of muscle coordination, or a loss of balance, are often ignored, delaying the diagnosis. The disease gradually takes its toll over a 10- to 25-year period.

Within a few years characteristic involuntary movement (chorea) of the body, limbs, and facial muscles appears. As HD progresses, speech becomes slurred and swallowing becomes difficult. The patients' cognitive abilities decline and there are distinct personality changes—depression and withdrawal, sometimes countered with euphoria. Eventually, nearly all patients must be institutionalized, and they usually die as a result of choking or infections.

Prevalence of HD

HD, once considered rare, is now recognized as one of the more common hereditary diseases. The Genetics Home Reference states in "Huntington Disease" (January 9, 2011, http://ghr.nlm.nih.gov/condition/huntington-disease) that HD affects 3 to 7 per 100,000 people of European ancestry. HD appears to be less common in other populations, including people of Japanese, Chinese, and African descent. In "Learning about Huntington's Disease"

FIGURE 5.21

Location of the Huntington's disease gene (HD)

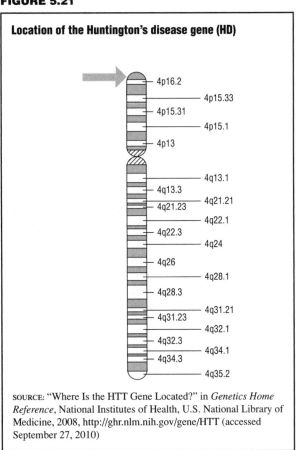

SOURCE: "Where Is the HTT Gene Located?" in *Genetics Home Reference*, National Institutes of Health, U.S. National Library of Medicine, 2008, http://ghr.nlm.nih.gov/gene/HTT (accessed September 27, 2010)

(December 28, 2010, http://www.genome.gov/10001215), the NHGRI reports that in the United States about 30,000 people have HD, an additional 35,000 people have some of the symptoms of HD, and 75,000 are carriers of the HD gene.

Prediction Test

In 1983 researchers identified a DNA marker that made it possible to offer a test to determine whether an individual has inherited the HD gene before symptoms appear. In some cases it is even possible to make a prenatal diagnosis on an unborn child. Many people, however, prefer not to know whether or not they carry the defective gene. Researchers are trying to determine if the exact number of excess triplets indicates when in life a person will be affected by the disease. Some scientists fear that the ability to tell people that they are going to develop an incurable disease and pinpoint when they will develop it will make genetic testing, already a difficult decision, even more complicated.

MUSCULAR DYSTROPHY

Muscular dystrophy (MD) is a term that applies to a group of more than 30 types of hereditary muscle-destroying disorders. More than a million Americans are affected by one of the forms of MD. Each variant of the disease is caused by defects in the genes that play important roles in the growth and development of muscles. Duchenne muscular dystrophy (DMD) is one of the most frequently occurring types of MD and is characterized by rapid progression of muscle degeneration that occurs early in life. DMD may also cause enlargement of some muscles, such as those in the calves, which is caused by fat and connective tissue replacing muscle. In all forms of MD the proteins produced by the defective genes are abnormal, causing the muscles to waste away. Unable to function properly, the muscle cells die and are replaced by fat and connective tissue. The symptoms of MD may not be noticed until as much as 50% of the muscle tissue has been affected.

DMD is X-linked, affects mostly males, and, according to the NHGRI, in "Learning about Duchenne Muscular Dystrophy" (December 20, 2010, http://www.genome .gov/19518854), strikes 1 out of 3,500 boys worldwide. The gene for DMD is located on the short arm of the X chromosome and encodes a large protein called dystrophin. (See Figure 5.22.) Dystrophin provides structural support for muscle cells, and without it the cell membrane becomes penetrable, allowing extracellular components into the cell. These additional components increase the intracellular pressure, causing the muscle cell to die.

With myotonic dystrophy the muscles contract but have diminishing ability to relax, and there is muscle weakening and wasting. Typically, the initial complaints are the loss of hand strength or tripping while walking or climbing stairs. Along with decreased muscle strength, myotonic dystrophy may cause mental deficiency, hair loss, and cataracts. The myotonic dystrophy gene is a

FIGURE 5.22

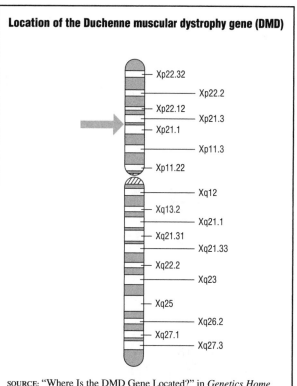

Location of the Duchenne muscular dystrophy gene (DMD)

SOURCE: "Where Is the DMD Gene Located?" in *Genetics Home Reference*, National Institutes of Health, U.S. National Library of Medicine, 2007, http://ghr.nlm.nih.gov/gene/DMD (accessed September 27, 2010)

protein kinase gene found on the long arm of chromosome 19. (See Figure 5.23.) The defect is a repeated set of three nucleotides—cytosine (C), thymine (T), and guanine (G)—in the gene. The symptoms of myotonic dystrophy frequently become more severe with each generation because mistakes in copying the gene from one generation to the next result in amplification of a genomic AGC/CTG triplet repeat, which is similar to the process that is observed in HD. Unaffected individuals have CTG repeats with between 3 and 37 iterations (repetitions) of the triplet. In contrast, people with the mild phenotype of myotonic dystrophy have between 40 and 170, and those with more serious forms of the disease have between 100 and 1,000 iterations.

All the various disorders labeled MD cause progressive weakening and wasting of muscle tissue. They vary, however, in terms of the usual age at the onset of symptoms, the rate of progression, and the initial group of muscles affected. The most common type, DMD, affects young boys, who show symptoms in early childhood and usually die from respiratory weakness or damage to the heart before adulthood. The gene is passed from the mother to her children. Females who inherit the defective gene generally do not manifest symptoms—they become carriers of the defective gene, and their children have a 50% chance of inheriting the disease. Other forms of MD appear later in life and are usually not fatal.

FIGURE 5.23

FIGURE 5.24

Chromosome 19 and the DM gene associated with myotonic dystrophy

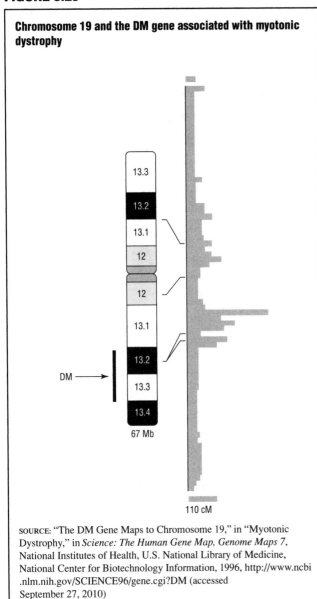

SOURCE: "The DM Gene Maps to Chromosome 19," in "Myotonic Dystrophy," in *Science: The Human Gene Map, Genome Maps 7*, National Institutes of Health, U.S. National Library of Medicine, National Center for Biotechnology Information, 1996, http://www.ncbi.nlm.nih.gov/SCIENCE96/gene.cgi?DM (accessed September 27, 2010)

Dystrophin and utrophin

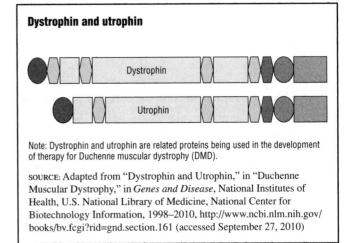

Note: Dystrophin and utrophin are related proteins being used in the development of therapy for Duchenne muscular dystrophy (DMD).

SOURCE: Adapted from "Dystrophin and Utrophin," in "Duchenne Muscular Dystrophy," in *Genes and Disease*, National Institutes of Health, U.S. National Library of Medicine, National Center for Biotechnology Information, 1998–2010, http://www.ncbi.nlm.nih.gov/books/bv.fcgi?rid=gnd.section.161 (accessed September 27, 2010)

In 1992 scientists discovered the defect in the gene that causes myotonic dystrophy. In people with this disorder, a segment of the gene is enlarged and unstable. This finding helps physicians to more accurately diagnose myotonic dystrophy. Researchers have since identified genes that are linked to other types of MD, including DMD, Becker MD, limb-girdle MD, and Emery-Dreifuss MD.

In 2005 Duygu Selcen and Andrew G. Engel of the Mayo Clinic identified a new form of MD that involves mutations in a protein called ZASP, which binds to cardiac (heart) and skeletal muscles, and reported their finding in "Mutations in ZASP Define a Novel Form of Muscular Dystrophy in Humans" (*Annals of Neurology*, vol. 57, no. 2, February 2005). The researchers detected ZASP mutations in 11 patients; in seven of the patients, they observed a dominant pattern of inheritance.

Treatment and Hope

There is no known cure for MD, but patients can be made more comfortable and functional by a combination of physical therapy, exercise programs, and orthopedic devices (special shoes, braces, or powered wheelchairs) that help them to maintain mobility and independence as long as possible.

Genetic research offers hope of finding effective treatments, and even cures, for these diseases. Gene therapy experiments designed to find a cure or a treatment for one or more of these types of MD are ongoing. Research teams have identified the crucial proteins produced by these genes, such as dystrophin, beta sarcoglycan, gamma sarcoglycan, and adhalin. One experimental treatment approach involves substituting a protein of comparable size, such as utrophin, for dystrophin to compensate for the loss of dystrophin. (See Figure 5.24.)

In "Progress in Muscular Dystrophy Research with Special Emphasis on Gene Therapy" (*Proceedings of the Japan Academy—Series B: Physical and Biological Sciences*, vol. 86, no. 7, 2010), Hideo Sugita and Shin'ichi Takeda report promising results of an experimental gene therapy called exon skipping that has been tested on small numbers of DMD patients. Exons are the areas of genes that have the instructions for synthesizing proteins such as dystrophin. Exon skipping aims to prevent the communication of faulty instructions in specific exons. The objective is to allow the remaining exons to be spliced together to correctly synthesize dystrophin. Sugita and Takeda find that by skipping exon 51 of the DMD gene, they can restore some dystrophin in the muscles of DMD patients.

PHENYLKETONURIA

Phenylketonuria (PKU) is an example of a disorder caused by a gene-environment interaction. As a result of the defect, the affected individual is unable to convert phenylalanine into tyrosine. Phenylalanine in the body accumulates in the blood and can reach toxic levels. (See Figure 5.4.) This toxicity may impair brain and nerve development and result in mental retardation, organ damage, and unusual posture. When it occurs during pregnancy, it may jeopardize the health and viability of the unborn child.

Originally, PKU was considered an autosomal recessive inherited error of metabolism that occurred when an individual received two defective copies, caused by mutations in both alleles of the phenylalanine hydroxylase gene found in the long arm of chromosome 12. (See Figure 5.25.) The environmental trigger—dietary phenylalanine—was not identified at first because phenylalanine is so prevalent in the diet, occurring in common foods such as milk and eggs and in the artificial sweetener aspartame. Recognition of dietary phenylalanine as a critical environmental trigger has enabled children born with PKU to lead normal lives when they are placed on low-phenylalanine diets, and mothers with the disease can bear healthy children.

The March of Dimes reports in "PKU (Phenylketonuria)" (March 2008, http://www.marchofdimes.com/baby/birthdefects_pku.html) that to identify people at risk of PKU, all newborns in the United States are screened at birth for high levels of phenylalanine in the blood. Approximately 1 out of 25,000 infants is diagnosed with PKU and, with proper diet, is likely to lead a healthy, normal life.

FIGURE 5.25

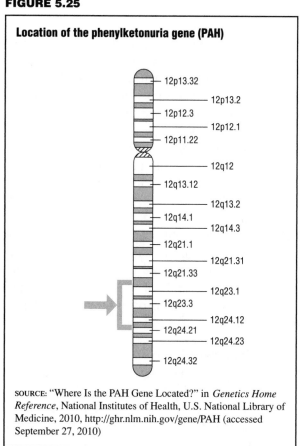

Location of the phenylketonuria gene (PAH)

12p13.32
12p13.2
12p12.3
12p12.1
12p11.22
12q12
12q13.12
12q13.2
12q14.1
12q14.3
12q21.1
12q21.31
12q21.33
12q23.1
12q23.3
12q24.12
12q24.21
12q24.23
12q24.32

SOURCE: "Where Is the PAH Gene Located?" in *Genetics Home Reference*, National Institutes of Health, U.S. National Library of Medicine, 2010, http://ghr.nlm.nih.gov/gene/PAH (accessed September 27, 2010)

SICKLE-CELL DISEASE

Sickle-cell disease (SCD) is a group of hereditary diseases, including sickle-cell anemia (SCA) and sickle B-thalassemia, in which the red blood cells contain an abnormal hemoglobin, called hemoglobin S (HbS). HbS is responsible for the premature destruction of red blood cells, or hemolysis. In addition, it causes the red blood cells to become deformed, actually taking on a sickle shape, particularly in parts of the body where the amount of oxygen is relatively low. These abnormally shaped cells cannot travel smoothly through the smaller blood vessels and capillaries. They tend to clog the vessels and prevent blood from reaching vital tissues. This blockage produces anoxia (lack of oxygen), which in turn causes more sickling and more damage.

SCA is an autosomal recessive disease caused by a point mutation in the hemoglobin beta gene (HBB) found on chromosome 11 at position 15.5. (See Figure 5.2.) A mutation in HBB results in the production of hemoglobin with an abnormal structure. Figure 5.26 shows how a point mutation in SCA causes the amino acid glutamine to be replaced by valine to produce abnormal hemoglobin called HbS. It also shows how when red blood cells with HbS are oxygen-deprived they become sickle shaped and may cause blockages that result in tissue death.

Symptoms of SCA

People with SCA have symptoms of anemia, including fatigue, weakness, fainting, and palpitations or an increased awareness of their heartbeat. These palpitations result from the heart's attempts to compensate for the anemia by pumping blood faster than normal.

In addition, patients experience occasional sickle-cell crises—attacks of pain in the bones and abdomen. Blood clots may also develop in the lungs, kidneys, brain, and other organs. A severe crisis or several acute crises can permanently damage various organs of the body. This damage can lead to death from heart failure, kidney failure, or stroke. The frequency of these crises varies from patient to patient. A sickle-cell crisis, however, occurs more often during infections and after an accident or injury.

Who Contracts SCD?

Both the sickle-cell trait and the disease exist almost exclusively in people of African, Native American, and Hispanic descent and in those from parts of Italy, Greece, Middle Eastern countries, and India. If one parent has the sickle-cell gene, then the couple's offspring will carry the trait; if both the mother and the father have the trait, then their children may be born with SCA. This trait is relatively common among African-Americans. People of African descent are advised to seek genetic counseling and testing for the trait before starting a family. The Genetics Home Reference notes in "Sickle Cell Disease" (January 9, 2011,

FIGURE 5.26

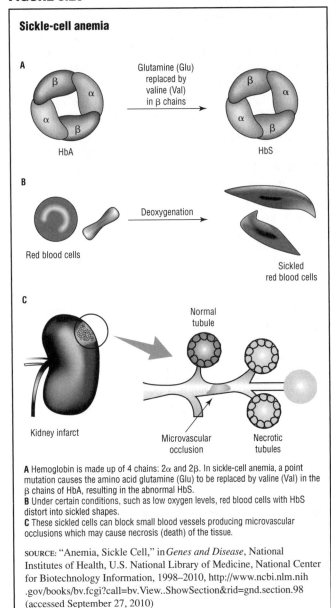

Sickle-cell anemia

A Hemoglobin is made up of 4 chains: 2α and 2β. In sickle-cell anemia, a point mutation causes the amino acid glutamine (Glu) to be replaced by valine (Val) in the β chains of HbA, resulting in the abnormal HbS.
B Under certain conditions, such as low oxygen levels, red blood cells with HbS distort into sickled shapes.
C These sickled cells can block small blood vessels producing microvascular occlusions which may cause necrosis (death) of the tissue.

SOURCE: "Anemia, Sickle Cell," in *Genes and Disease*, National Institutes of Health, U.S. National Library of Medicine, National Center for Biotechnology Information, 1998–2010, http://www.ncbi.nlm.nih.gov/books/bv.fcgi?call=bv.View..ShowSection&rid=gnd.section.98 (accessed September 27, 2010)

http://ghr.nlm.nih.gov/condition/sickle-cell-disease) that SCD is the most common inherited blood disorder in the United States, affecting 70,000 to 80,000 Americans, most of whom have African ancestry. SCA occurs in approximately 1 out of 500 African-American births and in 1 out of 1,000 to 1,400 Hispanic births. The occurrence of SCA in other groups is much lower.

Treatment of SCD

There is no universal cure for SCD, but the symptoms can be treated. Crises accompanied by extreme pain are the most common problems and can usually be treated with painkillers. Maintaining a healthy diet and receiving prompt treatment for any type of infection or injury is important. Special precautions are often necessary before any type of surgery, and for major surgery some patients receive transfusions to boost their levels of hemoglobin

(the oxygen-bearing, iron-containing protein in red blood cells). In early 1995 a medication that prevented the cells from clogging vessels and cutting off oxygen was approved by the U.S. Food and Drug Administration.

Many adults with SCD now take hydroxyurea, an anticancer drug that causes the body to produce red blood cells that resist sickling. Martin H. Steinberg et al. note in "Effect of Hydroxyurea on Mortality and Morbidity in Adult Sickle Cell Anemia: Risks and Benefits up to 9 Years of Treatment" (*Journal of the American Medical Association*, vol. 289, no. 13, April 2, 2003) that not only do patients on hydroxyurea have fewer crises but also they have a significant survival advantage when compared with SCD patients who do not take the medication. Subjects treated with it showed 40% lower mortality than those who did not take the medication.

Throughout the first decade of the 21st century scientists attempted to use gene therapy to replace the faulty genes that produce hemoglobin. Abiola Olowoyeye and Charles I. Okwundu conducted a review of the published research and reported their findings in "Gene Therapy for Sickle Cell Disease" (*Cochrane Collaboration Database Systematic Review*, no. 8, August 2010). The researchers find that because there have not been rigorous clinical trials comparing gene therapy to standard treatment, there is not yet evidence that gene therapy is effective for SCD.

TAY-SACHS DISEASE

Tay-Sachs disease (TSD) is caused by mutations in the HEXA gene, which is located on the long arm of chromosome 15 between positions 23 and 24. (See Figure 5.27.) It is a fatal genetic disorder in children that causes the progressive destruction of the central nervous system. It is caused by the absence of an important enzyme called hexosaminidase A (hex-A). Without hex-A, a fatty substance called GM2 ganglioside builds up abnormally in the cells, particularly in the brain's nerve cells. (Figure 5.28 shows how the absence of, or defect in, the hex-A protein prevents complete processing of GM2 ganglioside.) Eventually, these cells degenerate and die. This destructive process begins early in the development of a fetus, but the disease is not usually diagnosed until the baby is several months old. By the time a child with TSD is four or five years old, the nervous system is so badly damaged that the child dies.

How Is TSD Inherited?

TSD is an autosomal recessive genetic disorder caused by mutations in both alleles of the HEXA gene on chromosome 15. Both the mother and the father must be carriers of the defective TSD gene to produce a child with the disease.

FIGURE 5.27

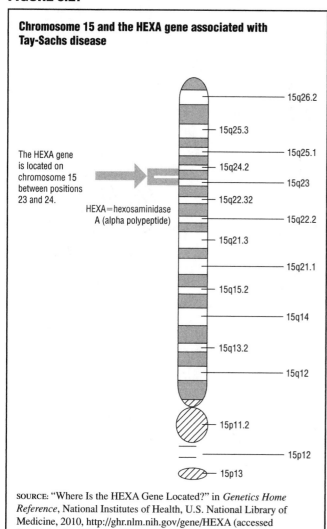

Chromosome 15 and the HEXA gene associated with Tay-Sachs disease

The HEXA gene is located on chromosome 15 between positions 23 and 24.

HEXA = hexosaminidase A (alpha polypeptide)

15q26.2
15q25.3
15q25.1
15q24.2
15q23
15q22.32
15q22.2
15q21.3
15q21.1
15q15.2
15q14
15q13.2
15q12
15p11.2
15p12
15p13

SOURCE: "Where Is the HEXA Gene Located?" in *Genetics Home Reference*, National Institutes of Health, U.S. National Library of Medicine, 2010, http://ghr.nlm.nih.gov/gene/HEXA (accessed September 27, 2010)

People who carry the gene for TSD are entirely unaffected and usually unaware that they have the potential to pass this disease onto their offspring. A blood test distinguishes Tay-Sachs carriers from noncarriers. Blood samples may be analyzed by enzyme assay or DNA studies. Enzyme assay measures the level of hex-A in the blood. Carriers have less hex-A than noncarriers. When only one parent is a carrier, the couple will not have a child with TSD. When both parents carry the recessive TSD gene, they have a one in four chance in every pregnancy of having a child with the disease. They also have a 50% chance of bearing a child who is a carrier. Prenatal diagnosis early in pregnancy can predict if the unborn child has TSD. If the fetus has the disease, the couple may choose to terminate the pregnancy.

Who Is at Risk?

Some genetic diseases, such as TSD, occur most frequently in a specific population. As the American neurologist Bernard Sachs (1858–1944) observed, individuals of east European (Ashkenazi) Jewish descent have the highest risk of being carriers of TSD. The National Tay-Sachs and Allied Diseases Association explains in "Tay-Sachs Disease" (2011, http://www.tay-sachs.org/taysachs_disease.php) that approximately 1 out of 27 Jews in the United States is a carrier of the TSD gene. French-Canadians and Cajuns also have the same carrier rate as Ashkenazi Jews. Among Irish-Americans, the carrier rate is approximately 1 out of 50. In the general population the carrier rate is 1 out of 250.

FIGURE 5.28

Molecular basis for Tay-Sachs disease

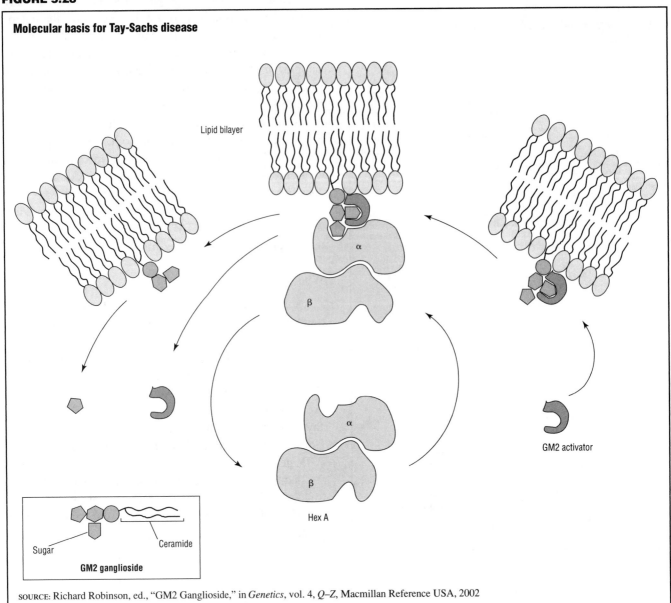

SOURCE: Richard Robinson, ed., "GM2 Ganglioside," in *Genetics*, vol. 4, *Q–Z*, Macmillan Reference USA, 2002

CHAPTER 6
GENETIC TESTING

When the consequences of the specific disorder in question are treatable, most people would agree that genetic testing makes sense. But what about disorders for which no preventive measures or treatments are available? What happens, for instance, when a young adult is found to have the mutation that causes a fatal disorder?

—Karen Norrgard, *Ethics of Genetic Testing: Medical Insurance and Genetic Discrimination* (2008)

Throughout the first decade of the 21st century the definitions of health and disease were transformed by advances in genetics. Genetic testing has enabled researchers and clinicians to detect inherited traits, diagnose heritable conditions, determine and quantify the likelihood that a heritable disease will develop, and identify genetic susceptibility to familial disorders. Many of the strides made in genetic diagnostics are direct results of the Human Genome Project, an international effort begun in 1990 by the U.S. Department of Energy and the National Institutes of Health, which mapped and sequenced the human genome in its entirety. The increasing availability of genetic testing has been one of the most immediate applications of this groundbreaking research.

A genetic test is the analysis of human deoxyribonucleic acid (DNA), ribonucleic acid (RNA), chromosomes, and proteins to detect heritable disease-related genotypes, mutations, phenotypes, or karyotypes (standard pictures of the chromosomes in a cell) for the purposes of diagnosis, treatment, and other clinical decision making. Most genetic testing is performed by drawing a blood sample and extracting DNA from white blood cells. Genetic tests may detect mutations at the chromosomal level, such as additional, absent, or rearranged chromosomal material, or even subtler abnormalities, such as a substitution in one of the bases that make up the DNA. There is a broad range of techniques that can be used for genetic testing. Genetic tests have diverse purposes, including screening for and diagnosis of genetic disease in newborns, children, and adults; the identification of future health risks; the prediction of drug responses; and the assessment of risks to future children.

There is a difference between genetic tests performed to screen for disease and testing conducted to establish a diagnosis. Diagnostic tests are intended to definitively determine whether a patient has a particular problem. They are generally complex tests and commonly require sophisticated analysis and interpretation. They may be expensive and are generally performed only on people believed to be at risk, such as patients who already have symptoms of a specific disease.

In contrast, screening is performed on healthy, asymptomatic (showing no symptoms of disease) people and often to the entire relevant population. A good screening test is relatively inexpensive, easy to use and interpret, and helps identify which individuals in the population are at higher risk of developing a specific disease. By definition, screening tests identify people who need further testing or those who should take special preventive measures or precautions. For example, people who are found to be especially susceptible to genetic conditions with specific environmental triggers are advised to avoid the environmental factors linked to developing the disease. Examples of genetic tests used to screen relevant populations include those that screen people of Ashkenazi Jewish heritage (the east European Jewish population primarily from Germany, Poland, and Russia, as opposed to the Sephardic Jewish population primarily from Spain, parts of France, Italy, and North Africa) for Tay-Sachs disease, African-Americans for sickle-cell disease, and the fetuses of expectant mothers over the age of 35 years for Down syndrome.

QUALITY AND UTILITY OF GENETIC TESTS

Like all diagnostic and screening tests, the quality and utility of genetic tests depend on their reliability, validity, sensitivity, specificity, positive predictive value, and negative predictive value. Reliability of testing refers to the test's ability to be repeated and to produce equivalent results in comparable circumstances. A reliable test is consistent and measures the same way each time it is used with the same patients in the same circumstances. For example, a well-calibrated balance scale is a reliable instrument for measuring body weight.

Validity is the accuracy of the test. It is the degree to which the test correctly identifies the presence of disease, blood level, or other quality or characteristic it is intended to detect. For example, if a person puts an object he or she knows weighs 10 pounds (4.5 kg) on a scale and the scale states it weighs 10 pounds, then the scale's results are valid. There are two components of validity: sensitivity and specificity.

Sensitivity is the test's ability to identify people who have the disease. Mathematically speaking, it is the percentage of people with the disease who test positive for the disease. Specificity is the test's ability to identify people who do not have the disease—it is the percentage of people without the disease who test negative for the disease. Ideally, diagnostic and screening tests should be highly sensitive and highly specific, thereby accurately classifying all people tested as either positive or negative. In practice, however, sensitivity and specificity are frequently inversely related—most tests with high levels of sensitivity have low specificity, and the reverse is also true.

The likelihood that a test result will be incorrect can be gauged based on the sensitivity and specificity of the test. For example, if a test's sensitivity is 95%, then when 100 patients with the disease are tested, 95 will test positive and five will test false negative—they have the disease but the test has failed to detect it. For example, disorders such as Charcot-Marie-Tooth disease (a group of inherited, slowly developing disorders that result from progressive damage to the nerves) can arise from mutations in one of many different genes, and because some of these genes have not yet been identified, they will not be detected, so a false negative result might be reported. By contrast, if a test is 90% specific, when 100 healthy, disease-free people are tested, 90 will receive negative test results and 10 will be given false-positive results, meaning they do not have the disease but the test has inaccurately classified them as being positive.

The positive predictive value is the percentage of people who actually have the disease of all those with positive test results. The negative predictive value measures the percent of all the people with negative test results who do not have the disease.

PREGNANCY, CHILDBIRTH, AND GENETIC TESTING

There are thousands of genetic diseases, such as sickle-cell anemia, cystic fibrosis, and Tay-Sachs disease, that may be passed from one generation to the next. Many tests have been developed to help screen parents at risk of passing on genetic disease to their children as well as to identify embryos, fetuses, and newborns who suffer from genetic diseases.

Carrier Identification

Carrier identification is the term for genetic testing to determine whether a healthy individual has a gene that may cause disease if passed on to his or her offspring. It is usually performed on people considered to be at higher than average risk, such as those of Ashkenazi Jewish descent, who have a 1 out of 27 chance of being Tay-Sachs carriers (in other populations the risk is 1 out of 250), according to the National Tay-Sachs and Allied Diseases Association, in "Tay-Sachs Disease" (2011, http://www.tay-sachs.org/taysachs_disease.php). Testing is necessary because many carriers have just one copy of a gene for an autosomal recessive trait and are unaffected by the trait or disorder. Only someone with two copies of the gene will actually have the disorder. So even though it is widely assumed that everyone is an unaffected carrier of at least one autosomal recessive gene, it only presents a problem in terms of inheritance when two parents have the same recessive disorder gene (or both are carriers). In this instance, the offspring would each have a one in four chance of receiving a defective copy of the gene from each parent and developing the disorder. Figure 6.1 shows the pattern of inheritance of an autosomal recessive disorder.

Carrier testing is offered to individuals who have family members with a genetic condition, people with family members who are identified carriers, and members of racial and ethnic groups known to be at high risk. Figure 6.2 indicates how carrier testing would be used for a family that is affected by cystic fibrosis and among African-Americans who may carry the gene for sickle-cell anemia.

Preimplantation Genetic Diagnosis

Preimplantation genetic diagnosis is a newer genetic test that enables parents undergoing in vitro fertilization (fertilization that takes place outside the body) to screen an embryo for specific genetic mutations when it is no larger than six or eight cells and before it is implanted in the uterus to grow and develop. Figure 6.3 explains how preimplantation genetic diagnosis is performed.

Prenatal Genetic Testing

Prenatal genetic testing enables physicians to diagnose diseases in the fetus. Most genetic tests examine blood or other tissue to detect abnormalities. An example

FIGURE 6.1

Autosomal recessive disorder inheritance pattern

○ Female: unaffected   □ Male: unaffected
● Female: affected   ■ Male: affected

Great grandparents

Grandparents

Parents

- Autosomal recessive conditions are usually seen in a single generaion of a family.
- Mothers and fathers are equally likely to transmit or inherit the disorder.

SOURCE: "Pedigree Illustrating Autosomal Recessive Inheritance Pattern" in *GeneTests Database: GeneReviews Illustrated Glossary*, National Institutes of Health, U.S. National Library of Medicine, National Human Genome Research Institute, 2004, http://www.genetests.org/servlet/access?id=8888892&qry=autosomal+recessive+disorder&submit=Search&db=genestar&site=gc&fcn=term&format=search&gtreport2=true&submit=Search+by+term&key=JLNoRAYoitCUQ (accessed September 29, 2010)

FIGURE 6.2

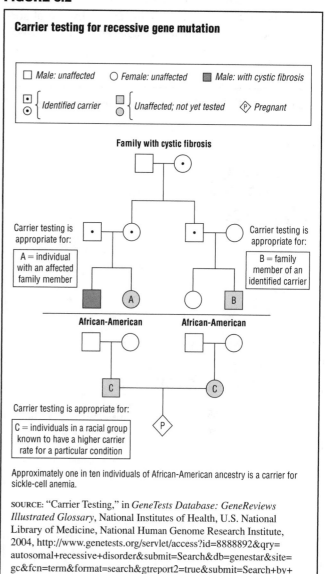

Carrier testing for recessive gene mutation

□ Male: unaffected   ○ Female: unaffected   ■ Male: with cystic fibrosis

⊡/⊙ Identified carrier   ▫/○ Unaffected; not yet tested   ◇ Pregnant

Family with cystic fibrosis

Carrier testing is appropriate for:

A = individual with an affected family member

Carrier testing is appropriate for:

B = family member of an identified carrier

African-American   African-American

Carrier testing is appropriate for:

C = individuals in a racial group known to have a higher carrier rate for a particular condition

Approximately one in ten individuals of African-American ancestry is a carrier for sickle-cell anemia.

SOURCE: "Carrier Testing," in *GeneTests Database: GeneReviews Illustrated Glossary*, National Institutes of Health, U.S. National Library of Medicine, National Human Genome Research Institute, 2004, http://www.genetests.org/servlet/access?id=8888892&qry=autosomal+recessive+disorder&submit=Search&db=genestar&site=gc&fcn=term&format=search&gtreport2=true&submit=Search+by+term&key=JLNoRAYoitCUQ (accessed September 29, 2010)

---

of a blood test is the triple-marker screen. This test measures levels of alpha fetoprotein (AFP), human chorionic gonadotropin (hCG), and unconjugated estriol and can identify some birth defects such as Down syndrome and neural tube defects. (Two of the most common neural tube defects are anencephaly [absence of most of the brain] and spina bifida [incomplete development of the back and spine].)

The fetal yolk sac and the fetal liver make AFP, which is continuously processed by the fetus and excreted into the amniotic fluid. A small amount crosses the placenta and can be found in maternal blood. Screening for AFP levels is based on maternal age, fetal gestation, and the number of fetuses the mother is carrying. Elevated levels of AFP are associated with conditions such as spina bifida and low levels are found with Down syndrome. Because AFP levels alone may not always adequately detect disorders, two other blood serum tests have been developed. The glycoprotein hCG is produced by the placenta. Normally, hCG is elevated at the time of implantation, but decreases at about eight weeks of gestation, and then drops again at approximately 12 weeks of gestation. Elevated levels of hCG are found with Down syndrome. The placenta also produces unconjugated estriol. As with AFP, lower unconjugated estriol maternal serum levels are also

found with Down syndrome. Triple-marker screen results are usually available within several days and women with abnormal results are often advised to undergo additional diagnostic testing such as chorionic villus sampling (CVS), amniocentesis, or percutaneous umbilical blood sampling (withdrawing blood from the umbilical cord).

CVS enables obstetricians and perinatologists (physicians specializing in the evaluation and care of high-risk expectant mothers and infants) to assess the progress of pregnancy during the first trimester (the first three months). A physician passes a small, flexible tube called a catheter through the cervix to extract chorionic villi tissue (cells that will become the placenta). Because the chorion and the fetus develop from trophoblasts, they are genetically identical (both contain the same DNA and chromosomes). The cells obtained via CVS are examined in the laboratory for indications of genetic disorders, such as cystic fibrosis, Down syndrome, Tay-Sachs disease, and thalassemia (a disorder characterized by the abnormal production of hemoglobin that can lead to growth failure, bone deformities, and enlarged liver and spleen), and the results of testing are available within seven to 14 days. Table 6.1 describes CVS and other prenatal diagnostic tests and specifies when during pregnancy they are performed.

Amniocentesis involves taking a sample of the fluid that surrounds the fetus in the uterus for chromosome analysis. An amniocentesis is usually performed at 15 to 20 weeks of gestation, although it can be done as early as 12 weeks. (See Table 6.1.) The physician inserts a hollow needle through the abdominal wall and the wall of the uterus to obtain approximately 0.7 ounces (20 ml) of amniotic fluid. Fetal karyotyping, DNA analysis, and biochemical testing may be performed on the isolated fetal cells. Like CVS, amniocentesis samples and analyzes cells that are derived from the baby to enable parents to learn of chromosomal abnormalities, as well as the gender of the unborn child, about two weeks after the test is performed.

Using samples of genetic material obtained from amniocentesis or CVS, physicians can detect disease in an unborn child. Down syndrome (also known as trisomy 21, because it is caused by an extra copy of chromosome 21) is the genetic disease most often identified using this technique. Down syndrome is rarely inherited; most cases result from an error in the formation of the ovum (egg) or sperm, leading to the inclusion of an extra chromosome 21 at conception. As with prenatal diagnosis for most inherited genetic diseases, this use of genetic testing is focused on reproductive decision making.

The most invasive prenatal procedure for genetic testing is periumbilical blood sampling. Table 6.1 describes the technique. Periumbilical blood sampling poses the greatest risk to an unborn child—1 out of 50 miscarriages occurs as a result of this procedure. It is used when a diagnosis must be made quickly. For example, when an expectant mother is exposed to an infectious agent with the potential to produce birth defects, it may be used to examine fetal blood for the presence of infection.

FIGURE 6.3

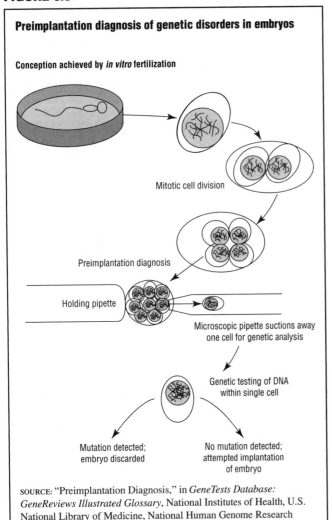

Preimplantation diagnosis of genetic disorders in embryos

Conception achieved by *in vitro* fertilization

Mitotic cell division

Preimplantation diagnosis

Holding pipette

Microscopic pipette suctions away one cell for genetic analysis

Genetic testing of DNA within single cell

Mutation detected; embryo discarded

No mutation detected; attempted implantation of embryo

SOURCE: "Preimplantation Diagnosis," in *GeneTests Database: GeneReviews Illustrated Glossary*, National Institutes of Health, U.S. National Library of Medicine, National Human Genome Research Institute, 2004, http://www.genetests.org/servlet/access?id=8888892&qry=autosomal+recessive+disorder&submit=Search&db=genestar&site=gc&fcn=term&format=search&gtreport2=true&submit=Search+by+term&key=JLNoRAYoitCUQ (accessed September 29, 2010)

TABLE 6.1

Prenatal diagnostic testing methods

Procedure	Technique (ultrasound guided)	Sample	Timing in gestation*
Chorionic villus sampling (CVS)	Needle inserted through mother's abdomen or catheter through cervix	Chorionic villus	10–12 weeks
Early amniocentesis	Needle inserted through mother's abdomen into amniotic sac	Amniotic fluid and/or amniocytes	<15 weeks
Amniocentesis	Needle inserted through mother's abdomen into amniotic sac	Amniotic fluid and/or amniocytes	15–20 weeks
Placental biopsy	Needle inserted through mother's abdomen into placenta	Placental tissue	>12 weeks
Periumbilical blood sampling (PUBS) (aka cordocentesis)	Needle inserted through mother's abdomen into fetal umbilical vein	Fetal blood	>18 weeks
Fetoscopy with fetal skin biopsy	Needle inserted through mother's abdomen, camera used to facilitate biopsy	Fetal skin	>18 weeks

*Menstrual weeks calculated either from the first day of the last normal menstrual period or by ultrasound measurements.

SOURCE: "Prenatal Diagnosis," in *GeneTests Database: GeneReviews Illustrated Glossary*, National Institutes of Health, U.S. National Library of Medicine, National Human Genome Research Institute, 2002, http://www.ncbi.nlm.nih.gov/bookshelf/br.fcgi?book=gene&part=glossary#IX-P (accessed September 29, 2010)

Until 2006 it was thought that women undergoing CVS were more likely to miscarry than those who had amniocentesis. However, Aaron B. Caughey, Linda M. Hopkins, and Mary E. Norton refute this notion in "Chorionic Villus Sampling Compared with Amniocentesis and the Difference in the Rate of Pregnancy Loss" (*Obstetrics and Gynecology*, vol. 108, no. 3, pt. 1, September 2006). The researchers analyzed the outcomes of 9,886 CVS and 30,893 amniocentesis procedures and found that CVS was no more likely than amniocentesis to lead to pregnancy loss. Caughey, Hopkins, and Norton attribute previously reported higher rates of miscarriage resulting from CVS to the fact that clinicians were not yet experienced at performing the newer procedure.

In "Update on Procedure-Related Risks for Prenatal Diagnosis Technique" (*Fetal Diagnosis and Therapy*, vol. 27, no. 1, 2010), Ann Tabor and Zarko Alfirevic observe that the introduction and widespread use of increasingly effective screening methods has resulted in a decreasing number of invasive prenatal diagnostic tests such as CVS and amniocentesis. They also reaffirm the observation that experienced clinicians have higher success rates and lower complication rates. A review of the literature reveals generally comparable rates of miscarriage for CVS and amniocentesis; however, amniocentesis performed prior to 15 weeks has a significantly higher rate of miscarriage than either CVS or mid-trimester amniocentesis. Tabor and Alfirevic assert that amniocentesis should not be performed before 15 weeks' gestation and that CVS should not be performed before 10 weeks' gestation because these invasive prenatal diagnostic tests may increase the risk of limb defects.

Genetic Testing of Newborns

The most common form of genetic testing is the screening of blood taken from newborn infants for genetic abnormalities. The Save Babies through Screening Foundation, Inc. (2010, http://www.savebabies.org/index.html) reports that more than 4 million newborns in the United States are screened every year for specific genetic disorders, such as phenylketonuria (PKU), and for other medical conditions that are only indirectly genetically linked, such as congenital hypothyroidism (underactive thyroid gland). PKU is an inherited error of metabolism resulting from a deficiency of an enzyme called phenylalanine hydroxylase. The lack of this enzyme can produce mental retardation, organ damage, and postural problems. Children born with PKU must pay close attention to their diet to lead a healthy, normal life.

Laboratory Techniques for Genetic Prenatal Testing

Genetic testing is performed on chromosomes, genes, or gene products to determine whether a mutation is causing or may cause a specific condition. Direct testing examines the DNA or RNA that makes up a gene. Linkage testing looks for disease-causing gene markers in family members from at least two generations. Biochemical testing assays certain enzymes or proteins, which are the products of genes. Cytogenetic testing examines the chromosomes.

Generally, a blood sample or buccal smear (cells from the mouth) is used for genetic tests. Other tissues used include skin cells from a biopsy, fetal cells, or stored tissue samples. (Table 6.1 lists the tissue samples that are used in each procedure.) Testing requires highly trained, certified technicians and laboratories because the procedures are complex and varied, the technology is new and evolving, and hereditary conditions are often rare. In the United States laboratories performing clinical genetic tests must be approved under the Clinical Laboratory Improvement Amendments (CLIA), passed by Congress in 1988 to establish the standards with which all laboratories that test human specimens must comply to receive certification. CLIA standards determine the qualifications of laboratory personnel, categorize the complexity of various tests, and oversee quality improvement and assurance. In 2009, 600 laboratories in the United States were performing genetic testing to detect and diagnose approximately 1,900 conditions. Figure 6.4 shows the tremendous growth of both laboratories and diseases for which testing was available from 1993 to 2009.

The polymerase chain reaction (PCR) technique permits rapid cloning and DNA analysis and allows selective amplification of specific DNA sequences. A PCR can be performed in hours and is a sensitive test that may be used to screen for altered genes, but it is limited by the size and length of the DNA sequences that can be cloned. Figure 6.5 explains how PCR is performed.

Fluorescence in situ hybridization (FISH), in which a fluorescent label is attached to a DNA probe that will bind to the complementary DNA strands, provides a unique opportunity to view specific genetic codes. The FISH technique may be used on cells and fluid obtained by CVS and amniocentesis as well as on maternal blood. With this technique, a single strand of DNA is used to create a probe that attaches at the specific gene location. (See Figure 6.6.) To separate the double-stranded DNA, heat or chemicals are used to break the chemical bonds of the DNA and obtain a single strand.

There are three kinds of chromosome-specific probes: repetitive probes, painting probes, and locus-specific probes. Repetitive probes produce intense signals by creating tandem repeats of base pairs. Painting probes are collections of specific DNA sequences that may extend along either part or all of an individual chromosome. These probe labels are most useful for identifying complex rearrangements of genetic material in structurally abnormal chromosomes. Probes that can hybridize to a single gene locus are called locus-specific probes. They can be used to identify a gene in a particular region of a chromosome. Locus-specific probes

FIGURE 6.4

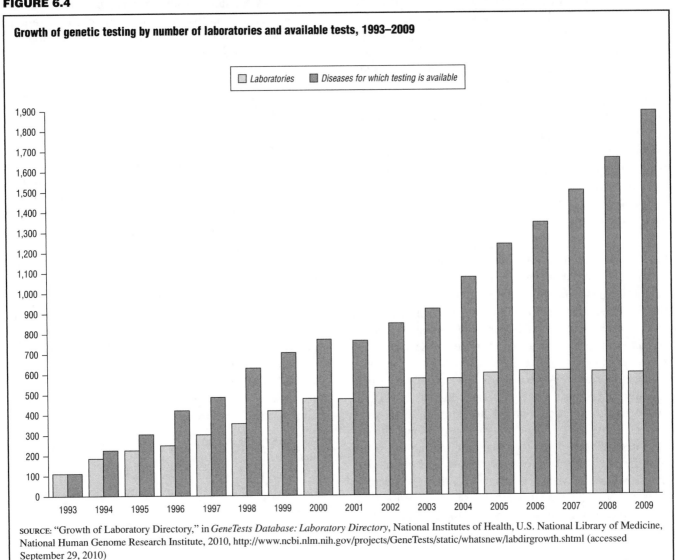

Growth of genetic testing by number of laboratories and available tests, 1993–2009

☐ Laboratories   ■ Diseases for which testing is available

SOURCE: "Growth of Laboratory Directory," in *GeneTests Database: Laboratory Directory*, National Institutes of Health, U.S. National Library of Medicine, National Human Genome Research Institute, 2010, http://www.ncbi.nlm.nih.gov/projects/GeneTests/static/whatsnew/labdirgrowth.shtml (accessed September 29, 2010)

are used to identify a deletion or duplication of genetic material. The FISH procedure allows a signal to be visualized that indicates the presence or absence of DNA. The entire process can be completed in less than eight hours using five to seven probes. For this reason, FISH is frequently used as a rapid screen for trisomies and genetic disorders.

Even though FISH is the most commonly used and most readily available prenatal diagnostic cytogenetic technique, it has limited ability to detect translocations, deletions, and inversions. Microdissection FISH (also used for prenatal diagnosis) is another method that is more sensitive to these alterations. Microdissection FISH constructs probes to define specific regions of the human chromosome. Figure 6.7 shows how the microdissection technique is used to identify structurally abnormal chromosomes.

Newer technologies allow for the identification of all human chromosomes, thereby expanding the FISH application to the entire genome. Multiplex FISH applies com-

binations of probes to color components of fluorescent dye during metaphase to visualize each chromosome. This type of FISH procedure evolved from the whole-chromosome painting probes and uses a combination of fluorochromes to identify each chromosome. Just five fluorophores are needed to decode the entire complement of human chromosomes. Multicolor spectral karyotyping uses computer imaging and Fourier spectroscopy to increase the analysis of genetic markers. (See Figure 6.8.) Even though FISH techniques are considered highly reliable, there are limitations to multiplex and multicolor FISH, in that inversion or subtle deletions may be overlooked. The greatest advantage of multiplex FISH is its application in cases of dysmorphic infants. Spectral karyotyping techniques can detect cryptic unbalanced translocations that cannot be identified by other techniques.

GENETIC TESTING FOR SICKLE-CELL ANEMIA. Sickle-cell anemia is an autosomal recessive disease that results when hemoglobin S is inherited from both parents. (When

FIGURE 6.5

Polymerase chain reaction (PCR)

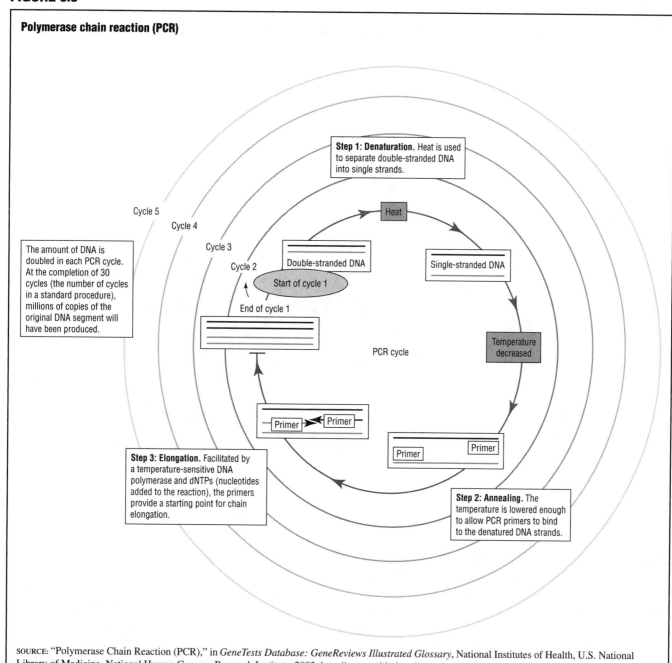

Step 1: Denaturation. Heat is used to separate double-stranded DNA into single strands.

Heat

Double-stranded DNA

Single-stranded DNA

Start of cycle 1

Cycle 2

Cycle 3

Cycle 4

Cycle 5

End of cycle 1

Temperature decreased

PCR cycle

The amount of DNA is doubled in each PCR cycle. At the completion of 30 cycles (the number of cycles in a standard procedure), millions of copies of the original DNA segment will have been produced.

Primer Primer

Primer Primer

Step 3: Elongation. Facilitated by a temperature-sensitive DNA polymerase and dNTPs (nucleotides added to the reaction), the primers provide a starting point for chain elongation.

Step 2: Annealing. The temperature is lowered enough to allow PCR primers to bind to the denatured DNA strands.

SOURCE: "Polymerase Chain Reaction (PCR)," in *GeneTests Database: GeneReviews Illustrated Glossary*, National Institutes of Health, U.S. National Library of Medicine, National Human Genome Research Institute, 2003, http://www.ncbi.nlm.nih.gov/bookshelf/br.fcgi?book=gene&part=glossary#IX-P (accessed September 29, 2010)

hemoglobin S is inherited from only one parent, the individual is a sickle-cell carrier.) Because normal and sickle hemoglobins differ at only one amino acid in the hemoglobin gene, a test called hemoglobin electrophoresis is used to establish the diagnosis.

Genetic testing for sickle-cell anemia involves restriction fragment length polymorphism (DNA sequence variant) that uses specific enzymes to cut the DNA. (See Figure 6.9.) These enzymes cut the DNA at a specific base sequence on the normal gene but not on a gene in which a mutation is present. As a result of this technique, there are longer fragments of sickle hemoglobin. Another technique

known as gel electrophoresis sorts the DNA fragments by size. Autoradiography renders the DNA fragments by generating an image after radioactive probes have labeled the DNA fragments that contain the specific gene sequence. The location of the fragments distinguishes carrier status (heterozygous) from sickle-cell anemia (homozygous), or normal blood.

Advanced Techniques Detect Fetal Gene Mutations

Ying Li et al. describe in "Detection of Paternally Inherited Fetal Point Mutations for Beta-Thalassemia Using Size-Fractionated Cell-Free DNA in Maternal Plasma"

FIGURE 6.6

Fluorescence in situ hybridization (FISH) is useful for gene mapping and for identifying abnormalities in chromosomes

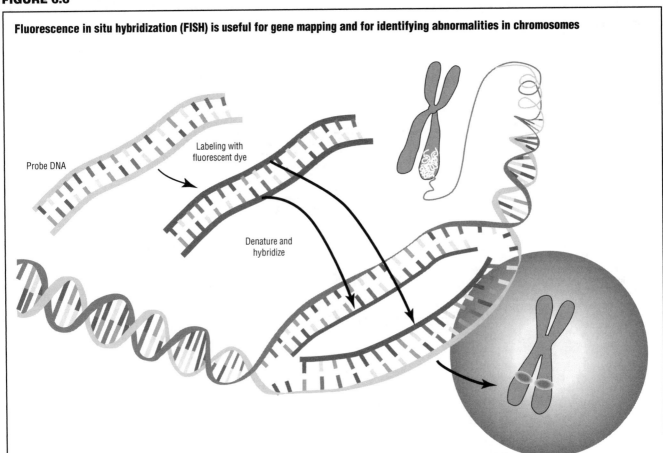

SOURCE: "Fluorescent In Situ Hybridization," in *GeneTests Database: GeneReviews Illustrated Glossary*, U.S. Department of Health and Human Services, National Institutes of Health, National Human Genome Research Institute, 2003, http://www.ncbi.nlm.nih.gov/bookshelf/br.fcgi?book=gene&part= glossary#IX-P (accessed September 29, 2010)

(*Journal of the American Medical Association*, vol. 293, no. 7, February 16, 2005) a technique that identifies fetal gene mutations such as beta-thalassemia from samples of maternal blood. The technique relies on the fact that circulatory fetal DNA sequences consist of fewer than 300 base pairs, whereas maternal DNA exceeds 500 base pairs. Li et al. used PCR amplification to select for paternally inherited DNA sequences. The presence of paternal mutant alleles for beta-thalassemia was then detected by allele-specific PCR. The researchers tested the technique on maternal blood samples from expectant mothers whose fetuses were at risk for beta-thalassemia because the father was a carrier for one of four beta-globin gene mutations. The results were verified by comparing the findings from CVS.

In "Prenatal Detection of Beta-Thalassemia CD17 (A→T) Mutation by Polymerase Chain Reaction/Ligase Detection Reaction/Capillary Electrophoresis for Fetal DNA in Maternal Plasma—a Case Report" (*Fetal Diagnosis and Therapy*, vol. 27, no. 1, 2010), Ping Yi et al. report successful use of PCR to amplify and detect a paternally inherited fetal CD17 (A→T) mutation in maternal blood plasma DNA. The results were consistent with those obtained by using PCR to examine amniotic fluid DNA. Yi et al. conclude that PCR used in conjunction with ligase detection reaction/capillary electrophoresis, which is another advanced laboratory technique, has a high level of sensitivity and effectively detects paternally inherited fetal point mutations in maternal plasma.

Another technique analyzes amniotic fluid using oligonucleotide hybridization and microarray data analysis. (Figure 6.10 shows how microarray technology is performed to distinguish normal tissue from a tumor. It is used to look at many genes simultaneously and allows detection of RNA and DNA molecules that are labeled with a dye such as biotin.) In "Global Gene Expression Analysis of the Living Human Fetus Using Cell-Free Messenger RNA in Amniotic Fluid" (*Journal of the American Medical Association*, vol. 293, no. 7, February 16, 2005), Paige B. Larrabee et al. note they found that gene expression patterns correlated with the gender of the fetus, gestational age, and disease status. The researchers assert that this technology may assist to advance human developmental research and in identifying new biomarkers for prenatal assessment.

FIGURE 6.7

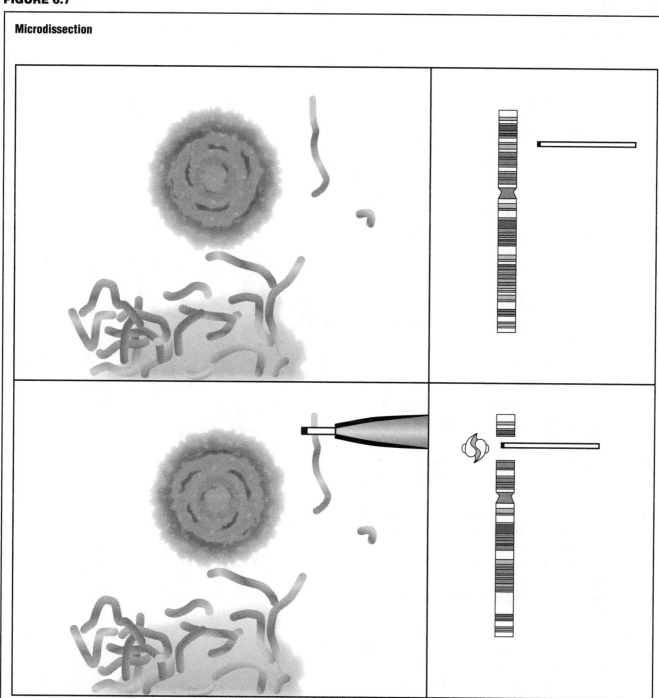

Microdissection

SOURCE: "How Does Microdissection Work?" in *Chromosome Microdissection Fact Sheet*, U.S. Department of Health and Human Services, National Institutes of Health, National Human Genome Research Institute, 2009, http://www.genome.gov/10000204 (accessed September 29, 2010)

Ana Bustamante-Aragones et al. report in "Early Noninvasive Prenatal Detection of a Fetal CRB1 Mutation Causing Leber Congenital Amaurosis" (*Molecular Vision*, vol. 14, August 4, 2008) a new technique, called denaturing High Performance Liquid Chromatography (dHPLC), that can detect a fetal mutation in the Crumbs homologue 1 (CRB1) gene from a sample of blood taken from an expectant mother. The CRB1 gene is associated with inherited vision disorders including retinitis pigmentosa and Leber congenital amaurosis. Besides

enabling diagnosis from a noninvasive test, dHPLC is a rapid, low-cost test.

In "Noninvasive Diagnosis of Fetal Aneuploidy by Shotgun Sequencing DNA from Maternal Blood" (*Proceedings of the National Academy of Sciences*, vol. 105, no. 42, October 21, 2008), H. Christina Fan et al. note the development of a genetic test for Down syndrome that relies on a blood sample taken from an expectant mother. The noninvasive test, which sequences the fetal DNA

FIGURE 6.8

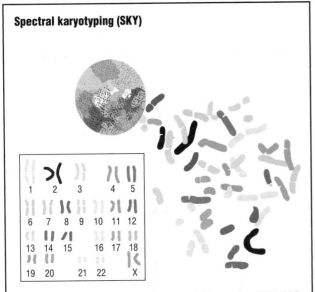

Spectral karyotyping (SKY)

SOURCE: "Spectral Karyotyping," in *Spectral Karyotyping (SKY)*, U.S. Department of Health and Human Services, National Institutes of Health, National Human Genome Research Institute, 2010, http://www.genome.gov/10000208 (accessed September 29, 2010)

floating in the mother's blood, may one day eliminate the need for more invasive prenatal diagnostics such as amniocentesis and CVS.

Christopher Cunniff and Louanne Hudgins observe in "Prenatal Genetic Screening and Diagnosis for Pediatricians" (*Current Opinion in Pediatrics*, vol. 22, no. 6, December 2010) that in recent years there have been rapid advances in prenatal diagnostics. In 2010 carrier screening for selected genetic disorders was offered routinely to pregnant women or those considering having children. Early screening (during the first trimester) for fetal chromosome abnormalities has improved the detection of these disorders. Cunniff and Hudgins note that professional societies, such as the American College of Obstetricians and Gynecologists and the American College of Medical Genetics, recommend that "invasive prenatal diagnostic testing be made available to all pregnant women, regardless of age or prenatal screening results."

GENETIC DIAGNOSIS IN CHILDREN AND ADULTS

Genetic testing can also be performed postnatally (after birth) to determine which children and adults are at increased risk of developing specific diseases. By January 2011, scientists could perform predictive genetic testing to identify which individuals were at risk for cystic fibrosis, Tay-Sachs disease, Huntington's disease, amyotrophic lateral sclerosis (a degenerative neurological condition commonly known as Lou Gehrig's disease), and several types of cancers, including some cases of breast, colon, and ovarian cancer.

FIGURE 6.9

Cutting DNA with restrictive enzymes

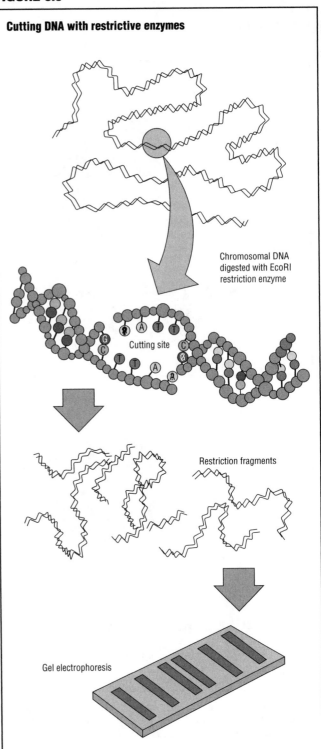

SOURCE: "Cutting DNA with Restrictive Enzymes," in *Human Genome Project Information Basic Genomics Image Gallery*, U.S. Department of Energy Office of Science, Office of Biological and Environmental Research, Human Genome Project, 1996, http://genomics.energy.gov/gallery/basic_genomics/detail.np/detail-06.html (accessed September 29, 2010)

About 1,900 genetic tests were available in 2009. (See Figure 6.4.) In 2010 that number continued to grow, but most of the tests were only used for research purposes because public health professionals do not consider it

FIGURE 6.10

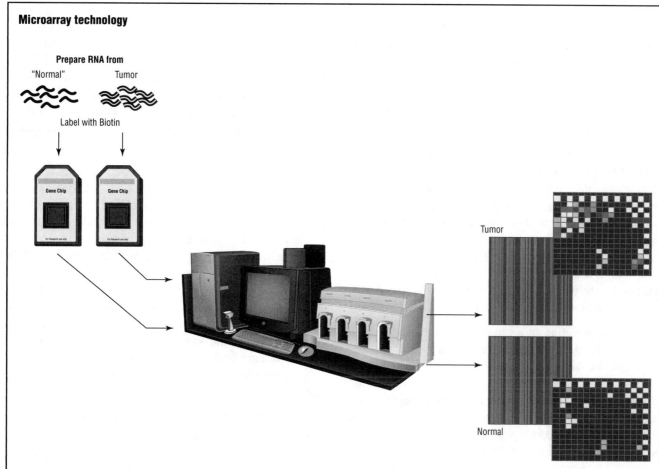

Microarray technology

Prepare RNA from

"Normal"   Tumor

Label with Biotin

Gene Chip

Gene Chip

Tumor

Normal

SOURCE: "Microarray Technology," in *Talking Glossary of Genetic Terms*, U.S. Department of Health and Human Services, National Institutes of Health, National Human Genome Research Institute, Division of Intramural Research, undated, http://www.genome.gov/Glossary/index.cfm?id=125 (accessed September 29, 2010)

practical to screen for conditions that are extremely rare, have only minor health consequences, or have no effective treatment. The most frequently performed genetic tests were those considered most useful in terms of their potential to screen populations for diseases that occur relatively frequently, have serious medical consequences (including death) if untreated, and have effective treatment.

A positive test result (the presence of a mutation—a defective or altered gene) from predictive genetic testing does not guarantee that the individual will develop the disease; it simply identifies the individual as genetically susceptible and at an increased risk for developing the disease. For example, a woman who tests positive for the BRCA1 gene has about an 80% chance of developing breast cancer before the age of 60 years. It is also important to note that, like other types of diagnostic medical testing, genetic tests are not 100% predictive—the results rely on the quality of laboratory procedures and the accuracy of interpretations. Furthermore, because tests vary in their sensitivity and specificity, there is always the possibility of false-positive and false-negative test results.

Researchers hope that positive test results will encourage people who are at higher than average risk of developing a disease to be especially vigilant about disease prevention and screening for early detection, when many diseases are most successfully treated. There is an expectation that genetic information will increasingly be used in routine population screening to determine individual susceptibility to common disorders such as heart disease, diabetes, and cancer. This type of screening will identify groups at risk so that primary prevention efforts such as diet and exercise or secondary prevention efforts such as early detection can be initiated.

Diagnostic Genetic Testing

Most genetic testing is performed on people who are asymptomatic. The objective of screening is to determine if people are carriers of a genetic disease or to identify their susceptibility or risk of developing a specific disease or disorder. There is, however, some testing performed on people with symptoms of a disease to clarify or establish the diagnosis and calculate the risk of developing the disease for other family members. This type of testing is known as diagnostic genetic testing or symptomatic genetic testing. It may also assist in directing treatment for symptomatic patients in whom a mutation in a single gene (or in a gene pair) accounts for a disorder. Cystic fibrosis and myotonic dystrophy are examples of disorders that may be confirmed or ruled out by diagnostic genetic testing and other methods (such as the sweat test for cystic fibrosis or a neurological evaluation for myotonic dystrophy).

Madhuri R. Hegde et al. describe in "Microarray-Based Mutation Detection in the Dystrophin Gene" (*Human Mutation*, vol. 29, no. 9, September 2008) a diagnostic genetic test that detects mutations in the dystrophin gene that cause muscular dystrophy. Using a microarray-based approach, the EmArray Dystrophin test detects 99% of mutations in the dystrophin gene, including deletions, duplications, and point mutations, and may be used to confirm the diagnosis of muscular dystrophy. It can also be used for prenatal and carrier testing.

One consideration in diagnostic genetic testing is the appropriate frequency of testing in view of rapidly expanding genetic knowledge and identification of genes that are linked to disease. Physicians frequently see symptomatic patients for whom there is neither a definitive diagnosis nor a genetic test. The as-yet-unanswered question is: Should such people be recalled for genetic testing each time a new test becomes available? Even though clinics and physicians who perform genetic testing counsel patients to maintain regular contact so they may learn about the availability of new tests, there is no uniform guideline or recommendation about the frequency of testing.

Genetic Tests Help Patients Choose Optimal Treatment

In 2005 the U.S. Food and Drug Administration (FDA) approved marketing of a genetic test that will help physicians make personalized drug treatment decisions for some patients. The Invader UGT1A1 Molecular Assay detects variations in a gene that affect how certain drugs are broken down and cleared by the body. Using this information, physicians can determine the optimal drug dosage for each patient and minimize the harsh and potentially life-threatening side effects of drug treatment. This test joins a growing list of genetic tests (called pharmacogenetic or pharmacogenomic tests) that are used to customize treatment decisions. Examples of these tests are the Roche AmpliChip, which is used to personalize the dosage of antidepressants, antipsychotics, beta-blockers, and some chemotherapy drugs, and TRUGENE HIV-1 Genotyping Kit, which is used to detect variations in the genome of the human immunodeficiency virus (HIV) that make the virus resistant to some antiretroviral drugs.

Anil Potti et al. report in "A Genomic Strategy to Refine Prognosis in Early-Stage Non-small-Cell Lung Cancer" (*New England Journal of Medicine*, vol. 355, no. 6, August 10, 2006) on a genetic test that predicts with up to 90% accuracy which early-stage lung cancer patients are likely to experience recurrence and therefore would benefit from chemotherapy (drug treatment for cancer). Currently, patients diagnosed with early-stage non-small-cell lung cancer undergo surgery followed by observation, without chemotherapy. About one-third of such patients experience a relapse. Potti et al. describe the test as "a fingerprint unique to the individual patient [that] predicts survival changes."

A test that helps determine which patients are more sensitive to warfarin, a commonly used anticoagulant that prevents dangerous blood clots, received FDA approval in 2007. The Nanosphere Verigene Warfarin Metabolism Nucleic Acid Test detects variants of two genes: CYP2C9, which helps the body metabolize warfarin, and VKORC1, which helps regulate the ability of the drug to prevent blood clots. People who have polymorphisms of these genes are generally more sensitive to warfarin than those without the variant and may require lower doses of the drug to achieve the same therapeutic benefit. Knowing which patients will metabolize the drug differently can help prevent the risk of a dose that will produce bleeding instead of the desired anticoagulation.

In 2007 the FDA approved Mammaprint, a genetic test that aims to predict whether a breast cancer patient will suffer a relapse. The test can help distinguish patients who would benefit from chemotherapy from those for whom surgery to remove the tumor is sufficient treatment. Two other tests intended to help guide cancer treatment choices were approved in 2008. The SPOT-Light HER2 CISH measures the number of copies of the HER2 gene in breast tumor tissue to identify patients who may benefit from treatment with the anticancer drug Herceptin, which targets HER2 protein production. The TOP2A FISH pharmDx determines whether patients have normal or variants of the TOP2A gene. It helps assess the risk of recurrence and long-term survival for patients with relatively high-risk breast cancer, such as premenopausal women or those with lymph node involvement.

The FDA notes in "Table of Pharmacogenomic Biomarkers in Drug Labels" (http://www.fda.gov/Drugs/ScienceResearch/ResearchAreas/Pharmacogenetics/ucm083378.htm) that by January 2011 it had identified nearly 80 pharmacogenomic biomarkers. These tests may be performed to identify patients who will respond to a particular drug and those who will not, to prevent drug toxicity, and to guide drug dosing to optimize efficacy and safety. For example, people with the CYP2D6 variant are called "ultra-rapid metabolizers of codeine" and may experience overdose symptoms from the normally prescribed dosages of this pain medication. According to the National Institute of General Medical Sciences, in "Personalized Medicines Fact Sheet" (September 2010, http://www.nigms.nih.gov/

Initiatives/PGRN/Background/FactSheet.htm), in 2010 the FDA modified the labels of more than 12 pharmaceutical drugs to include pharmacogenetic information about the drugs, which alerts physicians to the genetic basis of the response to the drugs. Examples of these include Purinethol, which may be prescribed for inflammatory bowel disease and childhood leukemia; Coumadin, which is used to prevent blood clots, heart attacks, and strokes; and Camptosar, a drug for colorectal cancer.

NEW GENETIC TEST PROMISES EARLY DETECTION OF COLON CANCER. In "New DNA Tests Aimed at Reducing Colon Cancer" (*New York Times*, October 28, 2010), Nicholas Wade reports that during the October 2010 meeting of the American Association for Cancer Research in Philadelphia, Pennsylvania, researchers reported on the development of two new genetic diagnostic tests for colon cancer. One test detects the presence of four gene variants that are linked to colon cancer in stool, and the other is a blood test that detects the presence of a single gene, Septin 9, which is not one of the four genes detected by the stool sample test. As of October 2010, both tests were in clinical trials, which will determine their detection rates in terms of sensitivity and specificity. The results of these trials are eagerly awaited because if proven effective, these genetic diagnostic tests could conceivably replace colonoscopy (visual examination of the colon that requires sedation and insertion of a long narrow tube into the rectum) as a screening test and may detect colon cancer earlier, when it is more amenable to treatment.

Population Screening

Population screening for heritable diseases is one potentially lifesaving application of molecular genetics technology. Prenatal screening has demonstrated benefits and gained widespread use; however, genetic screening has not yet become part of routine medical practice for adults. Geneticists have identified at least eight genes that might be candidates for use as population screening tests in adults in the United States. The genes are:

- HFE, for hereditary hemochromatosis (a disorder in which the body absorbs too much iron from food; rather than the excess iron being excreted, it is stored throughout the body, and the iron deposits damage the pancreas, liver, skin, and other tissues)

- Apolipoprotein E-4, which is linked to Alzheimer's disease

- CYP2D6, which is linked to ankylosing spondylitis (arthritis of the spine)

- BRCA1 and BRCA2, the genes for hereditary breast and ovarian cancer

- APC and MUTYH, genes that cause familial adenomatous polyposis, which is associated with precancerous growths in the colon

- Factor V Leiden, a genetic mutation in the factor V gene, causes the most common hereditary blood clotting disorder in the United States

As of January 2011, screening for variants in these genes had not entered into routine medical practice because there was considerable controversy about the predictive value of testing for these genes and how to monitor and care for people who test positive for them. In "Being More Realistic about the Public Health Impact of Genomic Medicine" (*PLoS Medicine*, vol. 7, no. 10, October 2010), Wayne D. Hall, Rebecca Mathews, and Katherine I. Morley observe that there are not yet genetic tests suitable for population-wide screening to predict genetic risk. The researchers assert that before predictive genomic tests are applied to population screening, they must be shown to:

- Predict disease risk more accurately than phenotypic information

- Identify diseases or conditions for which there are cost-effective interventions

- Be more cost effective than present-day population interventions

- Provide information that motivates desired risk-reducing behaviors or lifestyle changes

Hall, Mathews, and Morley also state that public health professionals are concerned about the potential misuse of genetic risk information by industries that wish to promote potentially harmful products such as tobacco, alcohol, or high-calorie foods with little nutritional value. They cite as an example the alcohol industry's contention that because alcohol-related problems only occur in a small minority of genetically susceptible drinkers, alcohol problems are better addressed by identifying problem drinkers rather than by adopting effective strategies for reducing population-wide alcohol consumption.

Hall, Mathews, and Morley conclude "that genetic screening of whole populations is unlikely to transform preventive health in the ways predicted 10 years ago. The integration of individual genomic risk prediction into public health disease prevention strategies will require good evidence that this approach improves on the cost-effectiveness of existing population level interventions. The utility and cost-effectiveness of predictive genomics, like any other new health technology, should be evaluated disease-by-disease and population-by-population."

Genetic Susceptibility

Susceptibility testing, also known as predictive testing, determines the likelihood that a healthy person with a family history of a disorder will develop the disease. Testing positive for a specific genetic mutation indicates an increased susceptibility to the disorder but does not establish a diagnosis. For example, a woman may choose to undergo testing to find out whether she has genetic mutations that would indicate a likelihood of developing hereditary cancer of the breast or ovary. If she tests positive for a genetic mutation, she may then decide to undergo some form of preventive treatment. Preventive measures may include increased surveillance, such as more frequent mammography; chemoprevention (prescription drug therapy that is intended to reduce risk); or surgical prophylaxis, such as mastectomy and/or oophorectomy (surgical removal of the breasts and ovaries, respectively).

Testing Children for Adult-Onset Disorders

In 2000 the American Academy of Pediatrics Committee on Genetics recommended genetic testing for people under the age of 18 years only when testing would offer immediate medical benefits or when there is a benefit to another family member and there is no anticipated harm to the person being tested. The committee considered genetic counseling before and after testing as essential components of the process.

That same year the American Academy of Pediatrics Committee on Bioethics and Newborn Screening Task Force recommended the inclusion of tests in the newborn-screening battery based on scientific evidence. The committee advocated informed consent for newborn screening. (As of January 2011, most states do not require informed consent.) However, it did not endorse carrier screening in people under 18 years of age, except in the case of a pregnant teenager. It also recommended against predictive testing for adult-onset disorders in people under the age of 18.

The American College of Medical Genetics, the American Society of Human Genetics (ASHG), and the World Health Organization have also weighed in on genetic testing of asymptomatic children, asserting that decisions about testing should emphasize the child's well-being. One issue involves the value of testing asymptomatic children for genetic mutations that are associated with adult-onset conditions such as Huntington's disease. Because no treatment can begin until the onset of the disease, and at present there is no treatment to alter the course of the disease, it may be ill advised to test for it. Another concern is testing for carrier status of autosomal recessive or X-linked conditions such as cystic fibrosis or Duchenne muscular dystrophy. Experts caution that children might confuse carrier status with actually having the condition, which in turn might provoke needless anxiety.

There are, however, circumstances in which genetic testing of children may be appropriate and useful. Examples are children with symptoms of suspected hereditary disorders or those at risk for cancers in which inheritance plays a primary role.

ETHICAL CONSIDERATIONS—CHOICES AND CHALLENGES

Since the 1980s rapid advances in genetic research have challenged scientists, health care professionals, ethicists, government regulators, legislators, and consumers to stay abreast of new developments. Understanding the scientific advances and their implications is critical for everyone involved in making informed decisions about the ways in which genetic research and information will affect the lives of current and future generations. U.S. citizens, scientists, ethicists, legislators, and regulators share this responsibility. According to Human Genome Project Information, in "Ethical, Legal, and Social Issues" (September 16, 2008, http://www.ornl.gov/sci/techresources/Human_Genome/elsi/elsi.shtml), the pivotal importance of these societal decisions was underscored by the allocation of 3% to 5% of the Human Genome Project budget for the study of ethical, legal, and social issues related to genetic research. As of January 2011, consideration of these issues had not produced simple or universally applicable answers to the many questions posed by the increasing availability of genetic information. Ongoing public discussion and debate is intended to inform, educate, and help people in every walk of life make personal decisions about their health and participate in decisions that concern others.

As researchers learn more about the genes that are responsible for a variety of illnesses, they can design more tests with increased accuracy and reliability to predict whether an individual is at risk of developing specific diseases. The ethical issues involved in genetic testing have turned out to be far more complicated than originally anticipated. Initially, physicians and researchers believed that a test to determine in advance who would develop or escape a disease would be welcomed by at-risk families, who would be able to plan more realistically about having children and going about their daily lives. Nevertheless, many people with family histories of a genetic disease have decided that not knowing is better than anticipating a grim future and an agonizing, slow death. They prefer to live with the hope that they will not develop the disease rather than having the certain knowledge that they will.

The discovery of genetic links and the development of tests to predict the likelihood or certainty of developing a disease raise ethical questions for people who carry a defective gene. Should women who are carriers of Huntington's disease or cystic fibrosis have children? Should a fetus with a defective gene be carried to term or aborted?

Historically, there have also been concerns about privacy and the confidentiality of medical records and the results of genetic testing and possible stigmatization. Some people are reluctant to be tested because they fear they may lose their health, life, and disability insurances, or even their job if they are found to be at risk for a disease. Genetic tests are sometimes costly, and some insurers agree to reimburse for testing only if they are informed of the results. The insurance companies feel they cannot risk selling policies to people they know will become disabled or die prematurely.

The fear of discrimination by insurance companies or employers who learn the results of genetic testing was often justified. An insurance carrier could charge a healthy person a higher rate or disqualify an individual based on test results, and an employer could choose not to hire or to deny an affected individual a promotion. Until 2008, such fears forced people to choose between having a genetic test that could provide lifesaving information and avoiding a test to retain health insurance coverage or save a job.

Legislation Bans Genetic Discrimination

Until May 2008, when the Genetic Information Non-discrimination Act (GINA) became law, Americans' fears of recrimination or undesired consequences of genetic testing were justified. Dissemination of genetic information was protected by an uneven array of state and federal regulations. Heralded as the first major civil rights bill of the new century, GINA prevents health insurers from denying coverage, adjusting premiums on the basis of genetic test results, or requesting that an individual undergo genetic testing. It prohibits employers from using genetic information to make hiring, firing, or promotion decisions. The law also sharply restricts an employer's right to request, require, or purchase workers' genetic information.

In June 2010 the Genetics and Public Policy Center of the Phoebe R. Berman Bioethics Institute at Johns Hopkins University, the National Coalition for Health Professional Education in Genetics, and the Genetic Alliance launched the interactive website "GINA & You" (http://www.GINA Help.org) to help consumers, clinicians, and educators understand the law that protects against the misuse of genetic information in health insurance and employment.

Unanticipated Information and Results of Genetic Testing

Sometimes genetic testing yields unanticipated information, such as paternity, or other unexpected results, such as the presence of a disorder that was not sought directly. Many health professionals consider disclosure of such information, particularly when it does not influence health or medical treatment decision making, as counter-productive and even potentially harmful.

The ASHG explains in "Mission & Vision" (2011, http://www.ashg.org/pages/about_mission.shtml) that it is the primary professional organization for human geneticists

in the United States. Its 8,000 members include researchers, academicians, clinicians, laboratory practice professionals, genetic counselors, nurses, and others involved in or with a special interest in human genetics. The ASHG recommends that family members not be informed of misattributed paternity unless the test requested was determination of paternity. The ASHG encourages all health professionals to educate, counsel, and obtain informed consents that include cautions regarding unexpected findings before performing genetic testing.

Advertising Genetic Testing Directly to Consumers

Pharmaceutical drugs have been advertised directly to consumers for more than two decades, but direct-to-consumer (DTC) advertising of genetic tests is a relatively recent phenomenon. The DTC genetic tests are marketed via print and electronic advertising and sold online. According to the National Library of Medicine's Genetics Home Reference, in "What Is Direct-to-Consumer Genetic Testing?" (January 17, 2011, http://ghr.nlm.nih.gov/handbook/testing/directtoconsumer), some home genetic test companies obtain a DNA sample by having their clients swab the inside of their cheek, whereas others require clients to visit a laboratory to have a blood sample drawn. Test results are generally available by telephone or online, and in some instances a genetic counselor or other health educator is available to discuss the test results.

Supporters of home genetic tests assert that the tests enhance public awareness of genetic diseases, enable consumers to assume an active role in managing their health, and offer a convenient way to explore genealogy and family history. In contrast, many health professionals and government agencies are concerned about how these tests are marketed and used.

The U.S. Government Accountability Office (GAO) explains in "Direct-to-Consumer Genetic Tests: Misleading Test Results Are Further Complicated by Deceptive Marketing and Other Questionable Practices" (July 22, 2010, http://www.gao.gov/new.items/d10847t.pdf) that in 2006 it looked at the marketing, advertising, and efficacy of genetic tests by purchasing 10 tests from four companies. The tests ranged from $299 to $999. The GAO chose five subjects and sent two DNA samples from each subject to each company: one sample included factual information about the subject and one included incorrect information, such as the wrong age, race, or ethnicity. The GAO compared the subjects' results and sought health advice by making undercover calls to the companies. Genetics experts were asked to determine whether the tests provided medically useful information.

According to the GAO:

- The subjects' test results were "misleading and of little or no practical use." Table 6.2 shows how one

TABLE 6.2

Contradictory risk predictions for prostate cancer and hypertension

Gender	Age	Condition	Company 1	Company 2	Company 3	Company 4
Male	48	Prostate cancer	Average	Average	Below average	Above average
		Hypertension	Average	Below average	Above average	Not tested

SOURCE: "Contradictory Risk Predictions for Prostate Cancer and Hypertension," in *Direct-To-Consumer Genetic Tests: Misleading Test Results Are Further Complicated by Deceptive Marketing and Other Questionable Practices*, U.S. General Accountability Office, July 22, 2010, http://www.gao.gov/highlights/d10847thigh.pdf (September 30, 2010)

male subject received widely varied and inconsistent predictions about his risk of developing prostate cancer and hypertension.

- The disease predictions that were made conflicted with the subjects' actual medical conditions.

- Three of the companies failed to provide the expert follow-up they had promised.

To evaluate advertising practices, the GAO contacted 15 companies, including the four that were tested, to inquire about supplement sales, test reliability, and privacy policies. The GAO notes 10 glaring instances of deceptive marketing, including:

- Four companies claimed that a consumer's DNA could be used to create a personalized supplement to cure diseases.

- Two companies asserted that their supplements could "repair damaged DNA" or "cure disease," even though there is no scientific evidence for such claims.

- Two companies asserted that their DNA analyses could predict the sports in which children would excel, even though there is no scientific basis for such claims.

- One company advised a subject that her test result meant she was "in the high risk of pretty much getting" breast cancer, a statement that alarmed health professionals because it suggests the test is diagnostic.

In "Risks of Presymptomatic Direct-to-Consumer Genetic Testing" (*New England Journal of Medicine*, vol. 363, no. 12, September 16, 2010), Justin P. Annes, Monica A. Giovanni, and Michael F. Murray point out that the risks and potential harms of home genetic testing include inaccurate results, population screening without consensus about how results should be interpreted, inappropriate follow-up, and the loss of consumer protections when the companies obtain test results outside of the traditional health care delivery system, such as physician-patient confidentiality. The researchers assert that these very real risks, some of which were documented in the GAO study, must be weighed against the potential benefits of genetic research and testing. Annes, Giovanni, and Murray exhort the government to assume an active role in "both

regulation and the funding of collaborative research that enables the optimal use of genetic information."

The FDA Moves to Regulate Genetic Testing

Table 6.3 displays the federal oversight and regulation of genetic tests in 2010. The chart was developed by the Johns Hopkins University's Genetics and Public Policy Center with support from the Pew Charitable Trusts. In *Guidance for Industry and FDA Staff: Pharmacogenetic Tests and Genetic Tests for Heritable Markers* (June 19, 2007, http://www.fda.gov/downloads/MedicalDevices/DeviceRegulationandGuidance/GuidanceDocuments/ucm071075.pdf), the FDA provides detailed and specific recommendations about many different aspects of genetic testing and how clinical evaluation studies should be conducted to determine efficacy and labeling. However, because these recommendations are nonbinding, the industry is not required to adhere to them and there is no way to determine the extent to which they have been adopted by manufacturers of genetic tests.

In view of the explosive growth in the number of DTC genetic tests and the frequency of questionable claims that companies make about them, such as determining cancer risk or response to a particular drug, Jeffrey Shuren (July 22, 2010, http://www.fda.gov/NewsEvents/Testimony/ucm219925.htm), the director of the FDA Center for Devices and Radiological Health, explained in his statement to the U.S. House of Representatives' Subcommittee on Oversight and Investigations that in 2010 the FDA had notified 20 diagnostic test manufacturing firms that offered DTC genetic tests that their tests should meet the "statutory definition of a medical device" on the basis of their claims about the test results and thus may require FDA approval.

This FDA action was prompted by the announcement by Pathway Genomics and Walgreens of their intent to market a testing kit to analyze genetic risk and provide information on more than 70 diseases or traits through the largest U.S. drugstore chain. According to Shuren, the FDA contacted two companies and indicated that clearance or approval by the agency was necessary for them to market their product. This notification caused Walgreens to postpone plans to sell the test in its more than 6,000 stores.

TABLE 6.3

Oversight and regulation of genetic tests, 2010

The federal goverment regulates a wide variety of medical products and services. Currently, the federal government has only limited oversight of genetic tests and the laboratories that perform them. Whether and how much the government regulates a genetic test depends on what methods and materials the laboratory uses to perform the test.

How genetic tests are regulated

The different scenarios for genetic testing are analogous to different ways to bake a cake.

Scenario A: Laboratory-developed

You make your own cocoa powder from cocoa beans, buy eggs, sugar, flour and butter from the local grocer, and put all the ingredients together using your grandma's recipe for chocolate cake. You bake the cake and sell it at the fair.

With genetic tests, companies can also make a test using available components "from scratch."

Who regulates it?

The Centers for Medicare and Medicaid Services (CMS) regulates genetic testing laboratories under the Clinical Laboratories Improvement Amendments (CLIA). Laboratories performing tests that provide information to guide diagnosis, treatment, prevention, or health assessment must be certified under CLIA and follow basic standards to ensure laboratory quality.

The Food and Drug Administration [FDA] regulates some components of home brew tests.

What is regulated?

Under CLIA, a laboratory must be certified. Certification requires demonstrating that the laboratory is clean, that the laboratory equipment is maintained, and that the test methods and procedures are appropriately organized and documented. A laboratory inspector comes once every two years to make sure everything is in order.

FDA regulates some "off the shelf" ingredients (analagous to the eggs, sugar, flour and butter for the cake) as"general purpose reagents," and requires that they be appropriately labeled and manufactured.

What is not regulated?

Ingredients made by the laboratory from scratch, analogous to the cocoa powder made from the cocoa beans, are not regulated by FDA at all.

For most genetic tests, CLIA does not impose specific requirements relating to proficiency testing (to assess whether the laboratory can get the "right answer" reliably), personnel, or quality control procedures.

Neither CLIA nor FDA assesses whether the test being performed detects a genetic alternation that relates to a disease or condition.

Scenario B: Laboratory-developed with analyte specific reagents

You don't want to make everything from scratch. In particular, you want to buy the "key ingredient" (the chocolate) from a gourmet shop.

Similarly, the "key ingredients" of a genetic test can be purchased from a manufacturer. These are called analyte specific reagents. They can be combined with the "general purpose reagents" (analogous to the eggs, sugar, flour and butter for the cake) to make a genetic test.

Who regulates it?

Regulation under CLIA for these tests is the same as in Scenario A.

In addition to regulating the general purpose reagents as in Scenario A, FDA also regulates analyte specific reagents.

What is regulated?

CLIA regulates the laboratory, as in Scenario A. FDA regulates both the general purpose reagents and the analyte specific reagents. ASRs must be labeled and manufactured appropriately and can be sold only to laboratories certified to perform particular types of tests.

What is not regulated?

For most genetic tests, CLIA does not impose specific requirements relating to proficiency testing (to assess whether the laboratory can get the "right answer" reliably), personnel, or quality control procedures.

Neither CLIA nor FDA assesses whether the test being performed detects a genetic alteration that relates to a disease or condition.

Scenario C: Test kits

You buy a cake mix at the store and "Just add water."

A small number of genetic tests are sold as complete "kits," like cake mixes. The kits come with ingredients pre-assembled, a label indicating what the test is for, and instructions for how to use the kit to perform the test. A laboratory can purchase the kit and use it to perform a genetic test.

Who regulates it?

Laboratories using test kits are regulated by CLIA as in Scenario A and B.

Test kits are regulated by FDA.

What is regulated?

CLIA regulates laboratories using the kit as in Scenario A.

Test kits must be reviewed by FDA before they can be sold. Test kit manufacturers must submit information to FDA showing that the test can be performed accurately and reliably using the instructions provided ("analytic validity") and that the kit tests for something relevant to a patient's health ("clinical validity"). FDA also regulates what claims the manufacturer can make about the kit. Unless FDA is satisfied that the manufacturer has provided adequate evidence that the kit is safe and effective, the kit cannot be sold.

Federal oversight

Federal agencies regulate some aspects of genetic testing. These agencies include the Food and Drug Administration (FDA), the Centers for Medicare and Medicaid Services (CMS), and the Centers for Disease Control and Prevention (CDC).

Genetic tests	Food and Drug Administration (FDA)	Centers for Medicare and Medicaid Services (CMS)	Centers for Disease Control and Prevention (CDC)
Laboratory developed genetic test	Regulates components called "general purpose reagents."	Oversees lab quality. No genetic testing "specialty area," meaning no specific, tailored requirements for genetic testing laboratories.	Advises CMS on CLIA issues.
Laboratory developed genetic test (with analyte specific reagent)	Regulates components called "analyte specific reagents" (ASRs).	Oversees lab quality. No genetic testing specialty area.	Advises CMS on CLIA issues.
Genetic test kit	Premarket review of safety and effectiveness of test kit.	Oversees lab quality. No genetic testing specialty area.	Advises CMS on CLIA issues.

SOURCE: Christina Ullman, "Genetic Tests," in *Genes & Society Graphics*, Genetics & Public Policy Center at Johns Hopkins University, 2008, http://www.dnapolicy.org/resources/Graphics_Genetic_Tests.pdf (accessed November 1, 2008). Courtesy of the Genetics & Public Policy Center with support from The Pew Charitable Trusts.

Shuren also noted that even though the FDA has the authority to regulate genetic tests because they are considered medical devices, in the past the agency had not exercised enforcement because there were not many tests offered and testing was marketed for the purposes of genealogy research as opposed to assessing disease risk. Shuren stated that "none of the genetic tests now offered directly to consumers has undergone premarket review by

FDA to ensure that the test results being provided to patients are accurate, reliable, and clinically meaningful."

Health professionals and policy makers continue to urge both the FDA and the Centers for Medicare and Medicaid Services, which regulates clinical labs, to strengthen the regulations on diagnostic genetic tests and establish ways to enforce adherence to these regulations. Stuart Hogarth considers in "Myths, Misconceptions, and Myopia: Searching for Clarity in the Debate about the Regulation of Consumer Genetics" (*Public Health Genomics*, vol. 13, no. 5, June 2010) whether genetic testing is in need of greater scrutiny and oversight. Hogarth identifies several myths or misconceptions that cloud the debate, specifically:

- Regulating genetic testing is equivalent to banning it—Hogarth explains that regulation entails gathering information, setting standards, and enforcing them and observes that many industries, such as air travel and restaurants, operate with regulations.

- Requiring a medical consultation to obtain genetic testing limits consumers' access to their genetic information—Hogarth observes that the testing companies already control access to the data by establishing standards on which data are reported, how the data are reported, and to whom they are reported. He asserts that health professionals or government agencies might be better custodians of these data than private companies.

- Banning DTC genetic testing is paternalistic and prevents self-care—Hogarth suggests that genetic tests be regulated as drugs are, with some available only by prescription and others sold over the counter.

- Online sales cannot be regulated—Hogarth concedes that Internet sales have created a market that probably cannot be universally or perfectly regulated. However, he avers that regulators can and do exert an effect on markets.

- Genetic tests should not be treated or regulated differently from other DTC tests. Hogarth contends that genetic tests are different because new research discoveries such as biomarkers can be rapidly commercialized and immediately sold on the Internet, with no period of review and adoption by the medical profession. Hogarth indicates that this is "an exceptional situation and it may require an exceptional policy response."

- It is too soon to act—Hogarth recounts that in 1997, "the U.S. Task Force on Genetic Testing warned that our capacity to respond to events would one day be outstripped by the rapid pace of commercialisation of new tests." He believes this point was reached in 2010.

- Consumer genetics has arrived and is the new reality—Hogarth observes that recent changes at some personal genomics companies and the demise of several other companies are evidence that the

commercial viability of consumer genetics has not yet been demonstrated. He believes there is still time to make intelligent, reasoned choices about how new genetic technologies should be deployed.

Personal Choices and Psychological Consequences of Genetic Testing

The results of genetic tests may be used to make decisions such as whether to have children or to end a pregnancy. Results that predict the likelihood that an individual will develop a disease may affect decisions about education, marriage, family, or career choices. The decision to undergo genetic testing and the results of the tests affect not only the individual tested but also his or her family members. For example, when an unaffected patient requests a genetic susceptibility test, another test of an affected relative may be required to accurately calculate probability. Family members may vary in their willingness to share genetic information and their desire to know about genetic risks.

There are psychological consequences of genetic testing and coming to terms with the results. People may be relieved or distressed when they learn the results of a genetic test. The results can change the way they feel about themselves and can influence their relationships with relatives. For example, family members who discover they are carriers for cystic fibrosis may feel isolated or estranged from siblings who have opted not to be tested. By contrast, those who find out they do not have a genetic mutation for Huntington's disease may feel guilty because they have been spared and other family members have not, or they may worry about assuming responsibility for family members who develop the disease.

The complexity of genetic testing and the uncertainty of many results pose an additional psychological challenge. The results of predictive genetic tests are often expressed in probabilities rather than in certainties, and, even for people with a high probability of developing a disease, there are often conflicting opinions about the most appropriate course of action. For example, a woman who tests positive for BRCA1 or BRCA2 has an increased lifetime risk of developing breast or ovarian cancer or both, but it does not mean that her risk of developing either or both is 100%. Depending on her personal circumstances and the medical advice she receives, she may opt to intensify screening to detect disease; use prescription medication that is intended to reduce her risk, such as Soltamox; or undergo a preventive surgical procedure, such as a mastectomy or oophorectomy.

In "Women's Constructions of the 'Right Time' to Consider Decisions about Risk-Reducing Mastectomy and Risk-Reducing Oophorectomy" (*BMC Women's Health*, vol. 10, no. 24, August 5, 2010), A. Fuchsia Howard et al. describe women's perspectives about actions they might take after learning they carry the BRCA1 or BRCA2 mutation. The researchers conducted in-depth interviews with 22 BRCA1- or BRCA2-positive women to learn about what they felt was the "right time" to consider risk-reducing surgical procedures.

Howard et al. find that women consider how the decisions fit into their life, especially in terms of changing their fertility or influencing prospective relationships or an existing marriage. The women were able to make decisions when they had sufficient time to think about them and were emotionally prepared to deal with the decisions and their consequences. Some women felt there were better options available, such as enhanced surveillance, and others felt that medical science was likely to produce less drastic alternatives in the foreseeable future. Howard et al. opine that these findings, of multiple concerns about risk-reducing procedures, may help explain the prolonged time between genetic testing and risk-reducing surgeries that have been observed in other studies.

Even a negative test result can be stressful, creating nearly as many questions as it does answers about disease risk. A negative test result for BRCA1 or BRCA2 in a woman with an affected family member (one who has the genetic mutation) means that even though she does not have the genetic mutation, her risk is the same as that of the general population. There are multiple genes as well as other factors associated with the risk of developing breast cancer that are not identified through genetic testing, such as the age at which a woman has her first child. In addition, most breast cancer is not believed to be hereditary, and most women diagnosed with breast cancer do not test positive for the BRCA1 or BRCA2 genetic mutation.

GENETIC COUNSELING

Generally, patients receive explicit counseling before undergoing genetic testing to ensure that they are able to make informed decisions about choosing to have the tests and the consequences of testing. Most genetic counseling is informative and nondirective—it is intended to offer enough information to allow families or individuals to determine the best courses of action for themselves but avoids making testing recommendations.

Patients undergoing tests to improve their care and treatment have different pretest counseling needs from those choosing susceptibility or predictive testing. In such instances genetic counselors do provide testing recommendations, particularly when a test offers an opportunity to prevent disease. As tests for genetic risk factors increasingly become routine in clinical medical practice, they are likely to be offered without formal pretest counseling. Genetic counselors urge physicians, nurses, and other health care professionals not to discount or rush through the process of obtaining informed consent to conduct a genetic test. They caution that potential psychological, social, and family implications should be acknowledged and addressed in advance of testing, including the risk of discrimination on the basis of genetic-risk status and the possibility that the predictive value of genetic information may be overestimated. The potentially life-changing consequences of genetic testing suggest that all health care professionals involved in the process should not only adhere to thoughtful informed consent procedures for genetic testing but also offer or make available genetic counseling when patients and families receive the test results.

CHAPTER 7
THE HUMAN GENOME PROJECT

When the full map of the human genome is known ... we shall have passed through a phase of human civilisation as significant as, if not more significant than, that which distinguished the age of Galileo from that of Copernicus, or that of Einstein from that of Newton. ... We have crossed a boundary of unprecedented importance. ... There is no going back. ... We are walking hopefully in the scientific foothills of a gigantic mountain range.

—Ian Lloyd, Official Record, British House of Commons, 1990

In 1953 James D. Watson (1928–) and Francis Crick (1916–2004) described the double helical structure of deoxyribonucleic acid (DNA). Their molecular DNA structure was published in "Molecular Structure of Nucleic Acids: A Structure for Deoxyribose Nucleic Acid" (*Nature*, vol. 171, no. 4356, April 25, 1953), an article that was only two pages long. Despite the brevity of this article, it ushered in a new age of discovery in genetics and laid the foundation for the sequencing of the human genome.

The word *genome* was derived from two words: *gene* and *chromosome*. In the 21st century the term *genome* is widely understood to be the entire complement of genetic material in the cell of an organism. A genome is composed of a series of four nitrogenous (containing nitrogen, a nonmetallic element that constitutes almost four-fifths of the air by volume) DNA bases: adenine (A), guanine (G), thymine (T), and cytosine (C). In each organism these bases are arranged in a specific order, or sequence, and this order constitutes the genetic code of the organism. In humans the genome is composed of approximately 3 billion bases. In 2001 a first draft sequence of the entire human genome was completed and made available to the public for study and research. The Human Genome Project (HGP) of the National Human Genome Research Institute, which is part of the National Institutes of Health (NIH), completed the full human genome sequence in April 2003.

LAYING THE GROUNDWORK FOR THE SEQUENCING OF THE HUMAN GENOME

During the 1960s and 1970s the techniques that would enable the study of molecular genetics were developed. In 1964 the American virologist Howard Temin (1934–1994) worked with ribonucleic acid (RNA) viruses and discovered that Crick's central tenet—that DNA makes RNA and that RNA makes protein—did not always hold true. In 1965 Temin described the process of reverse transcriptase—that genetic information in the form of RNA can be copied into DNA. The enzyme called reverse transcriptase used RNA as a template for the synthesis of a complementary DNA strand. Throughout the 1960s the American biochemists Robert W. Holley (1922–1993), Har Gobind Khorana (1922–), and Marshall Nirenberg (1927–2010), along with the American molecular geneticist Philip Leder (1934–), all contributed to deciphering the genetic code by determining the DNA sequence for each of the 20 most common amino acids. Holley, Khorana, and Nirenberg were awarded the 1968 Nobel Prize in Physiology or Medicine.

The American biochemist Paul Berg (1926–) created the first recombinant DNA in 1972, and his work paved the way for isolating and cloning genes. Recombinant DNA is formed by combining segments of DNA, frequently from different organisms. In 1975 the British molecular biologist Sir Edwin Southern (1938–) developed a method to isolate and analyze fragments of DNA that is still being used. Known as the Southern blot analysis, it is a procedure for separating DNA fragments by electrophoresis (a technique that separates molecules based on their size and charge) and identifying a specific fragment using a DNA probe. Figure 7.1 shows how the Southern blot analysis is performed. It is used in genetic research, forensic examinations of DNA evidence in legal proceedings, and clinical medical practice.

In 1977 the British biochemist Frederick Sanger (1918–), whose many accomplishments have been acknowledged by

FIGURE 7.1

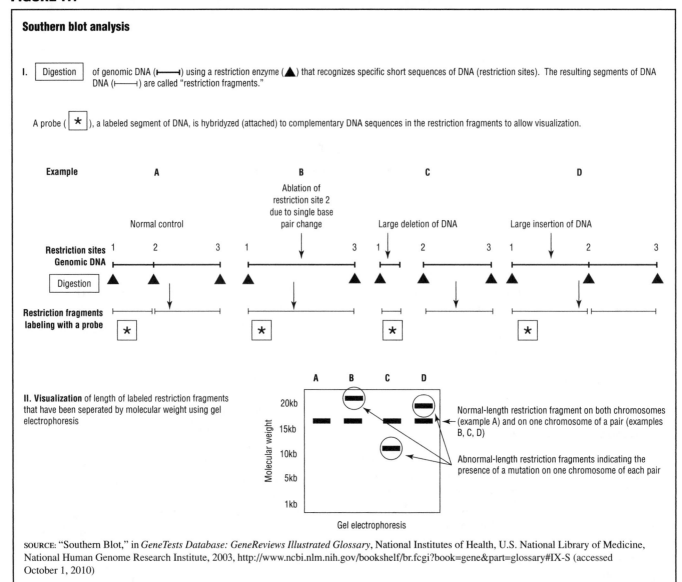

Southern blot analysis

I. Digestion of genomic DNA (⊢——⊣) using a restriction enzyme (▲) that recognizes specific short sequences of DNA (restriction sites). The resulting segments of DNA (⊢——⊣) are called "restriction fragments."

A probe (✱), a labeled segment of DNA, is hybridyzed (attached) to complementary DNA sequences in the restriction fragments to allow visualization.

Example

A — Normal control

B — Ablation of restriction site 2 due to single base pair change

C — Large deletion of DNA

D — Large insertion of DNA

II. Visualization of length of labeled restriction fragments that have been seperated by molecular weight using gel electrophoresis

Normal-length restriction fragment on both chromosomes (example A) and on one chromosome of a pair (examples B, C, D)

Abnormal-length restriction fragments indicating the presence of a mutation on one chromosome of each pair

Gel electrophoresis

SOURCE: "Southern Blot," in *GeneTests Database: GeneReviews Illustrated Glossary*, National Institutes of Health, U.S. National Library of Medicine, National Human Genome Research Institute, 2003, http://www.ncbi.nlm.nih.gov/bookshelf/br.fcgi?book=gene&part=glossary#IX-S (accessed October 1, 2010)

two Nobel Prizes in Chemistry in 1958 and 1980, and his colleagues developed techniques to determine the nucleic acid base sequence for long sections of DNA. In 1978 the American biologists Hamilton O. Smith (1931–) and Daniel Nathans (1928–1999) and the Swiss molecular geneticist Werner Arber (1929–) were awarded the Nobel Prize in Physiology or Medicine for an array of discoveries made during the 1960s, including the use of restriction enzymes, which ignited the biotechnology field. Restriction enzymes recognize and cut specific DNA sequences. That same year restriction fragment length polymorphisms (DNA sequence variants) were discovered. Figure 7.2 shows two single nucleotide polymorphisms—single base changes between homologous DNA fragments.

By using these new genetic techniques, researchers were able to identify several genes for serious human disorders during the 1980s. In 1982 the American molecular biologist James F. Gusella (1952–) and his colleagues at Harvard University began studying patients with Huntington's disease and determined that the gene for this degenerative, neuropsychiatric disorder was located on the short arm of chromosome 4. That same year a gene for neurofibromatosis type 1 was found on the long arm of chromosome 17. Neurofibromatoses are a group of genetic disorders that cause tumors (most of which are not malignant) to grow along various types of nerves and can affect the development of nonnervous tissues such as bones and skin. The disorder may also result in developmental abnormalities such as learning disabilities.

In 1985 the American biochemist Kary B. Mullis (1944–) and his colleagues at the Cetus Corporation in California pioneered the polymerase chain reaction, a fast, inexpensive technique that amplifies small fragments of DNA to make sufficient quantities available for DNA sequence analysis—that is, determining the

FIGURE 7.2

The identification of single nucleotide polymorphisms (SNPs) in soybean DNA

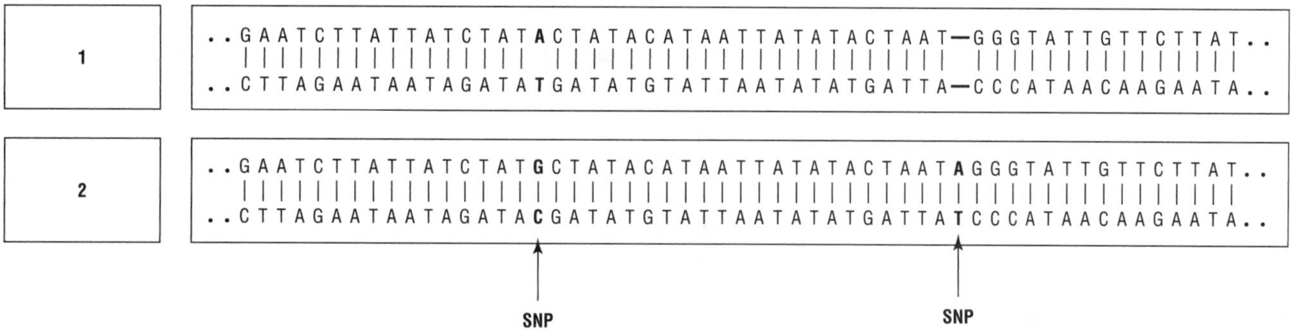

1
..GAATCTTATTATCTATACTATACATAATTATATACTAAT—GGGTATTGTTCTTAT..
..CTTAGAATAATAGATATGATATGTATTAATATATGATTA—CCCATAACAAGAATA..
2
..GAATCTTATTATCTATGCTATACATAATTATATACTAATAGGGTATTGTTCTTAT..
..CTTAGAATAATAGATACGATATGTATTAATATATGATTATCCCATAACAAGAATA..
SNP
SNP

SOURCE: Perry Cregan, "Single Nucleotide Polymorphisms," in *Soybeans: The Development of Genetic Maps*, U.S. Department of Agriculture, Beltsville Agricultural Research Center, Soybean Genomics and Improvement Laboratory, July 2003, http://bldg6.arsusda.gov/~pooley/soy/cregan/snp.html (accessed October 1, 2010)

exact order of the base pairs in a segment of DNA. Because it enabled researchers to make an unlimited number of copies of any piece of DNA, it was dubbed "molecular photocopying," and in 1993 Mullis was awarded the Nobel Prize in Chemistry for this tremendous breakthrough in gene analysis. By 1987 automated sequencers were developed, enabling even more rapid sequencing and analysis on large segments of DNA. Figure 7.3 shows the steps involved in a polymerase chain reaction.

In 1985 the Canadian molecular geneticist Lap-Chee Tsui (1950–) and his colleagues mapped the gene responsible for cystic fibrosis, the most common inherited fatal disease of children and young adults in the United States, to the long arm of chromosome 7. The gene for cystic fibrosis was discovered in 1989, and it was determined that three missing nucleic acid bases occurred in the altered gene of 70% of patients with cystic fibrosis.

The mutations associated with Duchenne muscular dystrophy were identified in 1987. This gene is located close to the gene for chronic granulomatous disease (an X-linked autosomal recessive disorder that, if left untreated, is fatal in childhood) on the short arm of the X chromosome. In 1990 the American geneticist Mary-Claire King (1946–) found the first evidence that a gene on chromosome 17 (now known as BRCA1) could be associated with an inherited predisposition to breast and ovarian cancer.

The discoveries and technological advances made by researchers during the 1970s and 1980s gave rise to modern clinical molecular genetics. The study of chromosome structure and function, called cytogenetics, produced methods to view distinct bands on each chromosome. Figure 7.4 is a cytogenetic map of human chromosomes.

FIGURE 7.3

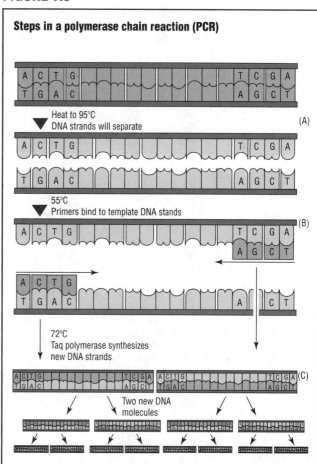

Steps in a polymerase chain reaction (PCR)

Heat to 95°C
DNA strands will separate
(A)
55°C
Primers bind to template DNA stands
(B)
72°C
Taq polymerase synthesizes
new DNA strands
(C)
Two new DNA
molecules

SOURCE: Marion Beal, "Polymerase Chain Reaction (PCR)," in "How Do Scientists Analyze Coral DNA?" in *Ocean Explorer: Explorations—Estuary to the Abyss*, U.S. Department of Commerce, National Oceanic and Atmospheric Administration, July 2005, http://oceanexplorer.noaa.gov/explorations/04etta/background/dna/dna.html (accessed October 1, 2010)

FIGURE 7.4

Cytogenetic map of human chromosomes

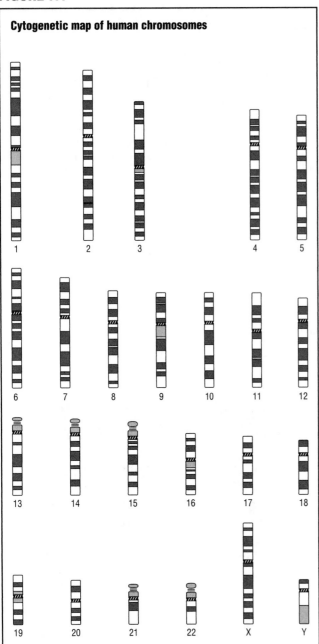

SOURCE: "Cytogenetic Map," in *Genomes and Maps*, U.S. Department of Health and Human Services, National Institutes of Health, National Human Genome Research Institute, Division of Intramural Research, 2007, http://www.ncbi.nlm.nih.gov/Class/MLACourse/Modules/Genomes/map_cytogenetic.html (accessed October 1, 2010)

TABLE 7.1

U.S. Human Genome Project funding, fiscal years 1988–2003

[$ millions]

Fiscal year	Department of Energy	National Institutes of Health	U.S. total
1988	10.7	17.2	27.9
1989	18.5	28.2	46.7
1990	27.2	59.5	86.7
1991	47.4	87.4	134.8
1992	59.4	104.8	164.2
1993	63.0	106.1	169.1
1994	63.3	127.0	190.3
1995	68.7	153.8	222.5
1996	73.9	169.3	243.2
1997	77.9	188.9	266.8
1998	85.5	218.3	303.8
1999	89.9	225.7	315.6
2000	88.9	271.7	360.6
2001	86.4	308.4	394.8
2002	90.1	346.7	434.3
2003	64.2	372.8	437

Note: These numbers do not include construction funds, which are a very small part of the budget.

SOURCE: "Human Genome Project Budget," in *Human Genome Project Information*, U.S. Department of Energy Office of Science, Office of Biological and Environmental Research, Human Genome Project, September 14, 2004, http://www.ornl.gov/sci/techresources/Human_Genome/project/budget.shtml (accessed October 1, 2010)

Cytogenetic studies are applied in three broad areas of medicine: congenital (from birth) disorders, prenatal diagnosis, and neoplastic diseases (cancer).

THE BIRTH OF THE HUMAN GENOME PROJECT

The first meetings to discuss the feasibility of sequencing the human genome were organized by Robert Louis Sinsheimer (1920–), a molecular biologist and chancellor of the University of California (UC), Santa Cruz, and were held on campus in 1985. The idea of sequencing the human genome generated excitement among the many well-known researchers in attendance—they considered the undertaking to be the "Holy Grail" of molecular biology. The following year Congress began to consider the feasibility of human genome research. However, Congress did not decide to fund the project until 1988, after it concluded that the establishment of administrative centers accountable to Congress could effectively manage key aspects of the project, such as databases, sharing of research findings and materials, and cultivating new technologies. The initial funding from Congress was $17.2 million to the NIH and $10.7 million to the U.S. Department of Energy's (DOE) Office of Health and Environmental Science, with progressive increases over the next several years. (See Table 7.1.)

The allocation of these funds incited an impassioned debate. Opponents argued that the financial and human resources devoted to the "big science" of the human genome project would divert research funds from vital scientific and biomedical research and that most of the sequence was of little biological interest and no medical utility. Other detractors warned that the sheer size of the human genome would impede completion of the project within a reasonable time frame without the creation of entirely new research methods and technologies. The project was launched despite considerable opposition, and most of these concerns were dispelled during the project's early years.

In 1988 Congress provided funding to the NIH and the DOE to "coordinate research and technical activities related to the human genome." The NIH established the

Office of Human Genome Research in September 1988. The following year the office was renamed the National Center for Human Genome Research (NCHGR). Watson served as its enthusiastic champion and director until April 1992. Larry Thompson of the National Human Genome Research Institute explains in "Background Paper on Internal Educational Activities at the National Human Genome Research Institute" (June 10, 2002, http://www.genome .gov/10005291) that following Watson's appointment, the NIH and the DOE committed 5% of the project's budget to address ethical, legal, and social issues that arose from the study of the human genome. This ambitious undertaking constituted the largest bioethics program, in terms of funding and human resources, in the world.

In "About the Human Genome Project" (August 19, 2008, http://www.ornl.gov/sci/techresources/Human _Genome/project/about.shtml), Human Genome Project Information describes the ambitious goals of the HGP when it began in 1990. The overarching HGP goals were to:

- Identify all the approximately 20,000 to 25,000 genes in human DNA

- Determine the sequences of the 3 billion chemical base pairs that make up human DNA

- Store HGP findings and other information in databases

- Improve tools for data analysis

- Transfer related technologies to the private sector

- Effectively address the ethical, legal, and social issues that might arise from the project

International genomic research was also under way in Italy, England, France, Germany, Japan, and other countries. For example, in 1987 the Italian National Research Council launched a genome research project, and the United Kingdom began its project in February 1989. An international group of geneticists founded in 1989 the Human Genome Organization (HUGO) in Switzerland. Many international collaborations had already been forged as individual scientists exchanged information in their quests for genetic links to disease. HUGO developed an international framework to coordinate research projects and prevent wasted resources through duplication, creating a culture of sharing data. In 1990 the European Commission initiated a two-year human genome project. The Soviet Union funded its genome research project that same year.

In April 1990 the initial planning stage of the HGP was completed with the publication of the joint research plan *Understanding Our Genetic Inheritance: The Human Genome Project—The First Five Years, Fiscal Years 1991–1995* (http://www.ornl.gov/sci/techresources/Human _Genome/project/5yrplan/summary.shtml). Just two years into the five-year plan, Watson resigned from his leadership position with the NCHGR because he vehemently disagreed with NIH decisions about the commercializa-

tion, propriety, and legality of patenting human gene sequences. Watson maintained that data from the HGP should be in the public domain and freely available to all scientists as well as to the public. In April 1993 Francis S. Collins (1950–) was named the director.

Many prominent researchers sided with Watson against the patenting and commercialization of HGP data. In 1996 scientists at leading research institutions throughout the world agreed to submit their findings and genome sequences to GenBank, a genome database maintained by the NIH. In a resounding and unanimous move, they required the publication of any submitted sequence data on the Internet within 24 hours of its receipt by GenBank. This action ensured that gene sequences were in the public domain and could not be patented.

Worming Away

Even though sequencing the human genome was the principal objective, the HGP also sought to sequence the genomes of other organisms. These other organisms served as models, enabling researchers to test and refine new methods and technologies that helped identify corresponding genes in the human genome. Table 7.2 is a list of some of the model organisms, including the roundworm, sequenced during the course of the HGP, along with the dates the sequences were published and the number of bases in each organism.

At Cambridge University, the British molecular biologist Sydney Brenner (1927–) was studying the nematode worm *Caenorhabditis elegans*. By 1989 Brenner and his colleagues had successfully produced a map of the entire *Caenorhabditis elegans* genome. The map consisted of multiple overlapping fragments of DNA, arranged in the correct order, and Brenner's research team printed the worm's genome on postcard-sized pieces of paper.

TABLE 7.2

Model organisms sequenced

Date sequenced[a]	Species	Total bases[b]
7/28/1995	*Haemophilus influenzae* (bacterium)	1,830,138
10/30/1995	*Mycoplasma genitalium* (bacterium)	580,073
5/29/1997	*Saccharomyces cerevisiae* (yeast)	12,069,247
9/5/1997	*Escherichia coli* (bacterium)	4,639,221
11/20/1997	*Bacillus subtillis* (bacterium)	4,214,814
12/31/1998	*Caenorhabditis elegans* (round worm)	97,283,371
		99,167,964[c]
3/24/2000	*Drosophila melanogaster* (fruit fly)	~137,000,000
12/14/2000	*Arabidopsis thaliana* (mustard plant)	~115,400,000
1/26/2001	*Oryza sativa* (rice)	~430,000,000
2/15/2001	*Homo sapiens* (human)	~3,200,000,000

[a]First publication date.
[b]Data excludes organelles or plasmids. These numbers should not be taken as absolute. Scientists are confirming the sequences; several laboratories were involved in the sequencing of a particular organism and have slightly different numbers; and there are some strain variations.
[c]The first number was originally published, and the second is a correction as of June 2000.

SOURCE: Richard Robinson, ed., "Model Organisms Sequenced," in *Genetics*, vol. 2, *E–I*, Macmillan Reference USA, 2002

Watson believed the genomes of smaller organisms would not only help refine research methods and the technology but also provide valuable sources of comparison once the human genome project was under way. The worm map convinced Watson that *Caenorhabditis elegans* should be the first multicellular organism to have its complete genome accurately sequenced. When the worm-sequencing project began in 1990, the first automatic sequencing machines had just become available from Applied Biosystems, Inc. The sequencing machines enabled the worm pilot project to meet its objective of sequencing 3 million bases in three years. Equally impor-

tant, the worm project demonstrated that the technology could scale up—that is, more machines and more technologists could produce more sequences faster.

A DECADE OF ACCOMPLISHMENTS: 1993 TO 2003

When the HGP began in September 1990, its projected completion date was 2005, at a cost of $3 billion for just the U.S. portion of the research. Ever-improving research techniques—including the use of restriction fragment length polymorphisms, which is described in detail in Figure 7.5, as well as the polymerase chain reaction,

FIGURE 7.5

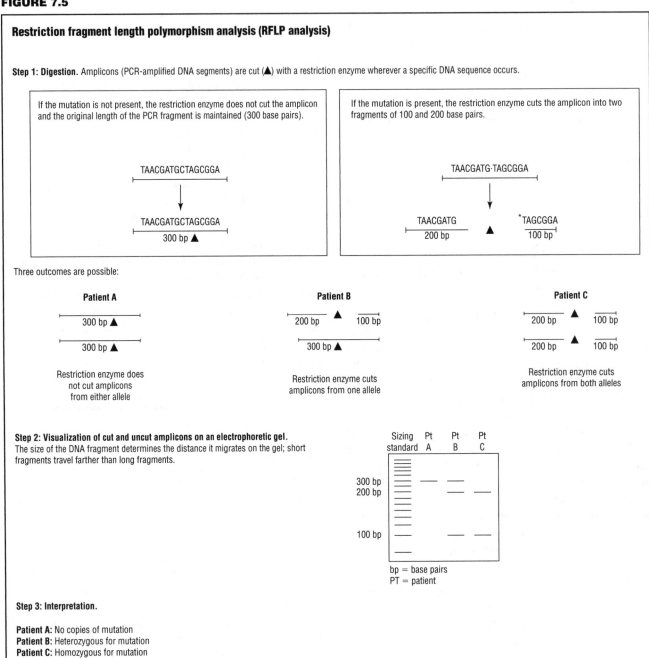

Restriction fragment length polymorphism analysis (RFLP analysis)

Step 1: Digestion. Amplicons (PCR-amplified DNA segments) are cut (▲) with a restriction enzyme wherever a specific DNA sequence occurs.

If the mutation is not present, the restriction enzyme does not cut the amplicon and the original length of the PCR fragment is maintained (300 base pairs).

If the mutation is present, the restriction enzyme cuts the amplicon into two fragments of 100 and 200 base pairs.

TAACGATGCTAGCGGA

TAACGATGCTAGCGGA
300 bp ▲

TAACGATG·TAGCGGA

TAACGATG   *TAGCGGA
200 bp   ▲   100 bp

Three outcomes are possible:

Patient A

300 bp ▲

300 bp ▲

Restriction enzyme does not cut amplicons from either allele

Patient B

200 bp ▲ 100 bp

300 bp ▲

Restriction enzyme cuts amplicons from one allele

Patient C

200 bp ▲ 100 bp

200 bp ▲ 100 bp

Restriction enzyme cuts amplicons from both alleles

Step 2: Visualization of cut and uncut amplicons on an electrophoretic gel. The size of the DNA fragment determines the distance it migrates on the gel; short fragments travel farther than long fragments.

Sizing standard   Pt A   Pt B   Pt C

300 bp
200 bp

100 bp

bp = base pairs
PT = patient

Step 3: Interpretation.

Patient A: No copies of mutation
Patient B: Heterozygous for mutation
Patient C: Homozygous for mutation

SOURCE: "Restriction Fragment Length Polymorphism Analysis (RFLP Analysis)," in *GeneTests Database: GeneReviews Illustrated Glossary*, National Institutes of Health, U.S. National Library of Medicine, National Human Genome Research Institute, 2002, http://www.genetests.org/servlet/access?qry=ALLTERMS&db=genestar&fcn=term&gtreport2=true&id=8888892&key=XlLW5MSdFPEu- (accessed October 1, 2010)

bacterial and yeast artificial chromosomes, and pulsed-field gel electrophoresis—accelerated the progress of the project. The HGP researchers finished the mapping two years earlier than scheduled, and U.S. scientists spent just $2.7 billion.

Even though the HGP was truly a collaborative, international effort, most of the sequencing work was performed at the Whitehead Institute for Medical Research in Massachusetts, the Baylor College of Medicine in Texas, the University of Washington, the Joint Genome Institute in California, and the Sanger Centre in the United Kingdom. Along with U.S government funding, the HGP was supported by the Wellcome Trust, a charitable foundation in the United Kingdom.

Public and Private Initiatives Compete

The NIH (March 25, 2010, http://www.nih.gov/about/almanac/organization/NHGRI.htm) notes that in 1993 the NCHGR established a Division of Intramural Research, which was charged with developing genome technology research of specific diseases. By 1996 eight NIH institutes and centers had also collaborated to create the Center for Inherited Disease Research to study the genetics of complex diseases. In 1997 the NCHGR gained full institute status at the NIH, becoming the National Human Genome Research Institute (NHGRI). The following year a third five-year plan, "New Goals for the U.S. Human Genome Project: 1998–2003," was published in *Science* (vol. 282, no. 5389, October 23, 1998).

The mid-1990s also saw the birth of a privately funded genomics effort led by the American geneticist J. Craig Venter (1946–). Venter had been working in a laboratory at the NIH when he decided to concentrate his sequencing efforts not on the genome itself, but on the gene products—that is, messenger RNAs produced by each cell. Besides the genes themselves, the genome is composed of a much larger amount of DNA whose function is as yet unknown. This portion of the genome is often referred to as "junk DNA," although it is possible that its true value has not yet been determined. Venter eventually left the NIH to establish The Institute for Genomic Research (TIGR), a private, nonprofit organization aimed at collecting and interpreting vast numbers of expressed sequence tags. (Expressed sequence tags are random fragments of complementary DNAs derived from the information in the RNAs, which contain all the information that is actually expressed in a given cell type.) In July 1995 TIGR published the first sequenced genome of the bacteria *Haemophilus influenzae*. (See Table 7.2.)

The international partners in the genome project met in Bermuda in February 1996 at a strategy meeting sponsored by the Wellcome Trust. There they created the "Bermuda Principles," a set of conditions that govern access to data, including the standard that sequence information be released into public databases within 24 hours. To adhere to this agreement, participating scientists were to deposit base sequences into one of three databases within 24 hours of sequencing completion. The data contained in the three databases were exchanged daily. Because these were public databases, access to the stored sequences was free and unrestricted. The agreement was extended to data on other organisms at a meeting later that same year.

In May 1998 Venter announced that he was allying with the Perkin-Elmer Corporation to form a new company that would compete directly with the public effort to sequence the entire human genome by 2001. The new company would become Celera Genomics. Venter planned to take a different approach from the one used by the HGP. Instead of the map-based, gene-by-gene approach taken by the HGP, his firm planned to break the genome into random lengths, and then sequence and reassemble it. This method saved time by eliminating the mapping phase, but it required robust computing capabilities to reassemble the human genome, which includes many repeated sequences. Venter's approach relied on the use of a supercomputer and 300 high-speed automatic sequencers manufactured by the Perkin-Elmer Corporation; it was the precursor to the large-scale genomics studies that have become standard.

The competition between public and private initiatives began in earnest. Within one week of the launch of the private initiative, the Wellcome Trust increased its funding to the Sanger Centre to step up the production of raw sequence. In response to this increased support, the Sanger Centre revised its objective by aiming to sequence a full one-third of the entire genome rather than just one-sixth. The race between the public and private projects was on, and the milestones of the sequencing project came ever more rapidly.

Venter's firm energized the publicly funded project and inspired intensified efforts. HGP investigators feared that if their efforts were perceived as slow and inefficient, the HGP would lose congressional support and funding. The threat of genomic information ending up solely in the hands of a private firm was simply unacceptable to the researchers. When Celera entered the race to sequence the human genome, it changed the landscape of the field. Celera resolved to make its data available only to paying customers and planned to patent some sequences before releasing them.

The publicly funded HGP released sequence information quickly both to provide the scientific community with timely data that were immediately usable and to place identified sequences beyond the reach of commercial companies wishing to patent them or charge for access to the data. The leadership of the HGP echoed the sentiments that had prompted Watson's resignation. It contended that patenting the human sequence was unethical and delayed the timely application of genomic information to medical disorders. Even though the HGP staunchly opposed privileged access to the data, to speed completion of the project it had agreed to grant Celera specific rights to the data generated by the collaborative effort between the two groups.

THE FIRST DRAFT OF THE COMPLETED HUMAN GENOME

In April 2000 Celera announced that it was prepared to present the first draft of the human genome. Not unexpectedly, scientists and the public eagerly anticipated this "first look" at the human genome. Even though geneticists and other scientists could better comprehend the mechanics and the future implications of this endeavor than the general public, the significance of this achievement was evident to professionals and lay people alike. The professional literature and the mass media had successfully communicated the importance of this achievement, and it was understood that knowledge of the human genome held the key to the singularity of the human species. Furthermore, it was widely assumed that this information would be the basis for unprecedented advances in medicine and biomedical technology.

In "Rival Demands Sink Genome Alliance Plans" (*Nature*, vol. 404, no. 6774, March 9, 2000), Natasha Loder explains that all hopes for continuing cooperation and collaboration between the HGP and Celera were dashed when a letter from the Wellcome Trust was released to the public. The letter described the HGP's concerns about Celera's intent to have its commercial rights extended to research applications and to publish data from the collaborative efforts under its name alone. Celera termed the release of the letter as unethical, and in a media interview Venter described it as "a low-life thing to do." The anger and bitterness between the two groups was not resolved and each vowed to publish the data separately.

In February 2001 the first working draft of the human genome was published in special issues of the journals *Nature* (vol. 409, no. 6822, February 15, 2001) and *Science* (vol. 291, no. 5507, February 16, 2001). *Nature* detailed the initial analysis of the descriptions of the sequence generated by the HGP, and *Science* contained the draft sequence reported by Celera. *Nature* said its choice to publish the HGP data reflected its traditional preference for publicly funded research. *Science* editors Barbara R. Jasny and Donald Kennedy lauded Venter's group in the editorial "The Human Genome," writing "his colleagues made it possible to celebrate this accomplishment far sooner than was believed possible."

One of several surprises from the first draft was that previous estimates of gene number appeared to have been wildly inaccurate. Most pregenome project estimates predicted that humans had as many as 60,000 to 150,000 genes. The first draft of the complete genome sequence indicated that the true number of genes required to make a human being was less than 40,000. (The current estimate is between 20,000 and 25,000 genes.) By comparison, yeast have 6,000 genes, fruit flies have 14,000, roundworms have 19,000, and the mustard weed plant has 26,000. Another surprise was the observation that humans share 99.9% of the nucleotide code in the human genome. Notably, human diversity at the genetic level is encoded by less than a 0.1% variation in DNA.

First Draft Is Headline News

In June 2000 the White House press release "President Clinton Announces the Completion of the First Survey of the Entire Human Genome" (http://www.ornl.gov/sci/techresources/Human_Genome/project/clinton1.shtml) predicted some of the anticipated medical outcomes of the project. These included the ability to:

- Alert patients that they are at risk for certain diseases. Once scientists discover which DNA sequence changes in a gene can cause disease, healthy people can be tested to see whether they risk developing conditions such as diabetes or prostate cancer later in life. In many cases, this advance warning can be a cue to start a vigilant screening program, to take preventive medicines, or to make diet or lifestyle changes that may prevent the disease.

- Reliably predict the course of disease. Diagnosing ailments more precisely will lead to more reliable predictions about the course of a disease. For example, a genetic fingerprint will allow doctors treating prostate cancer to predict how aggressive a tumor will be. New genetic information will help patients and doctors weigh the risks and benefits of different treatments.

- Precisely diagnose disease and ensure the most effective treatment is used. Genetic analysis allows us to classify diseases, such as colon cancer and skin cancer, into more defined categories. These improved classifications will eventually allow scientists to tailor drugs for patients whose individual response can be predicted by genetic fingerprinting. For example, cancer patients facing chemotherapy could receive a genetic fingerprint of their tumor that would predict which chemotherapy choices are most likely to be effective, leading to fewer side effects from the treatment and improved prognoses.

- Developing new treatments at the molecular level. Drug design, guided by an understanding of how genes work and knowledge of exactly what happens at the molecular level to cause disease, will lead to more effective therapies. In many cases, rather than trying to replace a gene, it may be more effective and simpler to replace a defective gene's protein product. Alternatively, it may be possible to administer a small molecule that would interact with the protein to change its behavior. This is the strategy behind a drug in development for chronic myelogenous leukemia, which targets the genetic flaw causing the disease. It attaches to the abnormal protein caused by the genetic flaw and blocks its activity. In preliminary tests, blood counts returned to normal in all patients treated with the drug.

That same month BBC News (June 26, 2000, http://news.bbc.co.uk/1/hi/sci/tech/807126.stm) compiled assessments of the achievement by the world's premier scientists and politicians. Venter spoke for many researchers when he said, "I think we will view this period as a very historic time, a new starting point." Michael Dexter of the Wellcome Trust echoed Venter's sentiments when he ventured, "This is the outstanding achievement not only of our lifetime, but in terms of human history. I say this, because the Human Genome Project does have the potential to impact on the life of every person on this planet." Randy Scott, the president of Incyte, another private firm involved in genomics research, predicted that "the availability of genome sequence is just the beginning. Scientists now want to understand the genes and the role they play in the prevention, diagnosis and treatment of disease." Mike Stratton of the Cancer Genome Project was equally optimistic when he said that "it would surprise me enormously if in 20 years the treatment of cancer had not been transformed." Sanger, the pioneer of DNA sequencing, expressed the collective awe of the scientific community when the HGP was completed earlier than anticipated. He admitted, "I never thought it would be done as quickly as this."

The U.S. media celebrated the achievement with a flood of press releases and features. Efforts were also made to explain this monumental accomplishment to the public and to educate students. The DOE Human Genome Program provided a wealth of information about the HGP and its findings on the Internet. Figure 7.6 is an example of the information the DOE made available to the public. Figure 7.7 presents some of the potential environmental benefits of the HGP, including reducing the impact of climate change and development of sustainable energy sources that may be realized in the future.

PUFFER FISH AND MOUSE GENOMES ARE SEQUENCED

In July 2002 the DOE Joint Genome Institute (JGI), operated by the Lawrence Berkeley National Laboratory, the Lawrence Livermore National Laboratory, and the Los Alamos National Laboratory, announced the draft

FIGURE 7.6

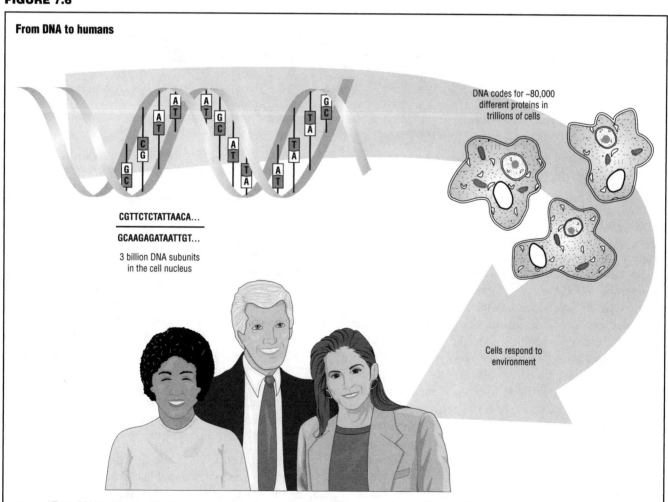

SOURCE: "From DNA to Humans," in *Human Genome Project Information Basic Genomics Image Gallery*, U.S. Department of Energy Office of Science, Office of Biological and Environmental Research, Human Genome Project, undated, http://genomics.energy.gov/gallery/basic_genomics/view.np/view-22.html (accessed October 1, 2010)

FIGURE 7.7

Projected national benefits of genomics research

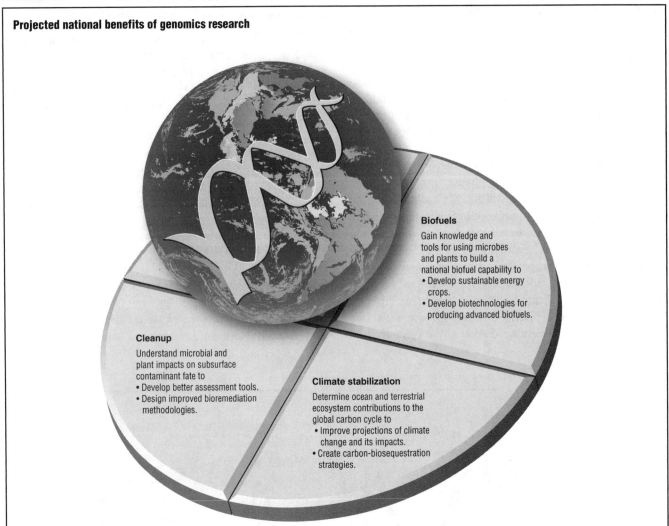

Biofuels

Gain knowledge and tools for using microbes and plants to build a national biofuel capability to
- Develop sustainable energy crops.
- Develop biotechnologies for producing advanced biofuels.

Cleanup

Understand microbial and plant impacts on subsurface contaminant fate to
- Develop better assessment tools.
- Design improved bioremediation methodologies.

Climate stabilization

Determine ocean and terrestrial ecosystem contributions to the global carbon cycle to
- Improve projections of climate change and its impacts.
- Create carbon-biosequestration strategies.

SOURCE: "Payoffs for the Nation," in *Human Genome Project Information Systems Biology Gallery*, U.S. Department of Energy Office of Science, Office of Biological and Environmental Research, Human Genome Project, March 2008, http://genomics.energy.gov/gallery/systems_biology/detail.np/detail-899 .html (accessed October 1, 2010)

sequencing, assembly, and analysis of the genome of the Japanese puffer fish *Fugu rubripes*. The Fugu Genome Project was initiated in 1989 in Cambridge, England, and in November 2000 the International Fugu Genome Consortium was formed, headed by the JGI. During 2001 the puffer fish genome was sequenced and assembled using the whole genome method pioneered by Celera. The puffer fish was the first vertebrate genome to be publicly sequenced and assembled in this manner and the first vertebrate genome published after the human genome. According to the JGI (2011, http://genome.jgi-psf.org/ Takru4/Takru4.home.html), puffer fish have the smallest known genomes among vertebrates (animals with bony backbones or cartilaginous spinal columns—fish, reptiles, birds, and mammals, including humans). The puffer fish sequence has about the same number of genes as the considerably larger human genome, but is more compact because it contains relatively little of the junk DNA present in the human genome sequence.

Comparison of the human and puffer fish genomes enabled investigators to predict the existence of nearly 1,000 previously unidentified human genes. Even though the function of these additional genes is as yet unknown, they contribute to the complete catalog of human genes. Ascertaining the existence and location of genes helped scientists begin to describe how they are regulated and function in the human body. Of the more than 30,000 puffer fish genes identified, the vast majority of human genes have counterparts in the puffer fish, with the most significant differences in genes of the immune system, metabolic regulation, and other physiological systems that are not alike in fish and mammals.

On December 5, 2002, the first draft of the sequence of the mouse genome was published in *Nature* (vol. 420, no. 6915). The mouse genome findings were deemed among the most important in terms of their comparability with humans. Mice and humans have about the same number of genes—approximately 20,000—and DNA base pairs—mice have 2.5 billion and humans have approximately 3 billion. More importantly, about 90% of genes associated with medical disorders in humans have counterparts in mice. This finding means that mice are especially well suited for studying diseases that afflict humans and for testing therapeutic treatments for disease.

THE HGP IS COMPLETED

After the publication of a first-draft human genome in 2001, researchers continued to fill in the blanks and produce a complete and accurate sequence. In January 2003 another milestone in the human genome sequencing effort was reported: the fourth human chromosome—chromosome 14, the largest one to date, with 87 million base pairs—had been sequenced. Roland Heilig et al. published their findings in "The DNA Sequence and Analysis of Human Chromosome 14" (*Nature*, vol. 421, no. 6923, February 6, 2003). They found two genes that are vital for immune responses on chromosome 14 and about 60 genes that, when defective, contribute to disorders such as spastic paraplegia and Alzheimer's disease.

In a remarkable coincidence that made the crowning achievement of the HGP even more poignant, the completion of the sequencing of the human genome occurred during the same year slated for celebrations of the 50th anniversary of the discovery of the DNA double helix. On April 14, 2003, the International Human Genome Sequencing Consortium, which is directed by the DOE and the NHGRI, announced in the press release "All Goals Achieved: New Vision for Genome Research Unveiled" (http://www.genome.gov/11006929) the successful completion of the HGP more than two years earlier than had been anticipated. The consortium included scientists at 20 sequencing centers in China, France, Germany, Great Britain, Japan, and the United States.

Nature, the same journal that had published the groundbreaking discoveries of Watson and Crick 50 years earlier, hailed the era of the genome in a special edition dated April 24, 2003 (vol. 422, no. 6934). As had been the practice since the inception of the HGP, the entirety of sequence data generated by the HGP was immediately entered into public databases and made freely available to the scientific community throughout the world, with no restrictions on its use or redistribution. The data are used by researchers in academic settings and industry, as well as by commercial firms that provide information services to biotechnologists. Figure 7.8 shows some of the landmarks in the sequenced human genome.

In "All Goals Achieved," the DOE and the NHGRI describe the international effort to sequence the 3 billion DNA base pairs in the human genome as "one of the most ambitious scientific undertakings of all time," comparing it to feats such as splitting the atom or traveling to the moon. Collins proudly declared that "the Human Genome

Project has been an amazing adventure into ourselves, to understand our own DNA instruction book, the shared inheritance of all humankind. All the project's goals have been completed successfully—well in advance of the original deadline and for a cost substantially less than the original estimates." In the same press release Eric Steven Lander (1957–), the director of the Whitehead Institute/Massachusetts Institute of Technology Center for Genome Research, predicted the postgenomic era when he asserted, "The Human Genome Project represents one of the remarkable achievements in the history of science. Its culmination this month signals the beginning of a new era in biomedical research. Biology is being transformed into an information science, able to take comprehensive global views of biological systems. With knowledge of all the components of the cells, we will be able to tackle biological problems at their most fundamental level."

Collins urged the scientific community not to rest on its laurels in the wake of this triumph, saying, "With this foundation of knowledge firmly in place, the medical advances promised from the project can now be significantly accelerated." The April 24, 2003, issue of *Nature* detailed the challenges researchers will face in the postgenomic era as they seek to employ the HGP data to treat disease and improve public health. Recommendations included collaborative efforts to produce:

- New tools to allow discovery in the not-too-distant future of the genetic contributions to frequently occurring diseases, including diabetes, heart disease, and mental illnesses such as schizophrenia.

- Improved methods for the early detection of disease and to enable timely treatment when it is likely to be effective.

- New technologies able to sequence the entire genome of any person affordably, ideally for less than $1,000.

- Wider access to tools and technologies of "chemical genomics" to enhance understanding of biological pathways and accelerate pharmaceutical and other treatment research.

Along with the special commemorative issue of *Nature*, the April 11, 2003, edition of *Science* (vol. 300, no. 5617) ran articles that described the HGP and detailed the multidisciplinary DOE plan dubbed "Genomes to Life," which aimed to use HGP data to understand the ways in which microbes can provide opportunities to develop clean energy, reduce climate change, and clean the environment.

THE HGP REVISES ITS ESTIMATE

In the October 2004 press release "International Human Genome Sequencing Consortium Describes Finished Human Genome Sequence" (http://www.genome.gov/12513430), the NHGRI reduced its estimate of the number of human genes from between 30,000 to 35,000 to between 20,000 and

FIGURE 7.8

Selected landmarks of the Human Genome Project

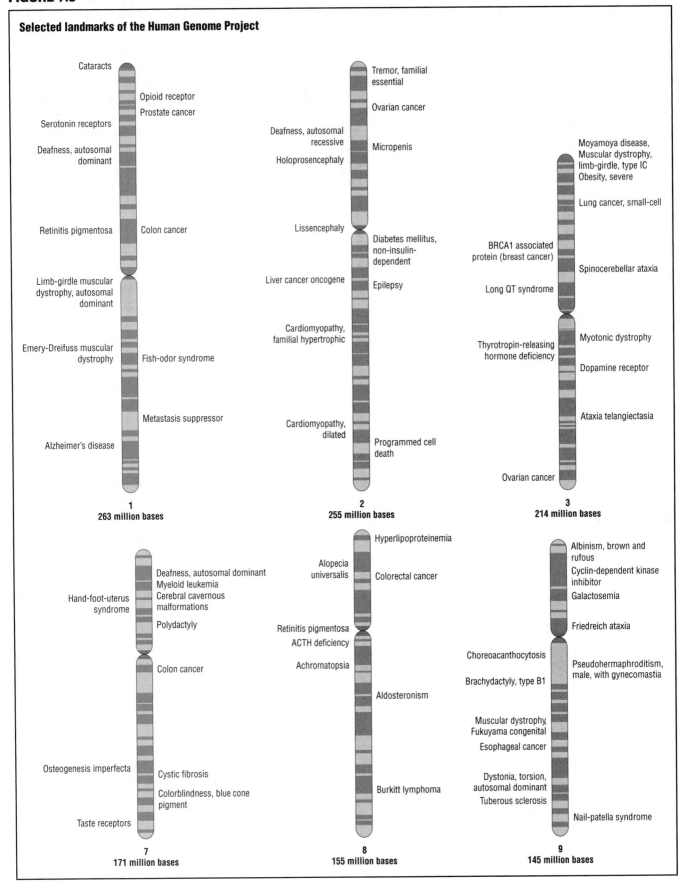

FIGURE 7.8

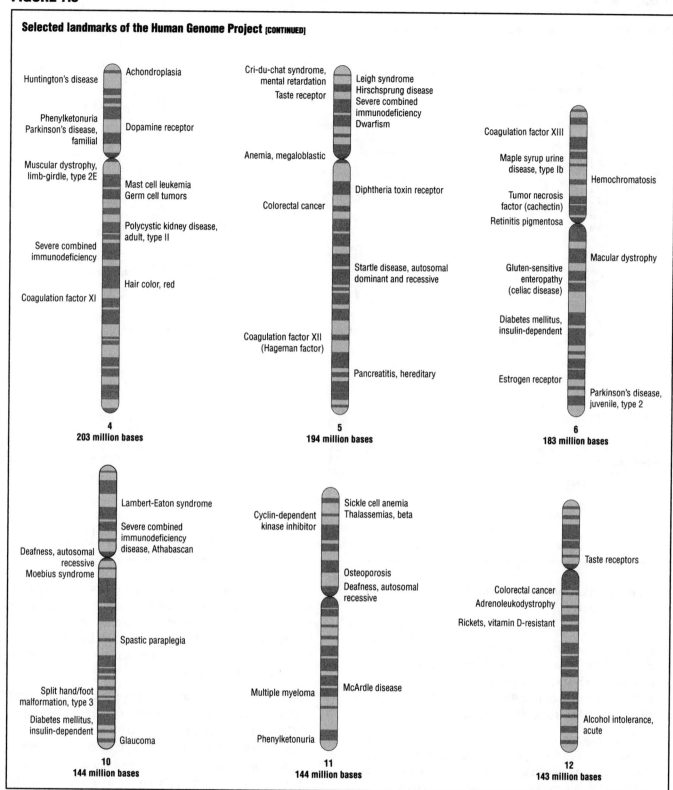

Selected landmarks of the Human Genome Project [CONTINUED]

4 — 203 million bases

- Huntington's disease
- Achondroplasia
- Phenylketonuria
- Parkinson's disease, familial
- Dopamine receptor
- Muscular dystrophy, limb-girdle, type 2E
- Mast cell leukemia
- Germ cell tumors
- Polycystic kidney disease, adult, type II
- Severe combined immunodeficiency
- Hair color, red
- Coagulation factor XI

5 — 194 million bases

- Cri-du-chat syndrome, mental retardation
- Taste receptor
- Leigh syndrome
- Hirschsprung disease
- Severe combined immunodeficiency
- Dwarfism
- Anemia, megaloblastic
- Diphtheria toxin receptor
- Colorectal cancer
- Startle disease, autosomal dominant and recessive
- Coagulation factor XII (Hageman factor)
- Pancreatitis, hereditary

6 — 183 million bases

- Coagulation factor XIII
- Maple syrup urine disease, type Ib
- Hemochromatosis
- Tumor necrosis factor (cachectin)
- Retinitis pigmentosa
- Macular dystrophy
- Gluten-sensitive enteropathy (celiac disease)
- Diabetes mellitus, insulin-dependent
- Estrogen receptor
- Parkinson's disease, juvenile, type 2

10 — 144 million bases

- Lambert-Eaton syndrome
- Severe combined immunodeficiency disease, Athabascan
- Deafness, autosomal recessive
- Moebius syndrome
- Spastic paraplegia
- Split hand/foot malformation, type 3
- Diabetes mellitus, insulin-dependent
- Glaucoma

11 — 144 million bases

- Cyclin-dependent kinase inhibitor
- Sickle cell anemia
- Thalassemias, beta
- Osteoporosis
- Deafness, autosomal recessive
- Multiple myeloma
- McArdle disease
- Phenylketonuria

12 — 143 million bases

- Taste receptors
- Colorectal cancer
- Adrenoleukodystrophy
- Rickets, vitamin D-resistant
- Alcohol intolerance, acute

25,000. The refined human genome sequence, published in "Finishing the Euchromatic Sequence of the Human Genome" (*Nature*, vol. 431, no. 7011, October 21, 2004), was the most complete version to date. According to HGP scientists, it covered 99% of the gene-containing parts of the human genome, identified nearly all known genes, and was 99.9% accurate.

DAWN OF THE POSTGENOMIC ERA

In "Potential Benefits of Human Genome Project Research" (October 9, 2009, http://www.ornl.gov/sci/techre sources/Human_Genome/project/benefits.shtml), Human Genome Project Information enumerates many of the potential benefits of HGP research. Besides its role in the practice of molecular medicine, other uses of HGP

FIGURE 7.8

Selected landmarks of the Human Genome Project [CONTINUED]

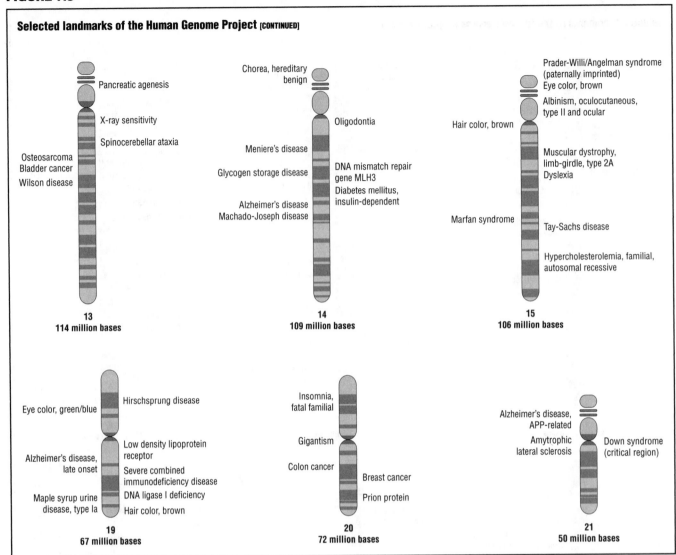

data and applications of human and other genomic research include:

- Microbial genomics—using bacteria to create new energy sources such as biofuels and safe, efficient toxic waste cleanup; enhancing understanding of how microbes cause disease; and protecting workers and the public from threats of biological and chemical terrorism and warfare. (See Figure 7.9.)

- Risk assessment—measuring the risks and health problems caused by exposure to radiation, carcinogens (cancer-causing agents), and mutagenic chemicals; and reducing the probability of heritable mutations.

- Archaeology, anthropology, evolution, and human migration—comparing the genomes of humans and other organisms such as mice has already identified similar genes associated with diseases and traits; improving the understanding of germline (cells that give rise to eggs or sperm) mutations; studying migration based on female genetic inheritance; examining mutations on the Y chromosome to trace lineage and migration of males;

and comparing the DNA sequences of entire genomes of different microbes to enhance the understanding of the relationships among the three domains of life: archaebacteria (cells that do not contain nuclei), eukaryotes (cells that contain nuclei), and prokaryotes (single-celled organisms without nuclei).

- DNA forensics—identifying crime victims, potential suspects, and catastrophe victims through examination of DNA; confirming paternity and other family relationships; clearing people wrongly accused of crimes; identifying and protecting endangered species; detecting bacterial and other environmental pollutants; matching organ donors and recipients for transplant programs; and determining pedigrees for animals and plants.

- Agriculture and livestock breeding—developing healthier and stronger crops and farm animals that can resist insects, disease, and drought; creating safer pesticides; growing more nutritious produce; incorporating vaccines into food products; and redeploying plants such as tobacco for use in environmental cleanup programs.

FIGURE 7.8

Selected landmarks of the Human Genome Project [CONTINUED]

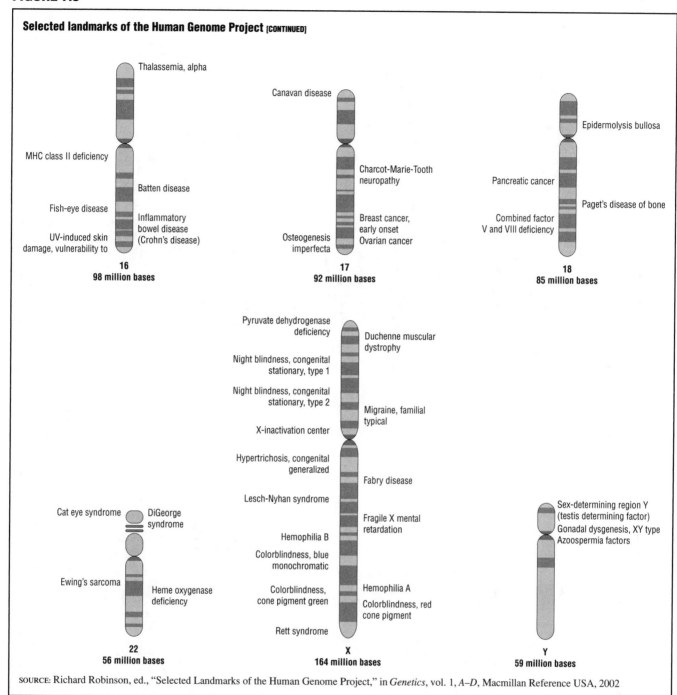

SOURCE: Richard Robinson, ed., "Selected Landmarks of the Human Genome Project," in *Genetics*, vol. 1, *A–D*, Macmillan Reference USA, 2002

Molecular Medicine

The HGP and the technological advances it has produced have moved the field of molecular medicine forward with extraordinary speed. Ivan C. Gerling, Solomon S. Solomon, and Michael Bryer-Ash assert in "Genomes, Transcriptomes, and Proteomes: Molecular Medicine and Its Impact on Medical Practice" (*Archives of Internal Medicine*, vol. 163, no. 2, January 27, 2003) that the HGP will not only influence the way science is conducted but will also advance the clinical practice of medicine.

The researchers credit the HGP for the technological advances that enable preclinical detection (recognition of disease before its earliest biochemical or visible expression). Gerling, Solomon, and Bryer-Ash foresee increasing accuracy and ease of preclinical detection, as well as the ability to predict disease based on three fundamental levels of biologic determination:

• The genomic DNA constitution of the individual (the genome) is unchanged from the moment of conception, except for some isolated, local mutations

FIGURE 7.9

Projected solutions to energy challenges arising from genomics research

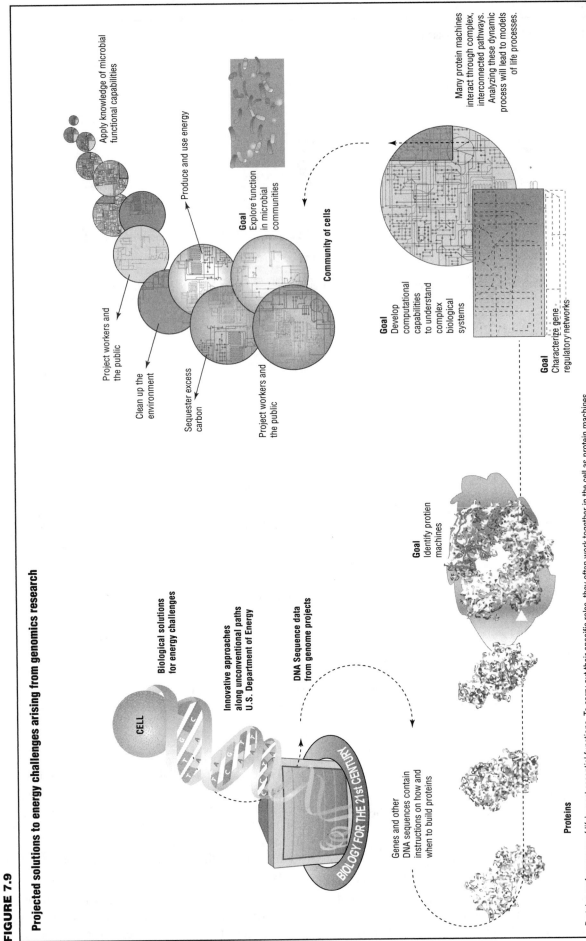

Proteins perform many of life's most essential functions. To carry out their specific roles, they often work together in the cell as protein machines.

SOURCE: "Genomics: GTL Program Pictorial," in *Human Genome Project Information Systems Biology Gallery*, U.S. Department of Energy Office of Science, Office of Biological and Environmental Research, Human Genome Project, http://genomics.energy.gov/gallery/systems_biology/detail-11.html (accessed October 1, 2010)

- The transcribed messenger RNA complement (the transcriptome)
- The full range of translated proteins (the proteome)

Gerling, Solomon, and Bryer-Ash posit that the environment influences gene expression and modifies gene products in ways that initiate, accelerate, or slow progress of disease-causing processes. This does not change the genome, but it does change the transcriptome and the proteome. Technological advances that occurred during the 1990s and the first few years of the first decade of the 21st century have provided the tools to perform the comprehensive molecular analyses needed to examine not only the genome but also the transcriptome and proteome. Using new technologies will dramatically increase understanding at the molecular level of the mechanisms of disease development.

Haplotype Mapping Project

In October 2002 an international effort to develop a haplotype map of the human genome was launched. A haplotype is a set of alleles (particular forms of genes) or markers on one of a pair of homologous chromosomes, and a haplotype map will show human genetic variation. The premise of the International HapMap Project was that within the human genome different genetic variants within a chromosomal region (haplotypes) occur together far more frequently than others. Based on common haplotype patterns (combinations of DNA sequence variants that are usually found together), the haplotype map simplifies the search for medically important DNA sequence variations and offers new understanding of human population structure and history.

Given that any two people are 99.9% identical genetically, understanding the 0.1% difference is important because it helps explain why one person may be more susceptible to a certain disease than another. Researchers can use the HapMap to compare the genetic variation patterns of a group of people known to have a specific disease with a group of people without the disease. Finding a certain pattern more often in people with the disease identifies a genomic region that may contain genes that contribute to the condition. Researchers hope that identifying single nucleotide polymorphisms (SNPs), which are specific positions in the genome sequence that are occupied by one nucleotide in some copies and by a different nucleotide in others, will enable them to identify the alleles that are associated with increased or decreased susceptibility to common diseases, such as asthma, heart disease, or psychiatric illness. Figure 7.10 shows that most SNP variation (about 85%) occurs within all populations.

Researchers hypothesize that differences between haplotypes may be associated with varying susceptibility to disease. As such, they indicate that mapping the hap-

FIGURE 7.10

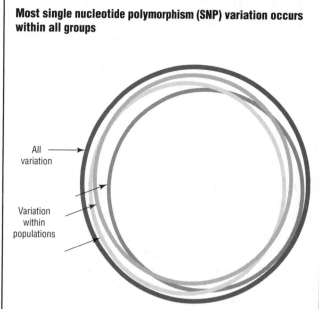

Most single nucleotide polymorphism (SNP) variation occurs within all groups

SOURCE: "Most SNP Variation Is within All Groups," in *Developing a Haplotype Map of the Human Genome for Finding Genes Related to Health and Disease*, U.S. Department of Health and Human Services, National Institutes of Health, National Human Genome Research Institute, Division of Extramural Research, 2001, http://www.genome.gov/10001665 (accessed October 1, 2010)

lotype structure of the human genome may be the key to identifying the genetic basis of many common disorders. The HapMap project serves as a resource for studying the genetic factors that contribute to variation in response to environmental factors, in susceptibility to infection, and in the identification of genetic variants that are associated with the effectiveness of, and adverse responses to, drugs and vaccines.

To create a haplotype map, researchers must have enough SNPs to be sure that regions containing disease alleles have been found and that regions not containing disease alleles can be excluded from further consideration. The HapMap enables researchers to study the genetic risk factors underlying a wide range of disorders. For any given disease, researchers may perform an association study by using the HapMap tag SNPs to compare the haplotype patterns of a group of people known to have the disease to a group of people without the disease. If the association study finds a specific haplotype more frequently in those with the disease, researchers scrutinize the precise genomic region in their search for the specific genetic variant.

By mid-2005 a draft of the HapMap, consisting of 1 million markers of genetic variation, was released. The first draft of the HapMap enabled researchers to analyze the human genome in ways that were not possible with the human DNA sequence alone. Second-generation data from the project were released in July 2006 and provided a denser map that enables scientists to narrow gene discovery more precisely to specific regions of the genome. The first- and second-generation HapMaps analyzed data from 270 people from four different populations: Yoruba in Ibadan, Nigeria; Japanese in Tokyo, Japan; Han Chinese in Beijing, China; and people with north and west European ancestry living in Utah.

In September 2010 the International HapMap 3 Consortium published a third generation of the HapMap in "Integrating Common and Rare Genetic Variation in Diverse Human Populations" (*Nature*, vol. 467, no. 7311). The third-generation HapMap is the largest survey of human genetic variation performed as of January 2011 and contains data from 1,184 people from a total of 11 different populations, the four original populations and seven additional populations: people with African ancestry living in the southwestern United States; people with Chinese ancestry in Denver, Colorado; Gujarati Indians in Houston, Texas; Luhya people in Webuye, Kenya; Maasai people in Kinyawa, Kenya; people with Mexican ancestry in Los Angeles, California; and Italians in Tuscany, Italy. Because the third-generation HapMap looked at 1.6 million SNPs in approximately 500 samples from the four original populations and over 650 samples from the seven new populations, it was able to detect rarer variants than those identified in the two previous HapMaps. Writing in *Nature*, the International HapMap 3 Consortium concludes, "This expanded public resource of genome variants in global populations supports deeper interrogation of genomic variation and its role in human disease, and serves as a step towards a high-resolution map of the landscape of human genetic variation."

The Future of Genomic Research

Established by the NHGRI in September 2003, the Encyclopedia of DNA Elements (ENCODE) Project (January 4, 2011, http://www.genome.gov/10005107) aims to develop efficient ways to identify and locate all the protein-coding genes, nonprotein-coding genes, and other sequence-based functional elements contained in the human DNA sequence. This ambitious undertaking has created an enormous resource for researchers seeking to use and apply the human sequence to predict disease risk and to develop new approaches to prevent and treat disease.

The ENCODE Project entails three phases: a three-year pilot project phase that began in September 2003, a second technology development phase that parallels phase 1, and a planned production phase. In "The ENCODE (ENCyclopedia Of DNA Elements) Project" (*Science*, vol. 306, no. 5696, October 22, 2004), the ENCODE researchers describe their plans to build a "parts list" of all sequence-based functional elements in the human DNA sequence. The researchers hope to identify as-yet-unrecognized functional elements. During the

pilot phase they developed and tested ways to efficiently identify functional elements. They focused on 44 DNA targets, which together cover 30 million base pairs (about 1% of the human genome). The target regions were strategically selected to provide a representative cross section of the entire human genome sequence. The conclusions from the pilot phase, which ended in 2006, were published in "Identification and Analysis of Functional Elements in 1% of the Human Genome by the ENCODE Pilot Project" (*Nature*, vol. 447, no. 7146, June 14, 2007). The ENCODE researchers describe the human genome as an "elegant but cryptic store of information" and report that their analyses provide a "rich source of functional information . . . of the human genome" and that the pilot phase constituted "an impressive platform for future genome exploration efforts." During the second phase the researchers worked to develop new technologies to apply to the ENCODE Project. In the press release "Researchers Expand Efforts to Explore Functional Landscape of the Human Genome" (October 9, 2007, http://www.genome.gov/26023194), the NHGRI states that it awarded grants totaling more than $80 million over the next four years to fund the production phase of the project.

Susan E. Celniker et al. observe in "Unlocking the Secrets of the Genome" (*Nature*, vol. 459, no. 7249, June 18, 2009) that free access to not only the human genome but also to the genomes of five model organisms—*Escherichia coli*, yeast (*Saccharomyces cerevisiae*), worm (*Caenorhabditis elegans*), fly (*Drosophila melanogaster*), and mouse (*Mus musculus*)—has provided new insights into how the information encoded in the genome produces complex multicultural organisms. The pilot and first phases of the ENCODE analysis produced new insights, pointed out the complexity of the biology, and raised new questions. In 2007 the NHGRI expanded the human ENCODE project to include the entire genome and initiated the model organism ENCODE (modENCODE) project to catalog the genomic elements of the worm and fly, which share similarities with other organisms, even humans. The advantage of cataloging these genomes is that they are small enough to be thoroughly examined using currently available technologies. Celniker et al. opine that the modENCODE results "will add value to the human ENCODE effort by illuminating the relationship between molecular and biological events. In the future, these data will provide a powerful platform for characterizing the functional networks that direct multicellular biology, thereby linking genomic data with the biological programs of higher organisms, including humans." Figure 7.11 shows the kind of genomic and other data the modENCODE project makes available online. The figure depicts how users of the electronic database can access the chromosome arm genomic sequence of a specific species, such as *D. melanogaster* (small fruit flies) and *C. elegans* (roundworms).

FIGURE 7.11

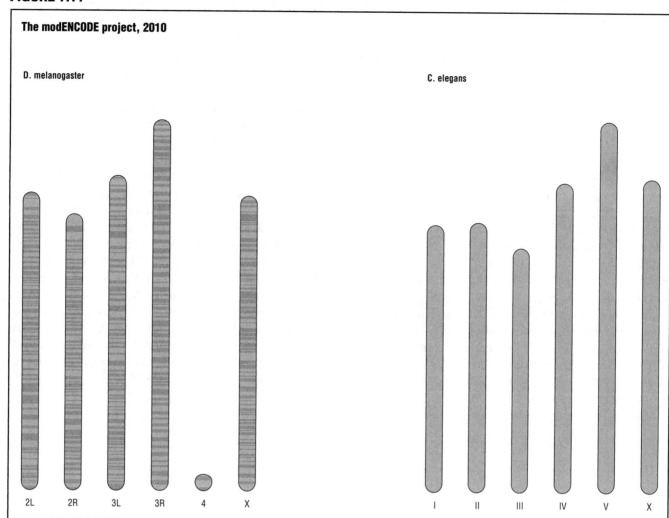

The modENCODE project, 2010

D. melanogaster

C. elegans

2L   2R   3L   3R   4   X   I   II   III   IV   V   X

SOURCE: Francis S. Collins, "The modENCODE Project," in *Perspectives on the Human Genome Project*, U.S. Department of Health and Human Services, National Institutes of Health, National Human Genome Research Institute, NHGRI Science Writers Conference, June 7, 2010, http://www.genome.gov/Pages/Newsroom/Webcasts/2010ScienceReportersWorkshop/Collins_NHGRIsciencewriters060710.pdf (accessed October 1, 2010)

Another NHGRI initiative is the creation of publicly available libraries, such as the NIH Chemical Genomics Center, which collects and catalogs data about chemical compounds for scientists who are engaged in developing chemical probes and charting biological pathways. These chemical compounds have many promising applications in genomic research. For example, their ability to enter cells readily makes them natural vehicles for pharmaceutical drug development and drug delivery system design. An endeavor of this size and scope requires significant financial and human resources, and the NHGRI is planning to use technologies such as robotic-enabled, high-output screening to create large libraries containing up to a million chemical compounds.

The United Kingdom's Wellcome Trust, along with Canadian funding organizations and the global pharmaceutical company GlaxoSmithKline, established the charitable organization Structural Genomics Consortium in April 2003 to round out international efforts in structural

genomics. Structural genomics is the systematic, high-volume generation of the three-dimensional structure of proteins. The goal of examining the structural genomics of any organism is the complete structural description of all proteins encoded by the genome of that organism. These descriptions are important for drug design, diagnosis, and treatment of disease. Like the HGP and the Chemical Genomics Center, the Structural Genomics Consortium is placing all the protein structures in public databases where scientists throughout the world may access them.

In "Dynamic Timeline" (February 4, 2010, http://www.genome.gov/25019887), the NHGRI illustrates some of the research challenges and accomplishments of the postgenomic era. Besides the HapMap project and the DOE's "Genomes to Life," it describes the 2006 launch of The Cancer Genome Atlas (TCGA, http://cancergenome.nih.gov/wwd/program/), a network of over 150 researchers at dozens of institutions across the United States that is mapping the genomic changes in more than 20 tumor types in an effort to create a resource for use in developing new strategies to prevent, diagnose, and treat cancer.

In October 2008 the TCGA reported the first results of its large-scale, comprehensive study of glioblastoma (GBM), the most common form of brain cancer. TCGA researchers described in "Comprehensive Genomic Characterization Defines Human Glioblastoma Genes and Core Pathways" (*Nature*, vol. 455, no. 7216, October 23, 2008) the discovery of new genetic mutations and other types of DNA alterations with potential implications for the diagnosis and treatment of GBM. In January 2010 TCGA researchers detailed in "Integrated Genomic Analysis Identifies Clinically Relevant Subtypes of Glioblastoma Characterized by Abnormalities in PDGFRA, IDH1, EGFR, and NF1" (*Cancer Cell*, vol. 17, no. 1, January 19, 2010) four distinct subtypes of GBM and reported that each subtype responded differently to chemotherapeutic and radiation therapies. Even though these research findings cannot be immediately translated into clinical practice, they may lead to more individualized treatments based on patients' genomic alterations. The TCGA indicates in "Scientific Publications" (2010, Scientific Publications, http://cancergenome.nih.gov/publications/scientific.asp)

that in 2010 alone its researchers published 21 articles describing their research findings in peer-reviewed science and medical journals.

According to the NHGRI, in "2006: Initiatives to Establish the Genetic and Environmental Causes of Common Diseases Launched" (April 10, 2006, http://www.genome.gov/25520287), two other disease-related initiatives were established in 2006. The NIH Genes and Environment Initiative (GEI) analyzes genetic variation among people with specific diseases and develops technology to effectively monitor the environmental exposures that interact with genetic variations to cause or trigger disease. The second initiative was a public-private partnership called Genetic Association Information Network (GAIN), which was based on the assumption that by working together, public and private scientists would progress faster than either could working alone. GAIN aimed to determine the genetic contributions to common diseases. GAIN researchers conducted genome-wide association studies that focused on psychiatric disorders, psoriasis (a common skin disorder), Crohn's disease (an inflammatory bowel disease), breast and prostate cancers, and diabetes. Funded by Pfizer Global Research and Development, this initiative concluded at the close of 2008.

CHAPTER 8

REPRODUCTIVE CLONING

Human reproductive cloning should not now be practiced. It is dangerous and likely to fail. The panel therefore unanimously supports the proposal that there should be a legally enforceable ban on the practice of human reproductive cloning. For this purpose, we define human reproductive cloning as the placement in a uterus of a human blastocyst derived by the technique that we call nuclear transplantation.

—Committee on Science, Engineering, and Public Policy Board on Life Sciences, *Scientific and Medical Aspects of Human Reproductive Cloning* (2002)

AAAS endorses a legally enforceable ban on efforts to implant a human cloned embryo for the purpose of reproduction. The scientific evidence documenting the serious health risks associated with reproductive cloning, as shown through animal studies, make it unconscionable to undertake this procedure. At the same time, we encourage continuing open and inclusive public dialogue, in which the scientific community is an active participant, on the scientific and ethical aspects of human cloning as our understanding of this technology advances.

—American Association for the Advancement of Science Statement on Human Cloning, February 14, 2002

The Human Genome Project defines three distinct types of cloning. The first is the use of highly specialized deoxyribonucleic acid (DNA) technology to produce multiple, exact copies of a single gene or other segment of DNA to obtain sufficient material to examine for research purposes. This process produces cloned collections of DNA known as clone libraries. The second kind of cloning involves the natural process of cell division to create identical copies of the entire cell. These copies are called a cell line. The third type of cloning, reproductive cloning, is the one that has received the most attention in the mass media. This is the process that generates complete, genetically identical organisms such as Dolly, the famous Scottish sheep cloned in 1996 and named after the entertainer Dolly Parton (1946–).

Cloning may also be described by the technology that is used to perform it. For example, the term *recombinant DNA technology* describes the technology and mechanism of DNA cloning. Also known as molecular cloning, or gene cloning, it involves the transfer of a specific DNA fragment of interest to researchers from one organism to a self-replicating genetic element of another species such as a bacterial plasmid. (See Figure 8.1.) The DNA under study may then be reproduced in a host cell. This technology has been in use since the 1970s and is a standard practice in molecular biology laboratories.

Figure 8.2 displays and describes the differences between different types of cloning. This graphic depiction was developed by the Johns Hopkins University's Genetics and Public Policy Center of the Phoebe R. Berman Bioethics Institute with support from the Pew Charitable Trusts. Figure 8.2 also presents some of the current and potential applications of therapeutic cloning, which produces stem cells that may be used to improve human health. This chapter focuses on cloning genes and reproductive cloning and the following chapter describes therapeutic cloning and stem cell research.

Just as GenBank is an online public repository of the human genome sequence, the Clone Registry database is a sort of "public library." Used by genome sequencing centers to record which clones have been selected for sequencing, which sequencing efforts are currently under way, and which are finished and represented by sequence entries in GenBank, the Clone Registry may be freely accessed by scientists worldwide. To effectively coordinate all of this information, a standardized system of naming clones is essential. The nomenclature used is shown in Figure 8.3.

CLONING GENES

Molecular cloning is performed to enable researchers to have many copies of genetic material available in the laboratory for the purpose of experimentation. Cloned

FIGURE 8.1

Bacterial plasmid

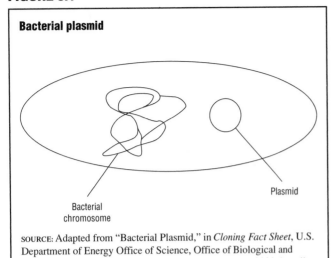

Plasmid

Bacterial
chromosome

SOURCE: Adapted from "Bacterial Plasmid," in *Cloning Fact Sheet*, U.S. Department of Energy Office of Science, Office of Biological and Environmental Research, Human Genome Project, May 2009, http://www.ornl.gov/sci/techresources/Human_Genome/elsi/cloning.shtml (accessed October 11, 2010)

genes allow researchers to examine encoded proteins and are used to sequence DNA. Gene cloning also allows researchers to isolate and experiment on the genes of an organism. This is particularly important in terms of human research; in instances where direct experimentation on humans might be dangerous or unethical, experimentation on cloned genes is often practical and feasible.

Cloned genes are also used to produce pharmaceutical drugs, insulin (a pancreatic hormone that regulates blood glucose levels), clotting factors, human growth hormone, and industrial enzymes. Before the widespread use of molecular cloning, these proteins were difficult and expensive to manufacture. For example, before the development of recombinant DNA technology, insulin used by people with diabetes was extracted and purified from cow and pig pancreases. Because the amino acid sequences of insulin from cows and pigs are slightly different from those in human insulin, some patients experienced adverse immune reactions to the nonhuman "foreign insulin." The recombinant human version of insulin is identical to human insulin so it does not produce an immune reaction.

Figure 8.4 shows how a gene is cloned. First, a DNA fragment containing the gene being studied is isolated from chromosomal DNA using restriction enzymes. It is joined with a plasmid (a small ring of DNA found in many bacteria that can carry foreign DNA) that has been cut with the same restriction enzymes. When the fragment of chromosomal DNA is joined with its cloning vector (cloning vectors, such as plasmids and yeast artificial chromosomes, introduce foreign DNA into host cells), it is called a recombinant DNA molecule. Once it has entered into the host cell, the recombinant DNA can be reproduced along with the host cell DNA.

Another molecular cloning technique that is simpler and less expensive than the recombinant cloning method is the polymerase chain reaction (PCR). PCR has also been dubbed "molecular photocopying" because it amplifies DNA without the use of a plasmid. Figure 6.5 in Chapter 6 shows how PCR is used to generate a virtually unlimited number of copies of a piece of DNA.

A collection of clones of chromosomal and vector DNA (a small piece of DNA containing regulatory and coding sequences of interest) is called a library. These libraries of clones containing partly overlapping regions are constructed to show that two particular clones are next to one another in the genome. Figure 8.5 shows how, by dividing the inserts into smaller fragments and determining which clones share the same DNA sequences, clone libraries are constructed.

ORGANISMAL OR REPRODUCTIVE CLONING

Another way to describe and classify cloning is by its purpose. Organismal or reproductive cloning is a technology used to produce a genetically identical organism—an animal with the same nuclear DNA as an existing animal.

The reproductive cloning technology used to create animals is called somatic cell nuclear transfer (SCNT). In SCNT scientists transfer genetic material from the nucleus of a donor adult cell to an enucleated egg (an egg from which the nucleus has been removed). This eliminates the need for fertilization of an egg by a sperm. The reconstructed egg containing the DNA from a donor cell is treated with chemicals or electric current to stimulate cell division. Once the cloned embryo reaches a suitable stage, it is transferred to the uterus of a surrogate (female host), where it continues to grow and develop until birth. Figure 8.6 shows the entire SCNT process that culminates in the transfer of the embryo into the surrogate mother and ultimately the birth of a cloned animal.

Organisms or animals generated using SCNT are not perfect or identical clones of the donor organism or "parent" animal. The clone's nuclear DNA is identical to the donor's, but some of the clone's genetic materials come from the mitochondria in the cytoplasm of the enucleated egg. Mitochondria, the organelles that serve as energy sources for the cell, contain their own short segments of DNA called mtDNA. Acquired mutations in the mtDNA contribute to differences between clones and their donors and are believed to influence the aging process.

Dolly the Sheep Paves the Way for Other Cloned Animals

In 1952 scientists transferred a cell from a frog embryo into an unfertilized egg, which then developed into a tadpole. This process became the prototype for cloning. Ever since, scientists have been cloning animals. During the 1980s the first mammals were also cloned

FIGURE 8.2

Types of cloning and use of stem cells

CLARIFYING CLONING

Reproductive cloning and research cloning involve the same technical processes until a cloned embryo is made. How that embryo is used determines the difference between reproductive and research cloning. Reproductive cloning results if that cloned embryo is used to start a pregnancy. Research cloning results if that cloned embryo is used as a source of embryonic stem cells.

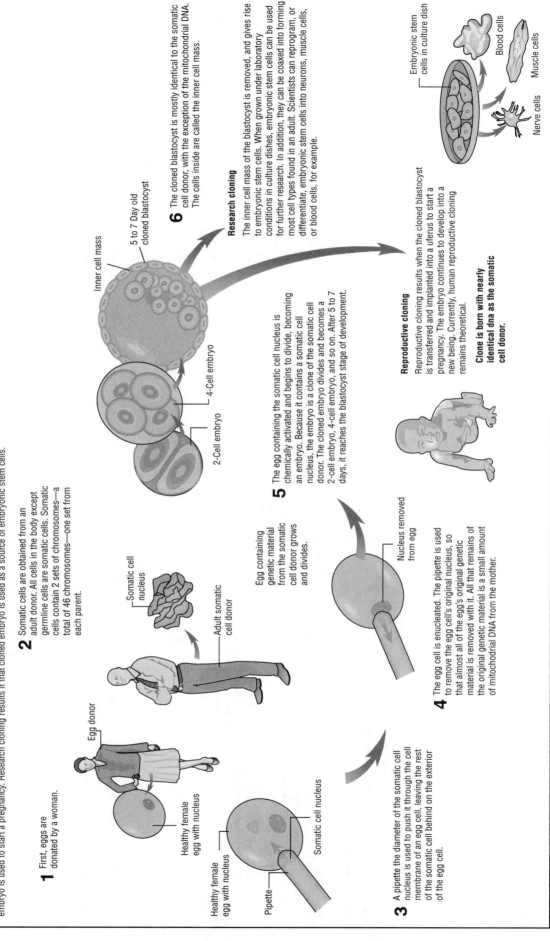

1 First, eggs are donated by a woman.

2 Somatic cells are obtained from an adult donor. All cells in the body except germline cells are somatic cells. Somatic cells contain 2 sets of chromosomes—a total of 46 chromosomes—one set from each parent.

3 A pipette the diameter of the somatic cell nucleus is used to push it through the cell membrane of an egg cell, leaving the rest of the somatic cell behind on the exterior of the egg cell.

4 The egg cell is enucleated. The pipette is used to remove the egg cell's original nucleus, so that almost all of the egg's original genetic material is removed with it. All that remains of the original genetic material is a small amount of mitochondrial DNA from the mother.

5 The egg containing the somatic cell nucleus is chemically activated and begins to divide, becoming an embryo. Because it contains a somatic cell nucleus, the embryo is a clone of the somatic cell donor. The cloned embryo divides and becomes a 2-cell embryo, 4-cell embryo, and so on. After 5 to 7 days, it reaches the blastocyst stage of development.

6 The cloned blastocyst is mostly identical to the somatic cell donor, with the exception of the mitochondrial DNA. The cells inside are called the inner cell mass.

Research cloning

The inner cell mass of the blastocyst is removed, and gives rise to embryonic stem cells. When grown under laboratory conditions in culture dishes, embryonic stem cells can be used for further research. In addition, they can be coaxed into forming most cell types found in an adult. Scientists can reprogram, or differentiate, embryonic stem cells into neurons, muscle cells, or blood cells, for example.

Reproductive cloning

Reproductive cloning results when the cloned blastocyst is transferred and implanted into a uterus to develop into a new being. The embryo continues to develop into a pregnancy. Currently, human reproductive cloning remains theoretical.

Clone is born with nearly identical dna as the somatic cell donor.

FIGURE 8.2

Types of cloning and use of stem cells [CONTINUED]

CURRENT AND POTENTIAL USES OF STEM CELLS

What are stem cells? Stem cells are a type of cell found in embryos as well as various tissues in the adult body. In adults, stem cells normally replace damaged or depleted cells. Scientists believe stem cells have the potential to treat degenerative diseases. These are just a few applications for stem cells:

Replace damaged cells
Adult stem cells such as those found in bone marrow currently are the only type of stem cell used to treat blood diseases such as leukemia and lymphoma. Scientists believe embryonic stem cells have the potential to replenish cells lost to age, damage, or disease because of their ability to develop into any cell type. Stem cells offer the possibility of a source of replacement cells to treat diseases and conditions. (See diagram at right.)

Learn more about human disease
Studying how stem cells transform into specialized cells will help scientists better understand these processes. This will help shed light on how certain medical conditions, such as cancer and birth defects, develop.

New drug development
Stem cells can be used to test the effectiveness or toxicity of new drugs.

Brain & nerve tissue
Stem cells may be used to repair brain and nerve tissue damaged by conditions such as Parkinson and Alzheimer disease or spinal cord injury.

Heart disease
Stem cells may be used to repair damaged heart muscle caused by heart disease.

Bone marrow replacement
Stem cells are currently used to restore bone marrow or blood cells for cancer patients.

Skin grafting
Stem cells in grafted skin are currently used to replace damaged skin for accident or burn victims.

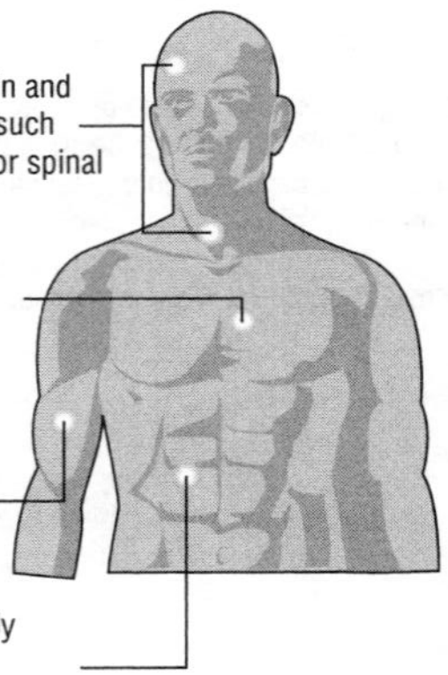

SOURCE: Christina Ullman, "Cloning," in *Genes & Society Graphics*, Genetics & Public Policy Center at Johns Hopkins University with support from the Pew Charitable Trusts, 2008, http://www.dnapolicy.org/resources/cloning_infographic_final.pdf (accessed October 11, 2010). Courtesy of the Genetics & Public Policy Center with support from The Pew Charitable Trusts.

FIGURE 8.3

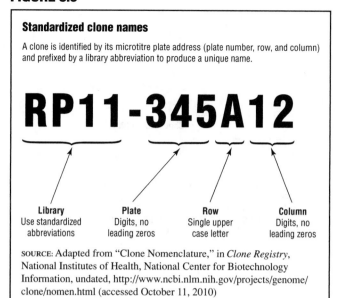

Standardized clone names

A clone is identified by its microtitre plate address (plate number, row, and column) and prefixed by a library abbreviation to produce a unique name.

RP11-345A12

Library
Use standardized abbreviations

Plate
Digits, no leading zeros

Row
Single upper case letter

Column
Digits, no leading zeros

SOURCE: Adapted from "Clone Nomenclature," in *Clone Registry*, National Institutes of Health, National Center for Biotechnology Information, undated, http://www.ncbi.nlm.nih.gov/projects/genome/clone/nomen.html (accessed October 11, 2010)

from embryonic cells. In 1996 cloning became headline news when, after more than 250 failed attempts, Ian Wilmut (1944–) and his colleagues at the Roslin Institute in Edinburgh, Scotland, announced they had successfully cloned a sheep, which they named Dolly. Dolly was the first mammal cloned from the cell of an adult animal, and since then researchers have used cells from adult animals and various modifications of nuclear transfer technology to clone a range of animals, including a gaur, sheep, goats, cows, horses, mules, oxen, deer, mice, rats, pigs, cats, dogs, and rabbits.

To create Dolly, the Roslin Institute researchers transplanted a nucleus from a mammary gland cell of a Finn Dorsett sheep into the enucleated egg of a Scottish blackface ewe and used electricity to stimulate cell division. The newly formed cell divided and was placed in the uterus of a blackface ewe to gestate. Born several months later, Dolly was a true clone—genetically identical to the Finn Dorsett mammary cells and not to the blackface ewe, which served as her surrogate mother. Her birth revolutionized the world's understanding of molecular biology, ignited worldwide discussion about the morality of generating new life through cloning, prompted legislation in dozens of countries, and launched an ongoing political debate in Congress.

Dolly was the object of intense media and public fascination. She proved to be a basically healthy clone and produced six lambs of her own through normal sexual means. Before her death by lethal injection in February 2003, Dolly had been suffering from lung cancer and arthritis. An autopsy (postmortem examination) of Dolly revealed that, other than her cancer and arthritis, which are common diseases in sheep, she was anatomically like other sheep.

FIGURE 8.4

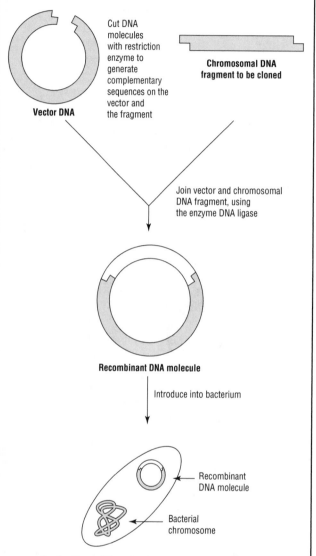

Cloning DNA in plasmids

By fragmenting DNA of any origin (human, animal, or plant) and inserting it in the DNA of rapidly reproducing foreign cells, billions of copies of a single gene or DNA segment can be produced in a very short time. DNA to be cloned is inserted into a plasmid (a small, self-replicating circular molecule of DNA) that is separate from chromosomal DNA. When the recombinant plasmid is introduced into bacteria, the newly inserted segment will be replicated along with the rest of the plasmid.

Cut DNA molecules with restriction enzyme to generate complementary sequences on the vector and the fragment

Vector DNA

Chromosomal DNA fragment to be cloned

Join vector and chromosomal DNA fragment, using the enzyme DNA ligase

Recombinant DNA molecule

Introduce into bacterium

Recombinant DNA molecule

Bacterial chromosome

SOURCE: Denise Casey, "Fig. 11a. Cloning DNA in Plasmids," in *Primer on Molecular Genetics*, U.S. Department of Energy Office of Science, Office of Biological and Environmental Research, Human Genome Project, 1992, http://www.ornl.gov/sci/techresources/Human_Genome/publicat/primer/fig11a.html (accessed October 11, 2010)

In February 1997 Don Paul Wolf (1939–) and his colleagues at the Oregon Regional Primate Center in Beaverton successfully cloned two rhesus monkeys using laboratory techniques that had previously produced frogs, cows, and mice. It was the first time that researchers used a nuclear transplant to generate monkeys. The monkeys were created using different donor blastocysts (early-stage embryos), so they were not clones of one another—each

FIGURE 8.5

Construction of an overlapping clone library

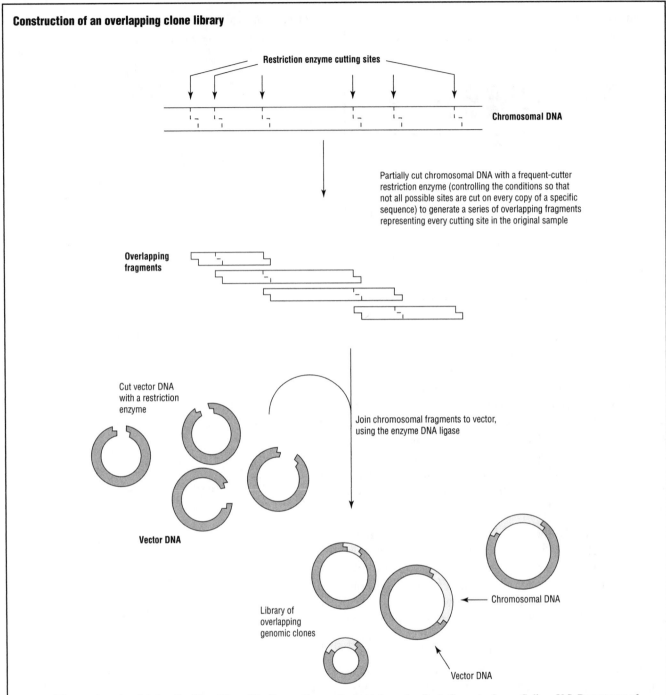

Restriction enzyme cutting sites

Chromosomal DNA

Partially cut chromosomal DNA with a frequent-cutter restriction enzyme (controlling the conditions so that not all possible sites are cut on every copy of a specific sequence) to generate a series of overlapping fragments representing every cutting site in the original sample

Overlapping fragments

Cut vector DNA with a restriction enzyme

Join chromosomal fragments to vector, using the enzyme DNA ligase

Vector DNA

Chromosomal DNA

Library of overlapping genomic clones

Vector DNA

SOURCE: "Construction of an Overlapping Clone Library," in *Human Genome Project Information Basic Genomics Image Gallery*, U.S. Department of Energy Office of Science, Office of Biological and Environmental Research, Human Genome Project, http://genomics.energy.gov/gallery/basic_genomics/det1ail.np/detail-05.html (accessed October 1, 2010)

monkey was a clone of the original blastocyst that had developed from a fertilized egg. Neither of the cloned monkeys survived past the embryonic stage.

An important distinction between the process that created Dolly and the one that produced the monkeys was that unspecialized embryonic cells were used to create the monkeys, whereas a specialized adult cell was used to create Dolly. The Oregon experiment was followed closely in the scientific and lay communities because, in terms of evolutionary biology and genetics, primates are closely related to humans.

In 2000 the Oregon researchers succeeded when one of four embryos that were created by splitting a blastocyst four ways and implanting the pieces into surrogate mothers survived. The survivor was named Tetra, from the Greek prefix for the number four. Researchers and the

FIGURE 8.6

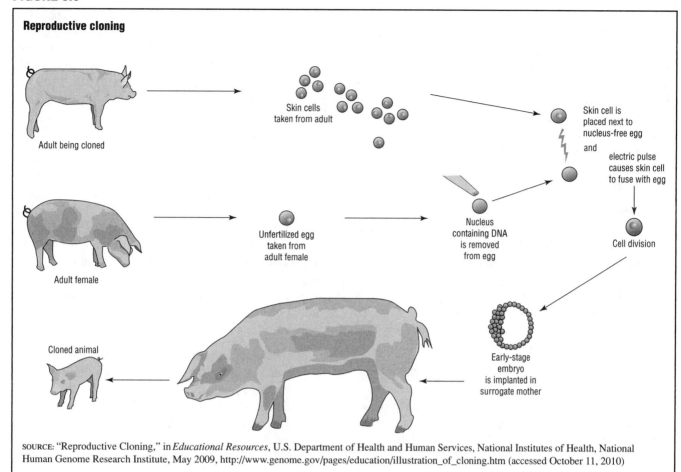

Reproductive cloning

Skin cells taken from adult

Adult being cloned

Skin cell is placed next to nucleus-free egg and electric pulse causes skin cell to fuse with egg

Adult female

Unfertilized egg taken from adult female

Nucleus containing DNA is removed from egg

Cell division

Cloned animal

Early-stage embryo is implanted in surrogate mother

SOURCE: "Reproductive Cloning," in *Educational Resources*, U.S. Department of Health and Human Services, National Institutes of Health, National Human Genome Research Institute, May 2009, http://www.genome.gov/pages/education/illustration_of_cloning.htm (accessed October 11, 2010)

public speculated that if monkeys could be cloned, it might become feasible to clone humans.

In May 2001 BresaGen Limited, an Australian biotechnology firm, announced the birth of that country's first cloned pig. The pig was cloned from cells that had been frozen in liquid nitrogen for more than two years, and the company employed technology that was different from the process used to clone Dolly the sheep. The most immediate benefit of this new technology was to improve livestock—cloning enables breeders to take some animals with superior genetics and rapidly produce more. Biomedical scientists were especially attentive to this research because of its potential for xenotransplantation (the use of animal organs for transplantation into humans). Pig organs that have been genetically modified so that they will not be rejected by the human immune system could prove to be a boon to medical transplantation.

That same year the first cat was cloned, and the following year rabbits were successfully cloned. In January 2003 researchers at Texas A&M University reported that cloned pigs behaved normally—as expected for a litter of pigs—but were not identical to the animals from which

they were cloned in terms of food preferences, temperament, and how they spent their time. The researchers explained the variation as arising from the environment and epigenetic (not involving DNA sequence change) factors, causing the DNA to line up differently in the clones. Epigenetic activity is defined as any gene-regulating action that does not involve changes to the DNA code and that persists through one or more generations, and it may explain why abnormalities such as fetal death occur more frequently in cloned species.

In May 2003 a cloned mule (the first successful clone of any member of the horse family) was born in Idaho. The clone was not just any mule, but the brother of the world's second-fastest racing mule. Named Idaho Gem, the cloned mule was created by researchers at the University of Idaho and Utah State University. The researchers attributed their success to changes in the culture medium they used to nurture the eggs and embryos.

In August 2003 scientists at the Laboratory of Reproductive Technology in Cremona, Italy, were the first to clone a horse. Cesare Galli et al. describe their cloning technique in "Pregnancy: A Cloned Horse Born to Its Dam Twin" (*Nature*, vol. 424, no. 6949, August 7, 2003).

The mule was cloned from cells that were extracted from a mule fetus, whereas the cloned horse's DNA came from her adult mother's skin cells. There were other differences as well. The University of Idaho and Utah State University researchers harvested fertile eggs from mares, removed the nucleus of each egg, and inserted DNA from cells of a mule fetus. The reconstructed eggs were then surgically implanted into the wombs of female horses. In contrast, Galli et al. harvested hundreds of eggs from mare carcasses, cultured the eggs, removed their DNA, and replaced it with DNA taken from either adult male or female horse skin cells.

In May 2004 the first bull was cloned from a previously cloned bull in a process known as serial somatic cell cloning or recloning. Before the bull, the only other successful recloning efforts involved mice. Chikara Kubota, X. Cindy Tian, and Xiangzhong Yang describe their cloning technique in "Serial Bull Cloning by Somatic Cell Nuclear Transfer" (*Nature Biotechnology*, vol. 22, no. 6, June 2004). Their effort was also cited in the *Guinness Book of World Records* as the largest clone in the world.

At the close of 2004 a South Korean research team reported cloning macaque monkey embryos, which would be used as a source of stem cells. Conservationists then focused research efforts on cloning rare and endangered species. In April 2005 Texas A&M University announced the first successfully cloned foal in the United States. That same month scientists at Seoul National University cloned a dog they named Snuppy. In May 2005 Embrapa, a Brazilian agricultural research corporation, reported the creation of two cloned calves from a Junquiera cow, which is an endangered species. In 2006 ferrets were cloned using somatic cell nuclear transfer.

In 2009 the first camel was cloned in Dubai, United Arab Emirates. In 2010 researchers in Spain reported cloning the first fighting bull. By the close of 2010, over 20 animal species had been successfully cloned using nuclear transfer and surrogate mothers, the same technique that was used to produce Dolly the sheep.

Cloning Endangered Species

Reproductive cloning technology may also be used to repopulate endangered species such as the African bongo antelope, the Sumatran tiger, and the giant panda, or animals that reproduce poorly in zoos or are difficult to breed. In January 2001 scientists at Advanced Cell Technology (ACT), a biotechnology company in Massachusetts, announced the birth of the first clone of an endangered animal, a baby bull gaur (a large wild ox from India and Southeast Asia). The gaur was cloned using the nuclei of frozen skin cells taken from an adult male gaur that had died eight years earlier. The skin cell nuclei were joined with enucleated cow eggs, one of which was implanted into a surrogate cow. The cloned gaur died from an infec-

tion within days of its birth. That same year scientists in Italy successfully cloned an endangered wild sheep. Cloning an endangered animal is different from cloning a more common animal because cloned animals need surrogate mothers to be carried to term. Furthermore, the transfer of embryos is risky, and researchers are reluctant to put an endangered animal through the rigors of surrogate motherhood, so they opt to use nonendangered domesticated animals whenever possible.

Cloning extinct animals is even more challenging than cloning living animals because the egg and the surrogate mother used to create and harbor the cloned embryo are not the same species as the clone. Furthermore, for most already extinct animal species such as the woolly mammoth (*Mammuthus primigenius*) or the smilodon (*Smilodon populator*), there is insufficient intact cellular and genetic material from which to generate clones. In the future, carefully preserving intact cellular material of imperiled species may allow for their preservation and propagation.

In 2003 ACT announced the birth of a healthy clone of a Javan banteng (an endangered cattlelike animal native to Asian jungles). The clone was created from a single skin cell that was taken from another banteng before it died in 1980. The skin cell was kept frozen until it was used to create the clone. The banteng embryo gestated in a standard beef cow in Iowa.

Born in April 2003, the cloned banteng developed normally, growing its characteristic horns and reaching an adult weight of about 1,800 pounds (816 kg). The banteng lived at the San Diego Zoo until it died in April 2010 at the age of seven, less than half of the anticipated lifespan of a uncloned banteng. Hunting and habitat destruction have reduced the number of banteng, which once lived in large numbers in the bamboo forests of Asia, by more than 75% from 1983 to 2003. In "Ecological-Economic Models of Sustainable Harvest for an Endangered but Exotic Megaherbivore in Northern Australia" (*Natural Resource Modeling*, vol. 20, no. 1, March 2007), Corey J. A. Bradshaw and Barry W. Brook of Charles Darwin University report that in 2007 a population of between 8,000 and 10,000 banteng lived on an isolated peninsula in northern Australia.

In August 2005 the Audubon Nature Institute in New Orleans, Louisiana, reported that two unrelated endangered African wildcat clones had given birth to eight babies. These births confirmed that clones of wild animals can breed naturally, which is vitally important for protecting endangered animals on the brink of extinction.

Rob Waters describes in "Animal Cloning: The Next Phase" (*Bloomberg BusinessWeek*, June 10, 2010) the Frozen Zoo, a laboratory repository of frozen skin cells and DNA from about 800 species that are housed at the

San Diego Zoo's Institute for Conservation Research. Even though the lab was established in 1972, the technology necessary to use the cells to clone animals was still under development in 2010. In June 2010 tissue from the Frozen Zoo was used to create stem cells of an endangered African monkey and the stem cells successfully developed into brain cells, giving rise to the hope that in the not-too-distant future it will be possible to clone endangered animals and save them from extinction.

Despite this progress, ethicists caution that there are moral considerations around the issue of animal cloning. Several cloned animals, including two endangered cattle and the banteng, died prematurely. Furthermore, ethicists question whether the risks (deformed and short-lived animals) outweigh the benefits if just a few endangered animals that ultimately live in zoos are created using cloning. According to Waters, Autumn M. Fiester of the University of Pennsylvania observed, "There has been a lot of suffering with these early deaths and malformations."

Reproductive Human Cloning

In December 2002 a religious sect known as the Raelians made news when their private biotechnology firm, Clonaid, announced that it had successfully delivered the world's first cloned baby. The announcement, which could not be independently verified or substantiated, generated unprecedented media coverage and was condemned in the scientific and lay communities. At least some of the media frenzy resulted from the beliefs of the Raelians—namely, the sect contends that humans were created by extraterrestrial beings.

Clonaid's announcement, which ultimately was viewed as a hoax, brought attention to the fact that several laboratories around the world had embarked on clandestine efforts to clone a human embryo. For example, in 2002 Panayiotis Zavos (1944–), a U.S. fertility specialist, claimed to be collaborating with about two dozen international researchers to produce human clones. Another doctor focusing on fertility issues, Severino Antinori (1945–), attracted media attention when he maintained that hundreds of infertile couples in Italy and thousands in the United States had already enrolled in his human cloning initiative. As of January 2011, neither these researchers nor anyone else had offered proof of successful reproductive human cloning.

Moral and Ethical Objections to Human Cloning

The difficulty and low success rate of much animal reproductive cloning (an average of just one or two viable offspring result from every 100 attempts) and the as-yet-incomplete understanding of reproductive cloning have prompted many scientists to deem it unethical to attempt to clone humans. Many attempts to clone mammals have failed, and about one-third of clones born alive suffer from anatomical, physiological, or developmental abnormalities that are often debilitating. Some cloned animals have died prematurely from infections and other complications at rates higher than conventionally bred animals, and some researchers anticipate comparable outcomes from human cloning. Furthermore, scientists cannot yet describe or characterize how cloning influences intellectual and emotional development. Even though the attributes of intelligence, temperament, and personality may not be as important for cattle or other primates, they are vital for humans. Without considering the myriad religious, social, and other ethical concerns, the presence of so many unanswered questions about the science of reproductive cloning has prompted many scientists to consider any attempts to clone humans as scientifically irresponsible, unacceptably risky, and morally unallowable.

People who oppose human cloning are as varied as the interests and institutions they support. Religious leaders, scientists, politicians, philosophers, and ethicists argue against the morality and acceptability of human cloning. Nearly all objections hinge, to various degrees, on the definition of human life, beliefs about its sanctity, and the potentially adverse consequences for families and society as a whole.

In an effort to stimulate consideration of and debate about this critical issue, the President's Council on Bioethics examined the principal moral and ethical objections to human cloning in *Human Cloning and Human Dignity: An Ethical Inquiry* (July 2002, http://bioethics .georgetown.edu/pcbe/reports/cloningreport/pcbe_cloning _report.pdf). The council's report distinguished between therapeutic and reproductive cloning and outlined key concerns by trying to respond to many as yet unresolved questions about the ethics, morality, and societal consequences of human cloning.

The council determined that the key moral and ethical objections to therapeutic cloning (cloning for biological research) center on the moral status of developing human life. Therapeutic cloning involves the deliberate production, use, and destruction of cloned human embryos. One objection to therapeutic cloning is that cloned embryos produced for research are no different from those that could be used in attempts to create cloned children. Another argument that has been made is that the ends do not justify the means—that research on any human embryo is morally unacceptable, even if this research promises cures for many dreaded diseases. Finally, there are concerns that acceptance of therapeutic cloning will lead society down a slippery slope to reproductive cloning, a prospect that is almost universally viewed as unethical and morally unacceptable.

The unacceptability of human reproductive cloning stems from the fact that it challenges the basic nature of human procreation by redefining having children as a form of manufacturing. Human embryos and children may then be viewed as products and commodities rather than as sacred and unique human beings. Furthermore, reproductive cloning might substantially change fundamental issues of human identity and individuality, and allowing parents unprecedented genetic control of their offspring may significantly alter family relationships across generations.

The council concluded that "the right to decide '*whether* to bear or beget a child' does not include a right to have a child *by whatever means*. Nor can this right be said to imply a corollary—the right to decide what *kind* of child one is going to have. ... Our society's commitment to freedom and parental authority by no means implies that all innovative procedures and practices should be allowed or accepted, no matter how bizarre or dangerous."

Nearly universal opposition to human cloning persisted in 2011. According to the Center for Genetics and Society, in "About Reproductive Cloning" (2011, http://www.geneticsandsociety.org/section.php?id=16), even though some researchers now believe that it may be possible to safely clone humans, many more researchers remain unconvinced. Other compelling arguments against human cloning focus on the psychological health of cloned children and the concern that if human reproductive cloning was permitted, it would also usher in the use of genetic manipulation techniques that could markedly alter the fabric of society and even change the definition of human life.

CHAPTER 9

THERAPEUTIC CLONING: STEM CELL RESEARCH

In addition to understanding normal human develop-ment more completely, human embryonic stem cells are providing key tools to help us study the origins of many devastating diseases that afflict babies and young chil-dren. Such research may even help to uncover targets for drug development.... Human embryonic stem cells remain the gold standard for pluripotency: to prohibit work on human embryonic stem cells will thus do severe collateral damage to the new and exciting research on iPS [induced pluripotent stem] cells.

—Francis S. Collins, Director of the National Institutes of Health, The Promise of Human Embryonic Stem Cell Research, Senate Subcommittee on Labor, September 16, 2010

Therapeutic cloning (also called embryo cloning) is the creation of embryos for use in biomedical research. The objective of therapeutic cloning is not to create clones but to obtain stem cells. Stem cells are "master cells" that are capable of differentiating into multiple other cell types. This potential is important to biomedical researchers because stem cells may be used to generate any type of specialized cell, such as nerve, muscle, blood, or brain cells. Many scientists believe that stem cells can not only provide a ready supply of replacement tissue but may also hold the key to developing more effective treat-ments for common disorders such as heart disease and cancer as well as for degenerative diseases such as Alz-heimer's and Parkinson's. Researchers believe it is pos-sible to induce stem cells to grow into complete organs.

Advocates of therapeutic cloning point to other treat-ment benefits such as using stem cells to generate bone marrow for transplants. They contend that scientists could use therapeutic cloning to manufacture perfectly matched bone marrow using the patient's own skin or other cells. This would eliminate the problem of rejection of foreign tissue that is associated with bone marrow transplant and other organ transplantation. Stem cells also have the potential to repair and restore damaged heart and nerve tissue. Furthermore, there is mounting evi-dence to suggest that stem cells from cloned embryos

have greater potential as medical treatments than stem cells harvested from unused embryos at fertility clinics, which are created by in vitro (outside a woman's body) fertilization and remain, even with the growing use of adult stem cells (which are generally limited to differ-entiating into different cell types of their tissue of origin), the major source of stem cells for research. Figure 9.1 shows how adult stem cells may be used to repair dam-aged heart muscle. These prospective benefits are among the most compelling arguments in favor of cloning to obtain embryonic stem cells.

EMBRYONIC STEM CELLS

Embryonic stem cells used in research are harvested from the blastocyst after it has divided for five days, during the earliest stage of embryonic development. Most embry-onic stem cells are derived from embryos that have been created in vitro and then donated for research. They are not from embryos conceived in a woman's body, nor are they embryos destined to gestate. Nonetheless, because many people regard human embryos as human beings or at least as potential human beings, they consider their destruction, or even using techniques to obtain stem cells that might imperil their future viability, as immoral or unethical.

Embryonic stem cells have several advantages over adult stem cells. Because embryonic stem cells are pluripotent, they can be differentiated to become any type of cell in the body, whereas adult stem cells only appear to be able to differentiate into cell types from their tissue of origin. Embry-onic stem cells are more plentiful and readily grown in cell cultures. In contrast, adult stem cells must be isolated from surrounding cells in the tissue and reliable techniques to grow them in cell cultures have not yet been developed.

The National Institutes of Health (NIH) explains in "Stem Cell Basics: What Are the Similarities and Differ-ences between Embryonic and Adult Stem Cells?" (January 20, 2011, http://stemcells.nih.gov/info/basics/basics5) that

FIGURE 9.1

Heart muscle repair with adult stem cells

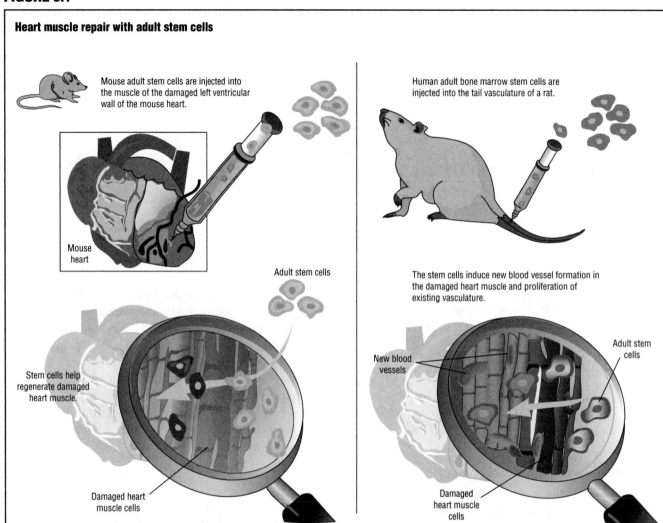

SOURCE: "Figure 4. Heart Muscle Repair with Adult Stem Cells," in *Stem Cell Basics*, U.S. Department of Health and Human Services, National Institutes of Health, 2008, http://stemcells.nih.gov/staticresources/images/figure4_lg.jpg (accessed October 11, 2010)

one potential advantage of adult stem cells is that because they are derived from a patient's own tissue, they may be less likely to be rejected following transplantation. It is not yet known whether tissues arising from embryonic stem cells will trigger the immune response that is associated with transplant rejection.

Therapeutic Cloning to Produce Stem Cells

This section considers selected milestones in therapeutic cloning (also known as somatic cell nuclear transfer) and stem cell research from 2001, when the first cloned human embryo was reported in the medical literature, to 2010, when the first patient in the United States was treated with stem cells.

In 2001 Jose B. Cibelli et al. of Advanced Cell Technology (ACT) reported in "Somatic Cell Nuclear Transfer in Humans: Pronuclear and Early Embryonic Development" (*e-biomed: The Journal of Regenerative Medicine*, vol. 2, November 26, 2001) that they created a cloned

human embryo, and, unlike groups that had claimed to have done this before, they published their results. In the press release "Advanced Cell Technology, Inc. (ACT) Today Announced Publication of Its Research on Human Somatic Cell Nuclear Transfer and Parthenogenesis" (November 25, 2001), the ACT boasted that this achievement offered "the first proof that reprogrammed human cells can supply tissue" and was a vital first step toward the objective of therapeutic cloning—using cloned embryos to harvest embryonic stem cells able to grow into replacement tissue perfectly matched to individual patients. To clone the human embryos, Cibelli et al. collected women's eggs and removed the genetic material from the eggs with a thin needle. A skin cell was inserted inside each of eight enucleated eggs, which were then chemically stimulated to divide. Just three of the eight eggs began dividing, and only one reached six cells before cell division ceased.

In 2003 a Chinese research team led by Ying Chen reported that it had made human embryonic stem cells by combining human skin cells with rabbit eggs. Their accomplishment was published in "Embryonic Stem Cells Generated by Nuclear Transfer of Human Somatic Nuclei into Rabbit Oocytes" (*Cell Research*, vol. 13, no. 4, August 2003). The researchers removed the rabbit eggs' deoxyribonucleic acid (DNA) and injected human skin cells inside them. The eggs then grew to form embryos containing human genetic material. After several days the embryos were dissected to extract their stem cells.

In 2004 Woo Suk Hwang et al. of Seoul National University reported in "Evidence of a Pluripotent Human Embryonic Stem Cell Line Derived from a Cloned Blastocyst" (*Science*, vol. 303, no. 5664, March 12, 2004) that they had successfully cloned healthy human embryos, removed embryonic stem cells, and grown them in mice. In January 2006, following a long investigation, Seoul National University concluded that the research reported in Hwang et al.'s *Science* article had been fabricated. As a result, the journal retracted this article along with another article, "Patient-Specific Embryonic Stem Cells Derived from Human SCNT Blastocysts" (*Science*, vol. 308, no. 5729, June 17, 2005), by Hwang et al. In May 2006 Hwang was charged with fraud, embezzlement, and violating South Korea's bioethics statutes. He was given a suspended jail sentence and served no jail time.

In 2005 Ian Wilmut (1944–) was granted a license by the British government to clone human embryos to generate stem cell lines to study motor neuron disease (MND). Wilmut and his colleagues focused on cloning embryos to generate stem cells that would in turn become motor neurons with MND-causing gene defects. By observing how stem cells grow into neurons, the researchers hoped to discover what causes the cells to degenerate. Their research involved comparing the stem cells with healthy and diseased cells from MND patients to gain a better understanding of the illness and to test potential drug treatments.

Human reproductive cloning remains illegal in the United Kingdom, but therapeutic cloning is allowed on an approved basis. The license granted to Wilmut and his colleagues was the second one granted by Britain's Human Fertilisation and Embryology Authority. However, Roger Highfield indicates in "Dolly Creator Prof Ian Wilmut Shuns Cloning" (*Telegraph* [London, England], November 17, 2007) that Wilmut, who had gained notoriety a decade earlier by creating Dolly, the sheep clone, announced in November 2007 that he was stopping his research on therapeutic cloning and turning his attention to research that focused on reprogramming adult cells into stem cells.

In 2006 Deepa M. Deshpande et al. restored movement to paralyzed rats using a new method that demonstrates the potential of embryonic stem cells to restore function to humans suffering from neurological disorders. They published their results in "Recovery from Paralysis in Adult Rats Using Embryonic Stem Cells" (*Annals of Neurology*, vol. 60, no. 1, July 2006). In 2008 other researchers reported successfully treating mice for Parkinson's disease using therapeutic cloning. In "Therapeutic Cloning in Individual Parkinsonian Mice" (*Nature Medicine*, vol. 14, no. 4, April 2008), Viviane Tabar et al. explain that they used skin cells from the tails of the mice to generate the missing nerve cells in Parkinson's disease. Even though clinical trials in humans are still years away, the results of this research represent an important advance in the quest for a cure for paralysis and other neurological disorders.

In 2006 Kevin A. D'Amour et al. of Novocell Inc. in San Diego, California, reported in "Production of Pancreatic Hormone-Expressing Endocrine Cells from Human Embryonic Stem Cells" (*Nature Biotechnology*, vol. 24, no. 11, October 2006) that they had developed a process to turn human embryonic stem cells into pancreatic cells that can produce insulin and other hormones.

Three studies—Volker Schächinger et al.'s "Intracoronary Bone Marrow–Derived Progenitor Cells in Acute Myocardial Infarction," Ketil Lunde et al.'s "Intracoronary Injection of Mononuclear Bone Marrow Cells in Acute Myocardial Infarction," and Birgit Assmus et al.'s "Transcoronary Transplantation of Progenitor Cells after Myocardial Infarction"—describing the use of stem cells in the treatment of heart disease were published in the *New England Journal of Medicine* (vol. 355, no. 12, September 21, 2006). The studies produced conflicting results: Schächinger et al. reported benefits for patients who had suffered myocardial infarction (heart attack). Lunde et al. found no benefit from stem cell treatment of such patients. Assmus et al. studied patients with chronic heart failure, who did show improvement after treatment. In the editorial "Cardiac Cell Therapy—Mixed Results from Mixed Cells" in the same issue of the journal, Anthony Rosenzweig writes that the three studies "provide a realistic perspective on this approach while leaving room for cautious optimism and underscoring the need for further study."

Lina Dahl et al. of Umeå University's Umeå Center for Molecular Medicine report in "Lhx2 Expression Promotes Self-Renewal of a Distinct Multipotential Hematopoietic Progenitor Cell in Embryonic Stem Cell-Derived Embryoid Bodies" (*PLoS One*, vol. 3, no. 4, April 23, 2008) that they used genetically modified embryonic stem cells to isolate the stem cell that produces blood cells. This finding not only promises enhanced ability to develop treatment for diseases of the blood such as anemia and leukemia but is also key to developing techniques to replace damaged or diseased tissue with new, healthy tissue.

For years scientists questioned whether cloned fetuses would have more mutations than naturally conceived fetuses. In 2009, after five years of research analyzing and comparing mutation rates in mouse fetuses, Patricia Murphey et al. reported in "Epigenetic Regulation of Genetic Integrity Is Reprogrammed during Cloning" (*Proceedings of the National Academy of Science*, vol. 106, no. 12, March 24, 2009) that the rate of spontaneous mutations was similar in fetuses produced by either cloning or natural conception, which means that cloned fetuses do not acquire mutations more rapidly than naturally conceived fetuses. The researchers even speculate that there "may be a relationship between pluripotency and enhanced maintenance of genetic integrity" for cloned fetuses.

One of the challenges of culturing human embryonic stem cells has been to grow them without adding other animal proteins to the culture, which then makes them unusable for treating humans. In "Long-term Self-Renewal of Human Pluripotent Stem Cells on Human Recombinant Laminin-511" (*Nature Biotechnology*, vol. 28, no. 6, June 2010), Sergey Rodin et al. report the first successful culture of human embryonic stem cells using a chemical environment that was free of animal proteins or other contaminants. The researchers cultured the human embryonic stem cells in a matrix of a single human protein called laminin-511.

Other research sheds new light on how embryonic stem cells behave and change. Maurice A. Canham et al. observe in "Functional Heterogeneity of Embryonic Stem Cells Revealed through Translational Amplification of an Early Endodermal Transcript" (*PLoS Biology*, vol. 8, no. 5, May 25, 2010) that embryonic stem cells change in response to certain environmental cues, which prompt them to become a specific tissue cell type. They not only can morph into precursors for adult cells but also are able to change into more primitive cells, such as those associated with the placenta (the lining of the uterus that surrounds the developing fetus). Understanding when embryonic stem cells commit to becoming a specific cell type is key to producing specific types of cells in the laboratory.

According to Jessica Reaves, in "Stem Cell Research Skirts Hurdles, but Raises Ethics Issues, Too" (*New York Times*, October 21, 2010), the first patient to be treated with human embryonic stem cells in the United States, a man with spinal cord injuries, began treatment in October 2010. The patient is participating in a Phase 1 trial that is being jointly run by Northwestern Memorial Hospital in Chicago, Illinois, and Geron Corporation. The two-year trial will involve approximately 10 patients and will focus on establishing the safety and tolerability of the treatment, which is called GRNOPC1. Researchers hope to demonstrate that this type of treatment is a safe,

effective way to regrow nerve cells to reverse paralysis resulting from spinal cord injury.

Research Promises Therapeutic Benefits without Cloning

In "Homologous Recombination in Human Embryonic Stem Cells" (*Nature Biotechnology*, vol. 21, no. 3, March 2003), Thomas P. Zwaka and James A. Thomson of the University of Wisconsin, Madison, report that they used human embryonic stem cells to splice out individual genes and substituted different genes in their place. Their accomplishment was heralded as a first step toward the goal of regenerating parts of the human body by transplanting either stem cells or tissues grown from stem cells into patients. Zwaka and Thomson used electrical charges and chemicals to make the cells' membranes permeable; the cells allowed the customized genes to enter, and they then found and replaced their counterparts in the cells' DNA.

The ability to make precise genetic changes in human stem cells could be used to boost their therapeutic potential or make them more compatible with patients' immune systems. Some researchers assert that the success of this bioengineering feat might eliminate the need to pursue the hotly debated practice of therapeutic cloning, but others caution that such research could heighten concerns among those who fear that stem cell technology will lead to the creation of "designer babies," which are bred for specific characteristics such as appearance, intelligence, or athletic prowess.

Karin Hübner et al. report in "Derivation of Oocytes from Mouse Embryonic Stem Cells" (*Science*, vol. 300. no. 5623, May 23, 2003) that they transformed ordinary mouse embryo cells into egg cells in laboratory dishes. The researchers selected from a population of stem cells the ones that bore certain genetic traits suggesting the potential to become eggs. They then isolated those cells in laboratory dishes. Eventually, the cells morphed into two kinds of cells, including young egg cells. The eggs matured normally and appeared to be healthy in terms of their appearance, size, and gene expression. When cultured for a few days, the eggs underwent spontaneous division and formed structures resembling embryos, a process called parthenogenesis. This finding implies that the eggs were fully functional and likely could be fertilized with sperm.

Once refined, this technology could be applied to produce egg cells in the laboratory that would enable scientists to engineer traits into animals and help conservationists rebuild populations of endangered species. It offers researchers the chance to observe mammalian egg cells as they mature, a process that occurs unseen within the ovary. The technology also offers an unparalleled opportunity to learn about meiosis, the process of cell division during which an egg or sperm disgorges half of its genes so that it can join with a gamete of the opposite sex. There are many potential medical benefits as well. For example, women who cannot make healthy eggs could use this technology to ensure healthy offspring.

Like many new technologies, transforming cells into eggs simultaneously resolves existing ethical issues and creates new ones. For example, because the embryonic stem cells spontaneously transformed themselves into eggs, this procedure overcomes many of the ethical objections to cloning, which involves creating offspring from a single parent. However, it also paves the way for the creation of "designer eggs" from scratch and, if performed with human cells, could redefine the biological definitions of mothers and fathers.

Sawako Ina et al. explain in "Expression of the Mouse Aven Gene during Spermatogenesis, Analyzed by Subtraction Screening Using Mvh-Knockout Mice" (*Gene Expression Patterns*, vol. 3, no. 5, October 2003) that they observed male mouse embryonic stem cells that developed spontaneously, with some cells actually becoming germ cells. When the researchers transplanted the germ cells into testicular tissue, the cells underwent meiosis and formed sperm cells. A possible medical application of this technology would be to assist couples who are infertile because the male cannot produce healthy sperm. One ethical issue is the prospect for two men to both be biological fathers of a child. Another issue is the potential to generate a human being who never had any parents using two laboratory-grown stem cells, one transformed into a sperm and the other into an egg. Many ethicists advise consideration of such issues before permitting human experimentation.

In "Growth Factors Essential for Self-Renewal and Expansion of Mouse Spermatogonial Stem Cells" (*Proceedings of the National Academy of Sciences*, vol. 101, no. 47, November 23, 2004), Hiroshi Kubota, Mary R. Avarbock, and Ralph L. Brinster indicate that they used cells from mice to grow sperm progenitor cells in a laboratory culture. Known as spermatogonial stem cells, the progenitor cells are incapable of fertilizing egg cells but give rise to cells that develop into sperm. The researchers transplanted the cells into infertile mice, which were then able to produce sperm and father offspring that were genetically related to the donor mice.

This breakthrough has many potential applications, including developing new treatments for male infertility and extending the reproductive lives of endangered species. Researchers will also attempt to genetically manipulate the sperm cells grown in a culture medium and then implant the cells into animals. In this way they could introduce new traits into laboratory animals and livestock, such as disease resistance. This culture technique offers researchers additional opportunities to investigate the potential of spermatogonial stem cells as a source for adult stem cells to replace diseased or injured tissue.

New Methods Obtain Stem Cells without Destroying Embryos

According to Young Chung et al., in "Embryonic and Extraembryonic Stem Cell Lines Derived from Single Mouse Blastomeres" (*Nature*, vol. 439, no. 7073, January 12, 2006), embryonic stem cell cultures can be derived from single cells of mouse embryos. Irina Klimanskaya et al. describe in "Human Embryonic Stem Cell Lines Derived from Single Blastomeres" (*Nature*, vol. 444, no. 7118, November 23, 2006) a technique for removing a single cell (called a blastomere) from a three-day-old embryo with eight to 10 cells and using a biochemical process to create embryonic stem cells from the blastomere. The method of removing a cell from the embryo is much like the technique that is used for pre-implantation genetic diagnosis, which is performed to screen the cell for genetic defects. Klimanskaya et al. note that human embryonic stem cell lines derived from a single blastomere were comparable to lines derived with conventional techniques. Even though the researchers assert that the new method "will make it far more difficult to oppose this research," opponents of stem cell research contend that the new technique is morally unacceptable because even a single cell removed from an early embryo may have the potential to produce a life.

Another technique reported in 2006 can obviate the need for embryonic stem cells. Erika Check notes in "Simple Recipe Gives Adult Cells Embryonic Powers" (*Nature*, vol. 442, no. 7098, July 6, 2006) that researchers in the United Kingdom discovered the gene, called nanog, that is the key to "reprogramming" adult cells back to an embryonic state. The reprogramming of adult cells using nanog may make it possible for scientists to generate cells that specialize and develop into every type of cell in the body without the controversial use of human embryonic stem cells.

According to the ACT, in the press release "Advanced Cell Technology Develops First Human Embryonic Stem Cell Line without Destroying an Embryo" (June 21, 2007, http://www.advancedcell.com/news-and-media/press-releases/advanced-cell-technology-develops-first-human-embryonic-stem-cell-line-without-destroying-an-embryo/), at the fifth annual meeting of the International Society for Stem Cell Research in Australia in 2007 it announced that it had successfully produced a human embryonic stem cell line without destroying an embryo. The technique, which involved removing blastomeres from embryos, was similar to that used for preimplantation genetic diagnosis. The stem cells were genetically normal and differentiated into various cell types including blood cells, neurons, heart cells, cartilage, and other cell types with significant therapeutic potential. In 2008 Young Chung et al. reported in "Human Embryonic Stem Cell Lines Generated without Embryo Destruction" (*Cell Stem Cell*, vol. 2, no. 2, February 7, 2008) the creation of fully functional liver cells from embryonic stem cells. That same year Shi-Jiang Lu et al. described success in "Biological Properties and Enucleation of Red Blood Cells from Human Embryonic Stem Cells" (*Blood*, vol. 112, no. 12, December 1, 2008) in transforming human embryonic stem cells into red blood cells. This is a tremendous stride because it may be possible to use this technique to produce blood for transfusion, which is often in short supply due to a lack of donors.

In "Breakthrough of the Year: Reprogramming Cells" (*Science*, vol. 322, no. 5909, December 19, 2008), Gretchen Vogel reports that scientists had successfully transformed human skin cells into stem cells. By observing this process, researchers are gaining insight into how cells grow, change, and transform. It is hoped that reprogrammed cells such as these may prove useful in helping physicians treat disease by using the patients' own cells.

The method that is used to transform the skin cells into stem cells is called induced pluripotent stem cells (IPSCs). IPSCs are created by genetically reprogramming adult cells, which involves disrupting their DNA, so that they have the same characteristics of embryonic stem cells. Because their DNA is disrupted, administering IPSCs to people is likely to increase their risk of developing cancer. Luigi Warren et al. report in "Highly Efficient Reprogramming to Pluripotency and Directed Differentiation of Human Cells with Synthetic Modified mRNA" (*Cell Stem Cell*, vol. 7, no. 5, November 5, 2010) a significant advance in the technique of creating IPSCs that does not involve genetic modification. If this technique proves effective, it has the capacity to produce IPSCs that can be used in patients, for transplants, and for other therapeutic purposes without the increased cancer risk.

In "Patient-Specific Induced Pluripotent Stem-Cell Models for Long-QT Syndrome" (*New England Journal of Medicine*, vol. 363 no. 15, October 7, 2010), Alessandra Moretti et al. report that they were able to generate heart muscle cells from human skin cells using IPSCs. Anthony Rosenzweig observes in the editorial "Illuminating the Potential of Pluripotent Stem Cells" (*New England Journal of Medicine*, vol. 363 no. 15, October 7, 2010) that "patient-specific cellular models derived from induced pluripotent stem cells provide an important additional tool for the study of human disease." However, Rosenzweig also cautions that there are many challenges to overcome and much additional work to do before these findings can be expected to benefit patients: "Working with induced pluripotent stem cells remains technically demanding, and the process of differentiation into specific cell types is incompletely understood, often inefficient, and somewhat variable."

OPINIONS SHAPE PUBLIC POLICY

Even though it is widely believed that stem cell research is the key to developing effective treatment for a range of serious and chronic diseases, public policy issues, especially funding for stem cell research, have been shaped by a heated moral controversy that pits scientific advancement against the sanctity of life. Opposition to stem cell research, and federal funding for it, arises from the concern that it may involve destroying human life contained in embryonic stem cells and the fear that embryos might be created solely for use in research, which could result in their destruction. President George W. Bush (1946–) was strongly opposed to embryonic stem cell research, characterizing it as a moral hazard, because he considered the destruction of human embryos to be morally wrong. He also expressed the belief that embryonic stem cell research might lead to human cloning.

According to the White House, in the press release "President Discusses Stem Cell Research" (August 9, 2001, http://georgewbush-whitehouse.archives.gov/news/releases/2001/08/20010809-2.html), Bush announced in August 2001 his decision to allow federal funds to be used for research on existing human embryonic stem cell lines as long as the derivation process (which begins with the removal of the inner cell mass from the blastocyst) had already been initiated and the embryo from which the stem cell line was derived no longer had the possibility of development as a human being. The president established the following criteria that research studies must meet to qualify for federal funding:

- The stem cells must have been drawn from an embryo created for reproductive purposes that was no longer needed for these purposes.

- Informed consent must have been obtained for the donation of the embryo and no financial inducements provided for donation of the embryo.

These criteria limited research to only 21 cell lines that had been created before August 9, 2001, and effectively banned federal funding for hundreds of embryonic stem cell lines, including some that carried genetic mutations for specific conditions and diseases that scientists wanted to study in the hope of finding cures. Many people argued that because embryonic stem cells are not, in fact, embryos, research on such cells should be allowed. Others, such as President Bush, believed that any research that creates or destroys human embryos at any stage, such as creating new embryos through cloning or destroying an embryo to create a stem cell line, should not be allowed. Despite Bush's ban of federal funding, such research remained legal as long as private funding was used.

In January 2002 the Panel on Scientific and Medical Aspects of Human Cloning was convened by the National Academy of Sciences; the National Academy of Engineering;

the Institute of Medicine Committee on Science, Engineering, and Public Policy; and the National Research Council, Division on Earth and Life Studies Board on Life Sciences. Following the panel, the report *Scientific and Medical Aspects of Human Reproductive Cloning* (2002, http://www.nap.edu/openbook.php?isbn=0309076374) was issued that called for a ban on human reproductive cloning.

The panel recommended a legally enforceable ban with substantial penalties as the best way to discourage human reproductive cloning experiments in both the public and private sectors. It cautioned that a voluntary measure might be ineffective because many of the technologies needed to accomplish human reproductive cloning are widely accessible in private clinics and in other organizations that are not subject to federal regulations.

The panel did not, however, conclude that the scientific and medical considerations that justify a ban on human reproductive cloning are applicable to nuclear transplantation to produce stem cells. In view of the potential to generate new treatments for life-threatening diseases and advance biomedical knowledge, the panel recommended that biomedical research using nuclear transplantation to produce stem cells be permitted. Finally, the panel encouraged ongoing national discussion and debate about the range of ethical, religious, and societal issues that are associated with human cloning research.

In February 2002 the American Association for the Advancement of Science, the world's largest general scientific organization, affirmed in "AAAS Resolution: Statement on Human Cloning" (2010, http://archives.aaas.org/docs/resolutions.php?doc_id=425) a legally enforceable ban on reproductive cloning, but stated its support for therapeutic cloning using nuclear transplantation methods under appropriate government oversight. Similarly, the American Medical Association, a national physicians' organization, addresses in "Human Cloning" (January 5, 2011, http://www.ama-assn.org/) many of the issues that are associated with human reproductive cloning. It also cautions that human cloning failures could jeopardize promising science and genetic research and prevent biomedical researchers and patients from realizing the potential benefits of therapeutic cell cloning.

According to the White House, in the press release "President Bush Calls on Senate to Back Human Cloning Ban" (April 10, 2002, http://georgewbush-whitehouse.archives.gov/news/releases/2002/04/20020410-4.html), in April 2002 President Bush called on the U.S. Senate to back legislation banning all types of human cloning. In his plea to the Senate, Bush said:

> Science has set before us decisions of immense consequence. We can pursue medical research with a clear sense of moral purpose or we can travel without an ethical compass into a world we could live to regret.

Science now presses forward the issue of human cloning. How we answer the question of human cloning will place us on one path or the other. ... Human cloning is deeply troubling to me, and to most Americans. Life is a creation, not a commodity.... Our children are gifts to be loved and protected, not products to be designed and manufactured. Allowing cloning would be taking a significant step toward a society in which human beings are grown for spare body parts, and children are engineered to custom specifications; and that's not acceptable. ... I believe all human cloning is wrong, and both forms of cloning ought to be banned, for the following reasons. First, anything other than a total ban on human cloning would be unethical. Research cloning would contradict the most fundamental principle of medical ethics, that no human life should be exploited or extinguished for the benefit of another.

On September 25, 2002, Elias A. Zerhouni (1951–; http://olpa.od.nih.gov/hearings/107/session2/testimonies/stemcelltest.asp), the director of the NIH, testified before the Senate Appropriations Subcommittee on Labor, Health and Human Services, and Education in favor of advancing the field of stem cell research. Zerhouni exhorted Congress to continue to fund both human embryonic stem cell research and adult stem cell research simultaneously to learn as much as possible about the potential of both types of cells to treat human disease. He observed that many studies that do not involve human subjects must be performed before any new therapy is tested on human patients. These preclinical studies include tests of the long-term survival and fate of transplanted cells, as well as tests on the safety, toxicity, and effectiveness of the cells in treating specific diseases in animals. Zerhouni promised that trials using human subjects, the clinical research phase, would begin only after the basic foundation had been established. Despite Zerhouni's impassioned plea and subsequent efforts by Congress to reverse limitations on embryonic stem cell research, President Bush vetoed such attempts, and, at the close of 2008, U.S. law continued to ban federal funding of any research that might harm human embryos.

Obama Administration Removes Barriers to Stem Cell Research

While campaigning for the office of president, Barack Obama (1961–) pledged to lift the ban on federal funding for embryonic stem cell research. On March 9, 2009, shortly after entering office in January, he did so by an executive order. Despite President Obama's reversal of President Bush's policy on embryonic stem cell research, for some research Congress would still need to overturn the Dickey-Wicker amendment, which has been included in the fiscal spending bills each year since 1996 (before stem cells had been successfully extracted from human embryos) and prohibits federal funding of any research that creates or destroys human embryos. Following President Obama's executive order, the NIH established policies and procedures enabling federally funded researchers to work with selected batches of human embryonic stem cells—largely stem cell lines that had been created with private, as opposed to federal, funds.

In June 2010 James Sherley of Boston Biomedical Research Institute and Theresa Deisher of AVM Biotechnology filed a suit seeking to stop federal funding of human embryonic stem cell research because they claimed such research destroys human embryos and diverts funds from researchers working with adult stem cells. On August 23, 2010, Judge Royce C. Lamberth (1943–) of the U.S. District Court issued a preliminary injunction that halted federal funding of embryonic stem cell research. The injunction also banned research on the stem cell lines that had been approved for research by President Bush in 2001.

The following day the Obama administration vowed to appeal the ruling, which would have suspended $54 million for 22 projects by the end of September 2010. The U.S. Department of Justice requested that the injunction be lifted, pending its appeal. Sheryl Gay Stolberg and Gardiner Harris note in "Stem Cell Ruling Will Be Appealed" (*New York Times*, August 24, 2010) that Francis S. Collins (1950–), the director of the NIH, asserted, "This decision has the potential to do serious damage to one of the most promising areas of biomedical research, just at the time when we were really gaining momentum."

On September 9, 2010, a federal appeals court suspended the district court's preliminary injunction, such that federal funding of stem cell research could proceed while the government's appeal was being considered. According to Pete Yost, in "Appeals Court Allows Stem-Cell Funding for Now" (*Washington Times*, September 9, 2010), Lisa Hughes, the president of the Coalition for the Advancement of Medical Research, expressed relief about the federal appeals court's decision but cautioned that "while this issue continues to be argued in the courts, we call on Congress to move swiftly to resolve this issue and secure the future of this important biomedical research." As of January 2011, the government's appeal was still being reviewed.

Legislation Aims to Completely Ban Human Cloning

In February 2003 the U.S. House of Representatives voted to outlaw all forms of human cloning. The legislation prohibited the creation of cloned human embryos for medical research as well as the creation of cloned babies. It contained strong sanctions, imposing a maximum penalty of $1 million in civil fines and as many as 10 years in prison for violations. The measure did not pass in the Senate, which was closely divided about whether therapeutic cloning should be prohibited along with reproductive cloning. In early February 2003 President Bush issued

a policy statement that strongly supported a total ban on cloning. In the Senate two bills were introduced: S. 245 was a complete ban intended to amend the Public Health Service Act to prohibit all human cloning, and S. 303 was a less sweeping measure that also prohibited cloning but protected stem cell research. S. 245 was referred to the Senate Committee on Health, Education, Labor, and Pensions and S. 303 was referred to the Senate Committee on the Judiciary. In March 2007 S. 812, the Human Cloning Ban and Stem Cell Research Protection Act of 2007, was introduced and referred to the Committee on the Judiciary. As of January 2011, none of these bills, nor any comparable proposed legislation, had emerged from the Senate committees.

Even though nearly all lawmakers concur that Congress should ban reproductive cloning, many disagree about whether legislation should also ban the creation of cloned human embryos that serve as sources of embryonic stem cells. Many legislators agree with scientists that stem cells derived from cloned human embryos have medical and therapeutic advantages over those derived from conventional embryos or adults. Those who oppose the legislation calling for a total ban assert that the aim of allowing research is to relieve the suffering of people with degenerative diseases. They say that the bill's sponsors are effectively thwarting advances in medical treatment and biomedical innovation.

Supporters of the total ban contend that Congress must send a clear message that cloning research and experimentation will not be tolerated. They consider cloning immoral and unethical, fear unintended consequences of cloning, and feel they speak for the public when they assert that it is not justifiable to create human embryos simply for the purpose of experimenting on them and then destroying them.

Even though President Obama lifted the ban on federal funding for all embryonic stem cell research on lines other than those President Bush approved (those created before August 9, 2001), his executive order did not address reproductive cloning. When he issued the executive order, President Obama expressed his strong support of embryonic stem cell research but stated that under no circumstances should the funding for embryonic stem cell research be used to explore human cloning.

State Human Cloning Laws

As of January 2011, 15 states had enacted legislation that addresses human cloning. According to the National Conference of State Legislatures, in "Human Cloning Laws" (January 2008, http://www.ncsl.org/default.aspx?tabid=14284), California was the first state to ban reproductive cloning in 1997. Since then, 12 other states—Arkansas, Connecticut, Indiana, Iowa, Maryland, Massachusetts, Michigan, Rhode Island, New Jersey, North Dakota, South

Dakota, and Virginia—have passed laws prohibiting reproductive cloning. Arizona's and Missouri's legislation addresses the use of public funds for cloning, and Maryland's prohibits the use of state stem cell research funds for reproductive cloning and possibly therapeutic cloning, depending on the interpretation of the statute. Louisiana also enacted legislation that prohibited reproductive cloning, but the law expired in July 2003. The laws of Arkansas, Indiana, Iowa, Michigan, North Dakota, and South Dakota also prohibit therapeutic cloning. Virginia's legislation may be interpreted as a complete ban on human cloning; however, it is unclear because the law does not define the term *human being*, which is used in the definition of human cloning. Rhode Island's law, which expired in July 2010, did not prohibit cloning for research, and California's and New Jersey's laws specifically permit cloning for the purpose of research.

California Leads the Way in Stem Cell Research

Even though some states specifically banned human cloning, state funding of stem cell research was supported by many individual states even when federal funds were not available. In 2002 the California state legislature passed a law encouraging therapeutic cloning. There were no provisions for funds in the law, but the move was interpreted as support for the research. In 2004 stem cell research advocates offered voters a sweeping ballot measure—Proposition 71—to make public funding available to support stem cell research and therapeutic cloning. Proposition 71 was championed by Robert Klein, a wealthy real estate developer and a father of a child with diabetes who might benefit from the research. It also received considerable financial support from the Microsoft founder Bill Gates (1955–) to finance campaign advertising and lobbying.

In November 2004 Californians approved Proposition 71, a ballot measure with the potential to make the state a leader in human embryonic stem cell research. Proposition 71 enabled the state to create a state agency called the California Institute for Regenerative Medicine (CIRM; http://www.cirm.ca.gov/). The proposition prohibits reproductive cloning but funds human cloning projects that are designed to create stem cells and allocates $3 billion over 10 years in research funds. Those supporting the legislation hoped that stem cell research would become the biggest, most important, and most profitable medical advancement of the 21st century. The legislation's supporters intended to use the funds to attract top researchers to the state, making California the epicenter of groundbreaking, lifesaving, and potentially lucrative medical research.

The CIRM was officially established in 2005. An independent oversight committee that consists of public officials governs the CIRM. These officials are appointed on the basis of their experience earned in California's leading public universities, nonprofit academic and research institutions, patient advocacy groups, and the biotechnology industry. The CIRM funds biomedical research based in California that focuses on developing diagnostics and therapies and on other vital research that will lead to life-saving medical treatments. By October 2010 the CIRM (http://www.cirm.ca.gov/InstitutionList) had approved 385 research grants totaling over $1.1 billion, making the CIRM the largest source of funding for embryonic and pluripotent stem cell research in the world.

Public Opinions about Stem Cell Research and Cloning

According to Gallup Organization poll data, in 2010 nearly six out of 10 (59%) Americans believed using stem cells derived from human embryos in medical research is morally acceptable. Table 9.1 reveals that the percentage of Americans that considers stem cell research morally acceptable declined from a high of 64% in 2007.

The percentage of Americans that deemed stem cell research morally acceptable in 2010 varied by political affiliation, with support highest among Democrats (68%) and Independents (62%), compared with Republicans

TABLE 9.1

Public opinion on the moral acceptability of embryonic stem cell research, 2002–10

	Morally acceptable	Morally wrong	Depends on situation (vol.)	Not a moral issue (vol.)	No opinion
2010 May 3–6	59	32	3	*	6
2009 May 7–10	57	36	2	1	4
2008 May 8–11	62	30	3	*	5
2007 May 10–13	64	30	2	*	4
2006 May 8–11	61	30	3	*	6
2005 May 2–5	60	33	3	—	4
2004 May 2–4	54	37	3	*	6
2003 May 5–7	54	38	3	—	6
2002 May 6–9	52	39	2	1	6

SOURCE: Lydia Saad, "Perceived Moral Acceptability of Behavior and Social Policies," in *Four Moral Issues Sharply Divide Americans*, The Gallup Organization, May 26, 2010, http://www.gallup.com/poll/137357/Four-Moral-Issues-Sharply-Divide-Americans.aspx (accessed October 14, 2010). Copyright © 2010 by The Gallup Organization. Reproduced by permission of The Gallup Organization.

TABLE 9.2

Public opinion on the moral acceptability of embryonic stem cell research, by political identification, May 2010

Percentage viewing matters as "morally acceptable," by party ID

Ranked by party difference (Democrat vs. Republican)

	Democrat	Independent	Republican	Difference
	%	%	%	pct. pts.
Gay or lesbian relations	61	61	35	26
Abortion	51	39	26	25
The death penalty	52	66	76	24
Embryonic stem cell research	68	62	47	21
Having a baby outside of marriage	61	59	41	20
Premarital sex	67	64	47	20
Buying/wearing clothing made of animal fur	54	61	67	13
Divorce	73	74	61	12
Doctor-assisted suicide	52	46	40	12
Cloning animals	34	32	27	7
Suicide	18	16	11	7
Gambling	64	60	59	5
Extramarital affairs	7	7	3	4
Medical testing on animals	58	57	62	4
Cloning humans	11	7	8	3
Polygamy	6	12	5	1

SOURCE: Lydia Saad, "Percentage Viewing Matters As 'Morally Acceptable,' by Party ID," in *Four Moral Issues Sharply Divide Americans*, The Gallup Organization, May 26, 2010, http://www.gallup.com/poll/137357/Four-Moral-Issues-Sharply-Divide-Americans.aspx (accessed October 14, 2010). Copyright © 2010 by The Gallup Organization. Reproduced by permission of The Gallup Organization.

TABLE 9.3

Public opinion on the moral acceptability of embryonic stem cell research, by gender, May 2010

Percentage viewing matters as "morally acceptable," by gender

Ranked by difference

	Men	Women	Difference
	%	%	pct. pts.
Buying/wearing clothing made of animal fur	73	48	25
Cloning animals	43	19	24
Medical testing on animals	69	49	20
Gambling	67	55	12
The death penalty	70	60	10
Cloning humans	13	4	9
Premarital sex	63	56	7
Doctor-assisted suicide	49	42	7
Abortion	41	36	5
Polygamy	10	5	5
Divorce	71	67	4
Embryonic stem cell research	61	57	4
Suicide	17	13	4
Having a baby outside of marriage	52	55	3
Gay or lesbian relations	53	51	2
Extramarital affairs	6	5	1

SOURCE: Lydia Saad, "Percentage Viewing Matters As 'Morally Acceptable,' by Gender," in *Four Moral Issues Sharply Divide Americans*, The Gallup Organization, May 26, 2010, http://www.gallup.com/poll/137357/Four-Moral-Issues-Sharply-Divide-Americans.aspx (accessed October 14, 2010). Copyright © 2010 by The Gallup Organization. Reproduced by permission of The Gallup Organization.

TABLE 9.4

Public opinion on the moral acceptability of cloning animals, 2001–10

	Morally acceptable	Morally wrong	Depends on situation (vol.)	Not a moral issue (vol.)	No opinion
2010 May 3–6	31	63	1	1	4
2009 May 7–10	34	63	1	1	2
2008 May 8–11	33	61	2	1	3
2007 May 10–13	36	59	2	1	3
2006 May 8–11	29	65	2	1	3
2005 May 2–5	35	61	1	*	3
2004 May 2–4	32	64	1	1	2
2003 May 5–7	29	68	1	*	2
2002 May 6–9	29	66	3	1	1
2001 May 10–14	31	63	2	1	3

SOURCE: Lydia Saad, "Perceived Moral Acceptability of Behavior and Social Policies," in *Four Moral Issues Sharply Divide Americans*, The Gallup Organization, May 26, 2010, http://www.gallup.com/poll/137357/Four-Moral-Issues-Sharply-Divide-Americans.aspx (accessed October 14, 2010). Copyright © 2010 by The Gallup Organization. Reproduced by permission of The Gallup Organization.

(47%). (See Table 9.2.) The Gallup poll also finds that more men (61%) than women (57%) view stem cell research as morally acceptable. (See Table 9.3.)

Even though there is considerable support for stem cell research, the overwhelming majority of Americans (88%) continue to feel that it is morally unacceptable to clone humans, and about two-thirds (63%) feel it is unacceptable to clone animals. (See Table 9.4 and Table 9.5) The percentages of Americans who feel cloning animals and humans is morally wrong has remained the same throughout the first decade of the 21st century.

TABLE 9.5

Public opinion on the moral acceptability of cloning humans, 2001–10

	Morally acceptable	Morally wrong	Depends on situation (vol.)	Not a moral issue (vol.)	No opinion
2010 May 3–6	9	88	*	1	3
2009 May 7–10	9	88	*	*	3
2008 May 8–11	11	85	1	*	2
2007 May 10–13	11	86	1	*	2
2006 May 8–11	8	88	1	*	3
2005 May 2–5	9	87	1	1	2
2004 May 2–4	9	88	1	*	2
2003 May 5–7	8	90	1	*	1
2002 May 6–9	7	90	2	*	1
2001 May 10–14	7	88	1	1	3

SOURCE: Lydia Saad, "Perceived Moral Acceptability of Behavior and Social Policies," in *Four Moral Issues Sharply Divide Americans*, The Gallup Organization, May 26, 2010, http://www.gallup.com/poll/137357/Four-Moral-Issues-Sharply-Divide-Americans.aspx (accessed October 14, 2010). Copyright © 2010 by The Gallup Organization. Reproduced by permission of The Gallup Organization.

CHAPTER 10
GENETIC ENGINEERING AND BIOTECHNOLOGY

Genetic engineers have looked at nature as a set of finished products to tweak and improve—a tomato that could be made into a slightly better tomato. But synthetic biologists imagine nature as a manufacturing platform: all living things are just crates of genetic cogs; we should be able to spill all those cogs out on the floor and rig them into whatever new machinery we want.

—Jon Mooallem, "Do-It-Yourself Genetic Engineering" (*New York Times*, February 10, 2010)

The dawn of the new millennium saw explosive advances in biotechnology. Technological breakthroughs offered scientists and physicians unprecedented opportunities to develop previously inconceivable solutions to pressing problems in agriculture, environmental science, and medicine. Simultaneously, researchers, politicians, ethicists, theologians, and the public were challenged to assess, analyze, and determine the feasibility of using new biotechnology in view of the opportunities, possibilities, risks, benefits, and diverse viewpoints about the safe, effective, and ethical applications of genetic research.

This chapter describes several examples of existing and proposed applications of genetic research and biotechnology, including uses that address pressing problems such as environmental pollution and world hunger as well as the role of genetic engineering in developing lifesaving medical therapeutics. It also considers industry and consumer viewpoints as well as recommendations scientists, policy makers, and ethicists have made regarding the wise, judicious, and equitable use of these technologies.

AGRICULTURAL APPLICATIONS OF GENETIC ENGINEERING

Genetically modified (GM) or transgenic crops (sometimes also called genetically engineered [GE] crops) contain one or more genes that have been artificially inserted instead of received through pollination (fertilization by the transfer of pollen from an anther to a stigma of a plant).

The inserted gene sequence, called the transgene, may be introduced to produce different results—either to overexpress or silence (direct a gene not to synthesize a specific protein) an existing plant gene, and it may come from another unrelated plant or from a completely different species. Transgenics is the science of inserting a foreign gene into an organism's genome. The ultimate product of this technology is a transgenic organism.

For example, the transgenic corn that produces its own insecticide contains a gene from a bacterium, and Macintosh apples with a gene from a moth that encodes an antimicrobial protein are resistant to a bacterial infection called fire blight. Even though all crops have been genetically modified from their original wild state by domestication, selection, and controlled breeding over long periods of time, the terms *transgenic crops*, *GE crops*, *GM crops*, and *biotech crops* usually refer to plants with transgenes (inserted gene sequences).

Genes are introduced into a crop plant to make it as useful and productive as possible by acting to protect the plant, improve the harvest, or enable the plant to perform a new function or acquire a new trait. Specific objectives of genetically modifying a plant include increasing its yield, improving its quality, or enhancing its resistance to pests or disease and its tolerance for heat, cold, or drought. Some of the GM traits that have been introduced into food crops are enhanced flavor, slowed ripening, reduced reliance on fertilizer, self-generating insecticide, and added nutrients. Examples of transgenic food crops include frost-resistant strawberries and tomatoes; slow-ripening bananas, melons, and pineapples; and insect-resistant and herbicide-tolerant corn, cotton, and soybeans. Figure 10.1 shows the increasing adoption of insect-resistant corn and cotton and herbicide-tolerant soybeans, cotton, and corn from 1996 to 2010.

Transgenic technology enables plant breeders to bring together in one plant useful genes from a wide

FIGURE 10.1

Adoption of genetically engineered crops, 1996–2010

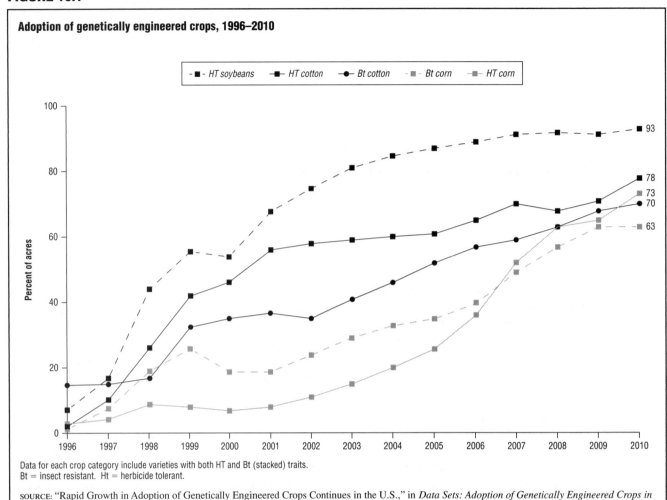

Data for each crop category include varieties with both HT and Bt (stacked) traits.
Bt = insect resistant. Ht = herbicide tolerant.

SOURCE: "Rapid Growth in Adoption of Genetically Engineered Crops Continues in the U.S.," in *Data Sets: Adoption of Genetically Engineered Crops in the U.S.*, United States Department of Agriculture, Economic Research Service, July 1, 2010, http://www.ers.usda.gov/Data/BiotechCrops/ (accessed October 15, 2010)

range of living sources, not just from within the crop species or from closely related plants. It provides a reliable means for identifying and isolating genes that control specific characteristics in one kind of organism and enables researchers to move copies of these genes into another organism that will then develop the chosen characteristics. This technology gives plant breeders the ability to generate more useful and productive crop varieties containing new combinations of genes, and it significantly expands the range of trait manipulation and enhancements well beyond the limitations of traditional cross-pollination and selection techniques.

Even though genetic modification of plants generates the same types of changes that are produced by conventional agricultural techniques, because it precisely alters a single gene the results are often more rapid and more complete. Traditional breeding techniques may require an entire generation or more to introduce or remove a single gene, and using conventional methods for breeding a polygenic trait into crops with multiyear generations could take several decades.

Creating Transgenic Crops

The first step in creating a transgenic plant is locating genes with the traits that growers, marketers, and consumers consider important. These are usually genes that increase productivity and yield and improve resistance to environmental stresses such as frost, heat, salt, and insects. Identifying the gene associated with a specific trait is necessary but not sufficient; researchers must determine how the gene is regulated, its other influences on the plant, and its interactions with other genes to express or silence various traits. Researchers must then isolate and clone the gene to have sufficient quantities to modify. Establishing the genomic sequence, called plant genomics, and the functions of genes of the most important crops is a priority of public- and private-sector plant genomic research projects.

Most genes that are introduced into plants come from bacteria; however, increasing understanding of plant genomics is anticipated to permit greater use of plant-derived genes to genetically engineer crops. In 2000 the first entire plant genome *Arabidopsis thaliana* was sequenced, which provided researchers with new insight into the genes that control specific traits in many other agricultural plants. There are several approaches to introducing genes into plant cells: vector- or carrier-mediated transformation, particle-mediated transformation, and direct deoxyribonucleic acid (DNA) insertion.

Vector-mediated transformation involves infecting plant cells with a virus or bacterium that during the process of infection inserts foreign DNA into the plant cell. The most convenient is through the soil bacterium *Agrobacterium tumefaciens*, which infects tomatoes, potatoes, cotton, and soybeans. This bacterium attacks cells by inserting its own DNA. When genes are added to the bacterium, they are transferred to the plant cell along with the other DNA.

Particle-mediated transformation involves gene transfer using a special particle tool known as a gene gun, which shoots tiny metal particles that contain DNA into the cell.

To perform direct DNA insertion or electroporation, cells are immersed in the DNA and electrically shocked to stimulate DNA uptake. The cell wall then opens for less than a second, allowing DNA to seep into the cell. (See Figure 10.2.) Following gene insertion, the cell incorporates the foreign DNA into its own chromosomes and undergoes normal cell division. The new cells ultimately form the organs and tissues of the "regenerated" plant. To ensure the systematic sequence of these steps, other genes may be added along with the gene that is associated with the desired trait. These helper genes are called promoters. They encourage the growth of cells that have integrated the inserted DNA, provide resistance to stresses (such as toxins present in the medium that is used to grow the cells), and may help regulate the functions of the gene that is linked to the desired trait.

To be certain that the new genes are in the organism, marker genes are sometimes inserted along with the gene for the desired trait. One common marker gene confers resistance to the antibiotic kanamycin. When this gene is used as a marker, investigators are able to confirm that the transfer was successful when the organism resists the antibiotic. The ultimate success of gene insertion is measured by whether the inserted gene functions properly by expressing, amplifying, or silencing the desired trait.

Are GM Crops Helpful or Harmful?

In *Agricultural Biotechnology (A Lot More than Just GM Crops)* (August 2010, http://www.isaaa.org/resources/publications/agricultural_biotechnology/download/Agricultural_Biotechnology.pdf), the International Service for the Acquisition of Agri-biotech Applications (ISAAA), a nonprofit organization that delivers new agricultural biotechnologies to developing countries, observes that the genetic modification of crops continues to be a rapidly adopted technology. (See Figure 10.1.) The ISAAA contends that GE crops consistently perform well and do not pose health or safety risks, stating, "To date, no scientifically valid demonstrations have shown that food safety issues of foods containing GE ingredients are greater than those from conventionally or organically produced foods." This assertion is tempered somewhat by the ISAAA observation that even though scientific testing and government regulatory efforts act to reduce the safety risks that are associated with conventionally grown, organic, and GM crops, assuring 100% safety is impossible.

The ISAAA asserts that GE crops have already helped ensure more sustainable agricultures and demonstrate the following benefits:

- Pesticide use for GE crops is lower than for conventionally grown crops.

- Farmers who have adopted GE crops realize economic benefits and enjoy time savings and easier agricultural practices. The ISAAA believes that resistance to GE crops stems almost exclusively from farmers' fear of rejection in the export market.

- GE crops decrease the environmental footprint of agriculture by significantly reducing the use of pesticides, diminish carbon dioxide emissions by reducing or eliminating plowing, and increase the efficiency of water and soil use.

- GE crops mitigate climate change and decrease greenhouse gases by reducing carbon dioxide emissions and pesticide use.

Opposition to GM crops takes several forms. Bioethicists contend that freedom of choice is a central tenet of ethical science and oppose what they deem to be interference with other forms of life. Environmentalists argue that transgenic technology poses the risk of altering delicately balanced ecosystems (biological communities and their environments) and causing unintended harm to other organisms. They are

FIGURE 10.2

A bacterial cell incorporating DNA in a process known as transformation

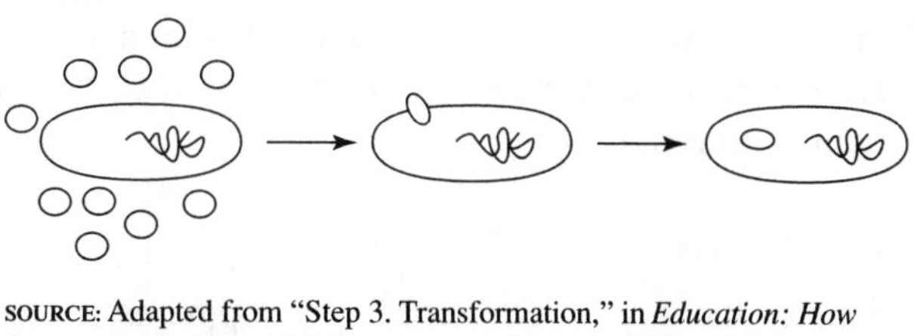

SOURCE: Adapted from "Step 3. Transformation," in *Education: How Sequencing Is Done*, U.S. Department of Energy Office of Science, Joint Genome Institute, September 2004, http://www.jgi.doe.gov/education/how/how_3.html (accessed October 15, 2010)

concerned that transgenic crops will replace traditional crop varieties, especially in developing countries, causing the loss of biological diversity.

In "Ethical Arguments Relevant to the Use of GM Crops" (*New Biotechnology*, vol. 27, no. 5, November 30, 2010), Albert Weale of the Nuffield Council on Bioethics in London, England, presents some of the ethical concerns that have been raised about GM crops, such as the "potential harm to human health, potential damage to the environment, negative impact on traditional farming practices and excessive corporate dominance." However, Weale concedes that the potential benefits of GM crops in developing countries, such as reducing hunger and poverty and developing profitable agriculture, create an ethical obligation to explore the use of GM crops and weigh the potential risks against the potential benefits of this technology.

Among the environmental concerns is the risk that pests may develop resistance to transgenics in much the same way that certain bacteria have become resistant to the antibiotics that once effectively eradicated them. Critics also fear the infiltration of transgenic crops beyond their intended areas and inadvertent gene transfer to species not targeted for transgenics. Dennis Obonyo et al. caution in "Challenges for GM Technologies: Evidence-Based Evaluation of the Potential Environmental Effects of GM Crops" (*GMOs for African Agriculture: Challenges and Opportunities* [July 2010, http://www.assaf.org.za/wp-content/uploads/PDF/ASSAf%20GMO%20African%20Agriculture%202010%20Web.pdf]) that the consequences of any potential gene transfer must be assessed along with the possibility of mutations that are triggered by GM technology.

According to Matin Qaim of the Georg-August-University of Göttingen, in "The Benefits of Genetically Modified Crops—and the Costs of Inefficient Regulation" (April 2, 2010, http://www.rff.org/Publications/WPC/Pages/The-Benefits-of-Genetically-Modified-Crops-and-the-Costs-of-Inefficient-Regulation.aspx), GM crops are growing on 10% of the earth's cultivatable land and as such are big business. Qaim observes that in developing countries the economic benefits of GM crops accrue to small farms and workers. Increased harvests translate into the need for more workers. In contrast, in developed countries biotechnology and seed companies realize tremendous profits. Opponents cite this economic concentration—the potential for companies that grow transgenic crops to drive out smaller farmers and create seed monopolies—as a risk of GM technology. Qaim asserts that complying with costly regulations also discourages small companies from using GM technologies, which further concentrates the economic gains in larger companies.

Health risks also concern those who object to widespread acceptance of transgenic crops. They call for the labeling of GM food to alert consumers that they are purchasing foods that contain GM organisms. According to the Center for Food Safety (CFS; 2010, http://www.centerforfoodsafety.org/genetical17.cfm), a nonprofit public interest and environmental advocacy membership organization that was established by the International Center for Technology Assessment for the purpose of challenging harmful food production technologies and promoting sustainable alternatives, as of 2010 approximately 45% of corn and three-quarters of all processed foods sold in the United States contained GM organisms, and there is no requirement that these foods be identified as transgenic or GM. Opponents cite safety issues such as possible allergies to transgenic foods and products because some transgenes may pose health risks when consumed. They also fear that there will be unforeseen and potentially harmful long-term adverse health consequences resulting from the consumption of foods containing foreign genes.

Despite fears about GM crops, skyrocketing food prices and global food shortages have helped reduce government and consumer resistance to GM crops. Andrew Pollack observes in "In Lean Times, Biotech Grains Are Less Taboo" (*New York Times*, April 21, 2008) that in Asia, the United States, and even Europe, where opposition to GM foods has been greatest, government officials are expediting approvals of imports of GM crops to prevent critical shortages of grain to feed livestock. Opponents of biotechnology contend that its advocates are using the global food crisis to advance their aims. However, Pollack notes that Michael Mack, the chief executive of Syngenta, a prominent agricultural chemical and biotechnology company, cautioned that the industry should not use the current crisis to push its agenda.

According to the article "EU Approves Six GM Maize Varieties for Import" (Reuters, July 28, 2010), in 2010 the European Union (EU) considered several proposals to allow trace amounts of GM materials in imported animal feed to prevent the shortages of animal feed supplies that occurred in 2009, when shipments of animal feed from the United States were prohibited because they contained unapproved GM materials. In July 2010 the EU approved six GM maize varieties, but this approval only eased the regulations for animal feed; the EU continued to maintain zero tolerance of GM foods for human consumption.

Terminator Technology

One of the most controversial developments in agricultural bioengineering is terminator technology (also known as genetic use restricted technology) because it is designed to genetically switch off a plant's ability to germinate a second time. Traditionally, farmers save seeds for the next harvest; however, the use of terminator technology effectively prevents this practice, forcing them to purchase a fresh supply of seeds each year.

The advocates of terminator technology are generally corporations and the organizations that represent them. They contend that the practice protects corporations from corrupt farmers. Controlling seed germination helps prevent growers from pirating the corporations' licensed or patented technology. If crops remained fertile, there is a chance that farmers could use any saved transgenic seed from a previous season. This would result in reduced profits for the companies that own the patents.

Opponents of terminator technology believe it threatens the livelihood of farmers in developing countries such as India, where many poorer farmers have been unable to compete and some have been forced out of business. Opponents considered it a victory when Monsanto Company, a major investor in this technology, decided not to market terminator technology. Barbara H. Peterson reports in "Monsanto's Terminator Making a Comeback? Enter the Zombie!" (*American Chronicle*, May 20, 2009) that despite a call from the United Nations Convention on Biological Diversity (CBD) in 2000 for a moratorium on terminator technology, and a reaffirmation of this moratorium in 2006, some companies continue to work on this technology and plan to introduce another restrictive technology, called zombie technology—plants that require the application of a proprietary chemical to trigger seed fertility, which would require farmers to pay for the chemical annually.

La Via Campesina, an international organization of small- and medium-sized producers and agricultural workers that focuses on conservation and sustainable advancement of cultural and ecological diversity and human rights, asserts that application and commercialization of terminator technology will restrict farmers' access to seeds and their rights to replant harvested seeds. In "La Via Campesina Call to Action—Help Stop Terminator's Return" (July 28, 2010, http://www.viacampesina.org/), La Via Campesina describes terminator technology as "a threat to food sovereignty and agrobiodiversity" and warns that "ending the moratorium on terminator will increase control of seed by transnational corporations," including Monsanto, DuPont, and ArborGen.

Still, even without terminator technology, under patent laws in the United States, in Canada, and in many other industrialized nations, it is illegal for farmers to reuse patented seed or to grow Monsanto's GM seed without signing a licensing agreement. This has the same effect on poor farmers as terminator technology; it renders them unable to compete.

Exorcist Technology

In 2004 the biotechnology industry introduced what it calls exorcist technology to some GE crops. This new technology requires chemical inducers to be applied to certain GE crops to help them shed their foreign DNA before they are harvested. The industry sees this technology as an effective way to counter anti-GE critics because the harvested crops will not contain foreign DNA. However, detractors assert that the intent of exorcist technology is to shift the responsibility from the biotechnology industry to farmers, because they have to apply the chemical inducers to remove the offensive transgenes.

Monsanto Takes Action against Farmers

In *Monsanto vs. U.S. Farmers* (January 12, 2005, http://www.centerforfoodsafety.org/pubs/CFSMOnsantovs FarmerReport1.13.05.pdf), the CFS reviews Monsanto's legal actions against U.S. farmers. The CFS finds that Monsanto engaged in investigations of farmers, out-of-court settlements, and litigation against farmers allegedly in breach of contract or engaged in patent infringement. It documents 90 Monsanto lawsuits in 25 states that involved 147 farmers and 39 small businesses or farm companies. In 2005 Monsanto had an annual budget of $10 million and 75 employees who were exclusively devoted to investigating and taking action against farmers.

The CFS indicates that by 2005 the largest judgment in favor of Monsanto was more than $3 million and that the total recorded judgments granted to Monsanto was nearly $15.3 million. Farmers paid an average of $412,259 for cases with recorded judgments. Some farmers were even forced to pay Monsanto's costs while they were under investigation.

Andrew Kimbrell, the executive director of the CFS, asserts in the press release "Monsanto Assault on U.S. Farmers Detailed in New Report" (January 13, 2005, http://www.centerforfoodsafety.org:80/press_release1.13 .05.cfm) that "these lawsuits and settlements are nothing less than corporate extortion of American farmers. Monsanto is polluting American farms with its GE crops, not properly informing farmers about these altered seeds, and then profiting from its own irresponsibility and negligence by suing innocent farmers. We are committed to stopping this corporate persecution of our farmers in its tracks."

In September 2006 the Public Patent Foundation (http://www.pubpat.org/monsantofiled.htm), a nonprofit legal services organization that represents the public's interests against the harms caused by the patent system, filed requests with the U.S. Patent and Trademark Office to revoke four patents owned by Monsanto "that the agricultural giant is using to harass, intimidate, sue—and in some cases literally bankrupt—American farmers." The

Public Patent Foundation notes in the press release "Monsanto Patents Asserted against American Farmers Rejected by Patent Office" (July 24, 2007, http://www.pubpat.org/monsantorejections.htm) that in July 2007 the Patent and Trademark Office issued complete rejections of the four patents.

According to the CFS, in *Monsanto vs. U.S. Farmers, November 2007 Update* (November 2007, http://www.centerforfoodsafety.org/pubs/Monsanto%20November%202007%20update.pdf), by October 2007 Monsanto had filed 112 lawsuits against farmers, alleging violations of its technology agreements or patents on GE seeds. Of these lawsuits, 57 ended with recorded damages totaling $21.6 million awarded to Monsanto; 24 ended in confidential settlements, meaning the damages awarded were unrecorded; 13 lawsuits were dismissed; and 18 were ongoing. The lawsuits involved 372 farmers and 49 small farm businesses and the average judgment was $385,418. The CFS estimates that farmers have paid as much as $160.6 million in settlements because of these seed piracy lawsuits.

SUPREME COURT RULES ON MONSANTO ROUNDUP READY ALFALFA SEEDS. In June 2010 the U.S. Supreme Court issued its first ruling on GM crops. The case involved the Monsanto GM product Roundup Ready alfalfa, which entered the market in 2005. In 2006 the CFS filed a federal lawsuit against Monsanto, asking that the sale and planting of the crop be suspended until an environmental impact study (EIS) could be completed to determine how the GM crop would affect nearby crops. In May 2007 a U.S. district court issued an injunction to stop the sale and planting of Roundup Ready alfalfa until the EIS report was completed. Monsanto appealed this decision to the U.S. Court of Appeals for the Ninth Circuit and the court of appeals ruled in September 2008 against Monsanto. Monsanto filed a petition with the U.S. Supreme Court in October 2009, claiming that the injunction imposed "unnecessary restrictions and costs on alfalfa hay and seed growers." In January 2010 the Supreme Court agreed to hear the case and in June Justice Samuel A. Alito Jr. (1950–) delivered the opinion of the court.

In *Monsanto Company v. Geertson Seed Farms* (No. 09-475), the court ruled 7–1 to reverse the lower court's ruling that suspended the planting of the Roundup Ready alfalfa. While at first glance this decision appears to be a clear win for Monsanto, the company did not win completely. The court only reversed the injunction; it upheld the balance of the lower court's ruling, meaning that the GM alfalfa will be subject to scrutiny and regulatory efforts before it can be planted commercially. Furthermore, the justices concurred that GM crops have the potential to harm the environment via cross-pollination. This observation may have an impact on future court decisions involving GM crops.

Pollack reports in "After Growth, Fortunes Turn for Monsanto" (*New York Times*, October 4, 2010) that Monsanto suffered another setback in September 2010, when its newest product—corn with eight inserted genes—failed to produce higher yields than the company's earlier, less costly GM corn, which contained just three inserted genes. Furthermore, the company's herbicide, Roundup, was experiencing a decrease in sales because of strong competition from generic, lower-cost versions that were made in China and was facing mounting pressure from its principal competitor, DuPont. The company's challenges were reflected in its stock, which plummeted 42% between January and October 2010.

In January 2011 the USDA announced that it was fully deregulating Roundup Ready alfalfa, paving the way for unrestricted planting. Organic farmers and others opposing the decision opined that by approving the GM alfalfa, the USDA had opened the door to unrestricted growing of other GM crops such as corn, soybeans, and cotton. Pollack reports in "U.S. Approves Genetically Modified Alfalfa" (*New York Times*, January 27, 2011) that Kimbrell asserted that the USDA had "caved to pressure from the biotech industry and Monsanto" and vowed to fight the ruling in court.

U.S. BIOTECHNOLOGY REGULATORY SYSTEM

The U.S. government operates a rigorous, coordinated regulatory process for determining the safety of agricultural products of modern biotechnology. The process ensures that all biotechnology products that are commercially grown, processed, sold, and consumed are as safe as their conventional counterparts. The U.S. Department of Agriculture (USDA) describes in *A Description of the U.S. Food Safety System* (March 3, 2000, http://www.fsis.usda.gov/OA/codex/system.htm) the government regulatory system as "transparent, science-based decision-making, and public participation." Within the USDA the agencies responsible for regulation are the Animal and Plant Health Inspection Service (APHIS) and the U.S. Food and Drug Administration (FDA). The U.S. Environmental Protection Agency (EPA), which is not a part of the USDA, also plays a role in regulating biotechnology. Policies, processes, and regulations are continuously reviewed, evaluated, and, when necessary, revised to meet the challenges of this evolving technology.

APHIS oversees U.S. agriculture, protecting against pests and diseases. It is the lead agency that regulates the field-testing of biotechnology-derived new plant varieties and certain microorganisms. As such, APHIS grants approval and licenses for veterinary biological substances including animal vaccines that may be the products of biotechnology.

The FDA ensures that foods derived from new bioengineered plant varieties are safe and nutritious—they

TABLE 10.1

Bioengineered foods approved by the U.S. Food and Drug Administration, 2005–10

BNF No.	Food	Intended effect	Designation	Letter date
119	Soybean	Resistance to lepidopteran insects	MON 87701 (MON-877Ø-2)	08/18/2010
112	Cotton	Resistance to lepidopteran insects	COT67B	02/13/2009
101	Plum	Resistance to plum pox virus	C5	01/16/2009
110	Soybean	Increased levels of monounsaturated fatty acid (oleic) and decreased levels of polyunsaturated fatty acids (linoleic and linolenic)	Event 305423 (DP-3Ø5423-1)	01/15/2009
100	Papaya	Resistance to papaya ringspot virus	X17-2	12/24/2008
113	Corn	Resistance to lepidopteran insects	MIR162	12/09/2008
109	Cotton	Tolerance to the herbicide glyphosate	GHB614	09/29/2008
111	Corn	Tolerance to the herbicide glyphosate	Event 98140	09/09/2008
108	Soybean	Tolerance to the herbicide glyphosate	Event 356043	09/21/2007
107	Corn	Resistance to lepidopteran insects	MON 89034	08/08/2007
95	Corn	Alpha-amylase expression	Event 3272	08/07/2007
99	Corn	Resistance to corn rootworm	MIR604	01/30/2007
104	Soybean	Tolerance to the herbicide glyphosate	MON 89788 (MON-89788-1)	01/19/2007
87	Corn	Increased lysine level for use in animal feed	REN-ØØØ38-3 or Maize Event LY038	10/05/2005
94	Cotton	Resistance to lepidopteran insects	Transformation Event COT102	07/08/2005
98	Cotton	Tolerance to the herbicide glyphosate	MON-88913-8	03/07/2005
97	Corn	Resistance to corn rootworm	MON 88017	01/12/2005

BNF = Biotechnology Notification File.

SOURCE: Adapted from "The FDA List of Completed Consultations on Bioengineered Foods," U.S. Food and Drug Administration, September 2010, http://www.accessdata.fda.gov/scripts/fcn/fcnNavigation.cfm?rpt=bioListing&displayAll=true (accessed October 15, 2010)

must meet or exceed the same high standards of safety that are applied to any food product. The FDA is also responsible for issuing and enforcing regulations to guarantee that all food and feed labels, including those related to biotechnology, are truthful and do not mislead consumers. Table 10.1 lists the genes, gene fragments, and GM products (with their intended effects) that were submitted to and approved by the FDA between 2005 and 2010.

The EPA approves new herbicidal and pesticidal substances. It issues permits for testing herbicides and biotechnology-derived plants that contain new pesticides. When the EPA makes a determination about whether to register a new pesticide, it considers human safety, environmental impact, its effectiveness on the target pest, and any consequences for other, nontarget species. The EPA establishes and enforces the guidelines that ensure the safe use of GE products that are classified as pesticides.

The Institute of Medicine Issues Reports

In 2004 the Institute of Medicine (IOM) of the National Academy of Sciences, an organization that was created by Congress in 1863 to advise the government about scientific and technical matters, published *Safety of Genetically Engineered Foods: Approaches to Assessing Unintended Health Effects* (http://books.nap.edu/openbook.php?record_id=10977). The USDA, the FDA, and the EPA commissioned the National Academy of Sciences to assess the potential for adverse health effects from GE foods compared with foods altered in other ways and to provide guidance on how to identify and evaluate the likelihood of these effects. Even though adverse health effects from GE foods have not been detected in the human population, the technique is relatively new and concerns about its safety remain.

The IOM urges federal agencies to assess the safety of genetically altered foods on a case-by-case basis to determine whether unintended changes in their composition have the potential to adversely affect human health. It also calls for greater scrutiny of foods containing new compounds or unusual amounts of naturally occurring substances.

The IOM presents a framework to guide federal agencies in selecting the course and intensity of safety assessment. A new GM food whose composition is similar to a commonly used conventional version may warrant little or no additional safety evaluation. If, however, an unknown substance has been detected in a food, more detailed analyses should be conducted to determine whether an allergen or toxin may be present. Similarly, foods with nutrient levels that fall outside the normal range should be assessed for their potential impact on consumers' diets and health.

The IOM is also charged with examining the safety of foods from cloned animals. It recommends that the safety evaluation of these foods should focus on the product itself rather than on the process that is used to create it, and advises that the evaluations compare foods from cloned animals with comparable food products from noncloned animals. Even though there is no evidence that foods from cloned animals pose an increased risk to consumers, the IOM cautions that cloned animals engineered to produce pharmaceuticals should not be permitted to enter the food chain.

In December 2008 the IOM held a workshop in which 10 authorities on food nanotechnology discussed the use of nanotechnology and nanomaterials in food products and the education of consumers about the application of nanotechnology in food products. In *Nanotechnology in Food Products: Workshop Summary* (2009, http://books.nap.edu/openbook.php?record_id=12633), the IOM describes the term *food nanotechnology* as "an emerging technology that enables researchers to manipulate matter at the atomic level ... to enhance food safety and make foods more nutritious and satisfying by enhancing their nutrition content and other characteristics." The workshop focused on the:

- Application of nanotechnology to food products
- Safety and efficacy of nanomaterials used with or in food products
- Education of consumers about the applications of nanotechnology to food products

The IOM summarizes the workshop findings in *Nanotechnology in Food Products*, which includes a review of the FDA's regulatory oversight of food products.

In June 2010 the IOM issued the consensus report *Enhancing Food Safety: The Role of the Food and Drug Administration* (http://books.nap.edu/openbook.php?record _id=12892) in response to a request from Congress to assess current U.S. food safety practices and recommend strategies to improve food safety. The IOM concludes that "the FDA lacks a comprehensive vision for food safety and ... should change its approach in order to properly protect the nation's food." Even though the bulk of the report focuses on efforts to prevent foodborne diseases and does not directly address GM foods, the IOM's recommendations include the development of a risk-based food safety system that can also be applied to GM food products. The IOM recommends the adoption of a risk-based approach that uses data and expertise to identify areas with the greatest potential for contamination along the food production, distribution, and handling chains. It calls for strengthening the FDA's capacity to gather, manage, and use data and reassessing its policies for sharing data with other agencies and organizations. The IOM also suggests increasing the FDA's authority by amending the Federal Food, Drug, and Cosmetic Act to strengthen the agency's role in preventive controls, risk-based inspection, mandatory recalls, and food import bans.

The FDA Institutes a New Regulatory Approach

In October 2010 the FDA issued the report *Advancing Regulatory Science for Public Health* (http://www.fda.gov/downloads/ScienceResearch/SpecialTopics/RegulatoryScience/UCM228444.pdf), which contained a summary of the

agency's efforts to protect the U.S. food supply. The FDA describes its intent to devote a substantial portion of its fiscal year 2011 budget to "Science to Address Emerging Technologies in FDA-regulated Products (eg: Nanotechnology and Expertise to Regulate New Animal Biotech Products)."

Criticisms of the U.S. Regulatory Approach

Opponents of GM foods do not believe there are sufficient government regulations in place to control U.S. production and distribution of these foods. They argue there has not been enough research or long-term experience with these foods, and as a result the health consequences of growing and eating such foods as well as the environmental impact are not yet known. Proponents claim the benefits of transgenic foods—improved flavor, increased nutritional value, longer shelf life, and greater yields—surpass any potential risks. They also discount the health risks, observing that nearly half the soybean crop and a quarter of all corn grown in the United States consists of transgenic varieties, meaning that Americans have been consuming transgenic food products for years and, as of January 2011, there were no reports of adverse health effects as a result.

Since the late 1990s there have been protests staged to oppose the widespread use and consumption of GM foods as well as attacks on facilities conducting research on transgenic crops and companies marketing GM products. Protesters have dubbed the transgenic crops "Frankenfoods," likening them to Frankenstein's monster, a manmade freak that was created through science. Greenpeace, an organization that opposes the creation of GM foods, and other activists staged a historic protest at the World Trade Organization (WTO) meeting in Seattle, Washington, in November 1999. Greenpeace also hung an anti-GM banner on Kellogg's Cereal City USA museum in Battle Creek, Michigan, in 2000. It hopes that such actions will move U.S. lawmakers to require the labeling of transgenic foods. Greenpeace is inspired by the example of European consumers, who demanded and have been granted product labeling that enables them to choose whether to purchase and consume GM foods. On the website "GM Contamination Register" (http://www.gmcontaminationregister.org/), Greenpeace documents "incidents of contamination arising from the intentional or accidental release of genetically modified (GM) organisms (which are also known as genetically engineered (GE) organisms)" and "illegal plantings of GM crops and the negative agricultural side-effects that have been reported." For example, the register indicates that in July 2010 Finland received shipments from Canada that were found to contain unauthorized linseed.

In 2010 Greenpeace agitated against GM crops via the video *Genetic Engineering: The World's Greatest*

Scam? (http://www.greenpeace.org/usa/en/multimedia/videos/Genetic-engineering-The-worlds-greatest-scam-/), in which the organization opines that "genetic engineering is a threat to food security, especially in a changing climate." The organization also advocates a worldwide ban on the release of any transgenic crop or seed and exhorts governments to halt both commercial and experimental growing of GE crops.

In August 2006 a federal court issued in *Center for Food Safety et al. v. Johanns et al.* (Civil No. 03-00621 JMS/LEK) the first ruling ever on molecular farming, the controversial practice of genetically altering food crops to produce experimental drugs and industrial compounds. The court ruled that the USDA violated the Endangered Species Act when it permitted the cultivation of drug-producing GE crops in Hawaii. That same month the CFS reported in the press release "Unapproved, Genetically Engineered Rice Found in Food Supply" (August 18, 2006, http://www.centerforfoodsafety.org/GE_Rice_8.18.06.cfm) that the USDA announced that an unapproved GE rice was found contaminating commercial long-grain rice supplies. The presence of GE rice, which was genetically altered to survive application of the powerful herbicide glufosinate, in the food supply is illegal because it has not undergone USDA review for potential environmental impacts and FDA review for possible harm to human health.

In September 2006 the CFS (http://www.centerforfoodsafety.org/PR9_14_06.cfm) filed a petition with the USDA to prevent after-the-fact approval of the illegal GE rice found in the world's food supply the previous month. The CFS (http://www.centerforfoodsafety.org/cloning_petitionPR10.12.06.cfm) and reproductive rights, animal welfare, and consumer protection organizations filed a petition with the FDA in October 2006 calling for a moratorium on the introduction of food products from cloned animals. The petition also asked the FDA to establish mandatory rules for premarket food safety and environmental review of cloned foods.

In January 2008 the CFS (http://www.centerforfoodsafety.org/CloningPR01_15_08.cfm) and other food industry, consumer, and animal welfare groups condemned the FDA's determination that milk and meat from cloned animals are safe for sale to the public and the administration's decision to allow the sale of such products without labeling that identified them as produced by cloning. In September 2008 the CFS announced in the press release "20 Leading Food Companies and Retailers Reject Ingredients from Cloned Animals in Their Products" (http://www.centerforfoodsafety.org/CloningPR9_3_08.cfm) that 20 leading U.S. food producers and retailers vowed not to use cloned animals in their foods. The CFS credited the companies' rejection of cloned animal products to "a growing industry trend of responding to consumer demand

for better food safety, environmental, and animal welfare standards." Lisa Bunin of the CFS asserted, "This rejection of food from clones sends a strong message to biotech firms that their products may not find a market. American consumers don't want to eat food from clones or their offspring, and these companies have realistically anticipated low market acceptance for this new and untested technology."

Jenifer Goodwin notes in "FDA Advisers Weigh Approval of Genetically Modified Salmon" (*Bloomsberg BusinessWeek*, September 20, 2010) that in September 2010 the CFS exhorted consumers to protest FDA approval of GE salmon, which if permitted would be the first GE animal product approved for human consumption. Even though the FDA Veterinary Advisory Committee indicates that the GM salmon "contained the same amount of nutrients and had 'no biologically relevant differences' from ordinary farmed Atlantic salmon," opponents fear that consuming the GM salmon might compromise human health and that approval of the salmon would open a floodgate of approvals for other GM livestock such as pigs.

U.S. Consumers' Opinions about Biotech Foods

Every year the International Food Information Council Foundation conducts a survey that tracks trends on public awareness and perceptions of various aspects of plant and animal biotechnology, confidence in the safety of the U.S. food supply, and U.S. consumer attitudes toward food labeling. The foundation notes in *Consumer Perceptions of Food Technology Survey* (August 26, 2010, http://www.foodinsight.org/Resources/Detail.aspx?topic=2010_Consumer_Perceptions_of_Food_Technology_Survey) that in 2010 consumers were fairly familiar with the term *biotechnology*. Seven out of 10 (69%) respondents said they had read or heard at least "a little" about the concept of biotech food. Almost one-third (32%) felt favorably toward plant biotechnology, and more than a quarter (29%) neither favored nor condemned biotech foods. About one-fifth (19%) said they feel somewhat or very unfavorably toward biotech foods.

The International Food Information Council Foundation also finds that consumer awareness of the presence of biotech foods in the supermarket has increased, from 23% in 2008 to 28% in 2010. One interesting finding is that when biotechnology is presented as a way to increase crops to meet food demands, approval rates rise, with more than half (51%) of the respondents saying they support farmers' use of biotechnology. The survey asked consumers how likely they would be to purchase biotech wheat products such as bread, crackers, cookies, cereal, or pasta products containing wheat that was grown using plant biotechnology, if the products were produced using "sustainable practices to feed more people using less resources (such as land and pesticides)." Twenty-six percent of the survey respondents said they are "very likely," 54% said they are "somewhat likely," 14% said they are "not too likely," and 5% said they are "not at all likely" to purchase biotech wheat products.

The majority of respondents said they are somewhat or very likely to buy biotech foods that offer health benefits such as higher levels of healthy fats (76%), have reduced trans fats (74%), and taste better/fresher (67%). More than three-quarters (77%) said they are likely to purchase biotech foods to reduce pesticide use, and 73% said they are likely to purchase biotech wheat products to conserve land and water and reduce the use of pesticides.

Comparable proportions of respondents felt somewhat or very favorable (29%) toward animal biotechnology, 24% were neither favorable nor unfavorable, and 27% were somewhat or very unfavorably inclined toward it. The lack of information may influence consumers' feelings. More than half (55%) of respondents who indicated unfavorable feelings selected "I don't have enough information" about animal biotechnology as their primary reason, and 39% chose "I don't understand the benefits of using biotechnology with animals."

Improved understanding of animal biotechnology increased favorable ratings. For example, when it was explained that "animal biotechnology can increase farm efficiency; that is, increase the amount of food produced while decreasing the amount of resources needed, such as animal feed (i.e., corn, water, etc.)," 53% of the survey respondents said this information had a positive impact on their impression. Similarly, about two-thirds (65%) of respondents reacted positively to the statement that "animal biotechnology can improve the quality and safety of our food (for example, through improved animal health or improved nutritional quality of the food produced)."

When genomics was defined for survey respondents as "a form of animal biotechnology that uses knowledge about the genetic makeup of farm animals to aid in producing better offspring for improved meat, milk, and egg quality," 44% of respondents gave "very favorable" or "somewhat favorable" ratings to it. A comparable proportion (41%) reported a "very favorable" or "somewhat favorable" impression of genetic engineering when it was described as "a form of animal biotechnology that allows for the transfer of beneficial traits from one animal to another in a precise way that allows for improved nutritional content or less environmental impact."

According to the International Food Information Council Foundation, U.S. consumers appear to be satisfied with the information currently available on food labels. The overwhelming majority (82%) cannot think of any additional information they would like on the label. Just 18% said they would like to see additional information on the label and a scant 3% named information about biotechnology. Furthermore, about two-thirds (63%) of survey respondents expressed satisfaction with FDA labeling of biotech foods, which only introduces labeling changes in the event that the use of biotechnology changes the nutritional content or composition of a food. Trust in the FDA is high, as reflected by the finding that 68% of respondents indicated a willingness to purchase meat, milk, and eggs from biotech animals because the FDA deems them safe.

AN INTERNATIONAL FOOD FIGHT

Cartagena Protocol on Biosafety

In January 2000, after participating in the negotiation of the International BioSafety Protocol, a treaty developed by the CBD in which all signatories concurred that GM crops are significantly different from traditional crops, the United States refused to join other countries in signing. Known as the Cartagena Protocol on Biosafety (http://www.cbd.int/biosafety/), the treaty aims to ensure the safe transfer, handling, and use of living modified organisms—such as GE plants, animals, and microbes—across international borders. The protocol is also intended to prevent adverse effects on the conservation and sustainable use of biodiversity without unnecessarily disrupting world food trade. By participating in the treaty negotiations, the U.S. government formally acknowledged that GM crops are not, as many U.S. federal agencies had previously maintained, substantially equivalent to traditional crops.

The Cartagena Protocol on Biosafety provides countries the opportunity to obtain information before new bioengineered organisms are imported. It acknowledges each country's right to regulate bioengineered organisms, subject to existing international obligations. It also aims to improve the capacity of developing countries to protect biodiversity. The protocol does not, however, cover processed food products or address the safety of GM food for consumption. Instead, the protocol is intended to protect the environment from the potential effects of introducing bioengineered products (referred to in the treaty as living modified organisms).

In September 2003 the Cartagena Protocol on Biosafety entered into force, empowering the more than 100 countries that had ratified the protocol to bar imports of live GM organisms that they believe carry environmental or health risks. If the government of the importing country has concerns about safety, it can ask the exporting country to provide a risk assessment. The protocol also established a central online database of information on the potential risks of several types of GM organisms.

Even though the protocol was acknowledged as an important first step in the protection of biodiversity, environmental advocacy groups including Greenpeace cautioned that there are still important unresolved issues for the international community, such as the amount of information that is required on shipments of GM crops. More important, Stephen Leahy reports in "Bioterror Fears Dim Biotech Potential" (February 28, 2006, http://ipsnews.net/news.asp?idnews=32325) that in 2006 Argentina, Canada, and the United States—the countries that produce about 90% of the GE crops in the world—had not ratified the protocol. As of January 2011, these three countries had still not ratified the protocol.

In 2009 the CBD conducted a survey to obtain information about participating countries' socioeconomic experience and considerations arising from implementation of the protocol and published the results in *Summary of Results of the Survey on the Application of and Experience in the Use of Socio-economic Considerations in Decision-Making on Living Modified Organisms* (January 31, 2010, http://www.cbd.int/doc/meetings/bs/bscmcb-06/information/bscmcb-06-inf-02-en.pdf). Besides financial challenges, the countries named "political challenges, political will, inconsistency between organizations, coordination between government departments, public concern about the environment, World Trade Organization regulations and distributional challenges" as key socioeconomic considerations.

The CBD indicates in the communiqué "Somalia Becomes the 160th Party to the Cartagena Protocol on Biosafety" (2010, http://www.cbd.int/doc/press/2010/pr-2010-08-06-somalia-en.pdf) that as of August 2010, 160 countries had ratified the protocol and that many of the protocol's signatories were embarking on developing legal, administrative, and other measures to facilitate its implementation. According to the CBD, in the press release "A United Nations Conference on the Safe Use of Modern Biotechnology Opens Tomorrow" (October 10, 2010, http://www.cbd.int/doc/press/2010/pr-2010-10-10-pre-mop-en.pdf), the governing body of the Cartagena Protocol on Biosafety met in October 2010 to "discuss and adopt further decisions to ensure the safe transfer, handling and use of living modified organisms (LMOs) resulting from modern biotechnology." The article "Can the Cartagena Protocol Help Us Make Better Choices about GMOs?" (GENET-news, October 15, 2010) explains that conference participants from developing countries expressed the concern that the lack of money prevents them from implementing the protocol. As of October 2010, 90 countries were awaiting funds to institute their biosafety programs.

Europe Bans Biotech Foods and Halts U.S. Trade

U.S. supermarket shelves are stocked with GM foods, but historically European consumers and farmers have nearly driven GM foods and crops out of the EU market, which is the largest market in the world. Europeans were so wary of GM foods that, beginning in 1998, the EU's European Commission and member states unofficially began a suspension on imports of biotech foods. At the onset of the ban, European Commission officials told the United States they would resume the approval process once companies submitting applications agreed to abide by newly proposed revisions to the approval procedures before they become law. Even though applicant companies agreed to comply voluntarily with the new procedures, the European Commission approval process did not resume as promised.

In June 1999 EU members called for an official moratorium on new approvals of agricultural biotech products. The EU Environmental Council contended that the administration of new approvals should be linked to new labeling rules for biotech foods. Ministers from Denmark, France, Greece, Italy, and Luxembourg vowed to suspend approvals until new rules were established. A year later the EU environmental ministers moved to continue the moratorium at least until the European Commission prepared proposals for labeling and for tracing minute amounts of biotech products in foods such as vegetable and corn oils. The European Commission assured the United States that it would develop its proposals by the end of 2000 and promptly resume the approval process.

The labeling and traceability requirements were not presented until July 2001, when the European Commission promised to lift the moratorium within weeks. Frustrated by the many delays and effects of these sanctions, the USDA determined that the EU moratorium on agricultural biotech products violated international law. The USDA argued in the fact sheet "Agricultural Biotechnology: WTO Case on Biotechnology" (September 2006, http://www.fas.usda.gov/itp/wto/eubiotech/factsheets/2006-09-28-biotech-wtocase.pdf) that:

- The EU "has pursued policies that undermine the development and use of agricultural biotechnology."

- During the late 1990s six member states (Austria, France, Germany, Greece, Italy, and Luxembourg) violated European law when they "banned imports of biotech corn and rapeseed approved by the European Union" and the European Commission did not challenge the bans.

- In 1998 member states began to block EU regulatory approval for new agricultural biotech products. The moratorium prohibited most U.S. corn exports from entering Europe and violated EU law.

- The ban also breached WTO rules, which do not require automatic approval of biotech foods. The WTO requires that measures regulating imports be based on "scientific principles and evidence" and that countries operate regulatory approval procedures without "undue delay."

The USDA asked the EU to apply a scientific, rules-based review and approval process to agricultural biotech product applications. Furthermore, the USDA contended that it was not attempting to force European consumers to accept biotech foods or products. Instead, it denounced Europe's actions, which it considered whimsical rather than based on scientific, health, or environmental evidence, as depriving consumers of choice.

In 2004 Europe's moratorium on GM foods came to an end when the European Commission agreed to import worm-resistant GM corn known as BT-11 that was developed by the Swiss firm Syngenta. In 2006 the WTO condemned the long approval process for GE products entering the European market as unscientific and determined that the process amounted to a trade embargo against agriculture producers in the United States, Canada, and Argentina. EU officials declined to revise the approval process, and European consumers continued to view GE products with suspicion. However, by 2008 there were signs that the EU might soften its stance on GM crops.

In "Europe Warms to GM Crops as Possible Solution to Food Crisis" (*Independent* [London, England], June 21, 2008), Andrew Grice and Vanessa Mock report that in June 2008 the EU began studying to determine if significant expansion of GM crops will curb escalating global food prices. The research was prompted by concern that blocking GM food production will result in higher food prices in Europe and in the rest of the world.

Leo Cendrowicz notes in "Is Europe Finally Ready for Genetically Modified Foods?" (*Time*, March 9, 2010) that beginning in 2010 the EU permitted European farmers to grow GM potatoes. The GM potatoes are intended for use in industries such as paper manufacturing. Supporters of this move observe that GM potatoes are designed to require less water, energy, and chemical pesticides. Detractors are concerned that because the GM potatoes harbor a gene that confers resistance to specific antibiotics, this resistance might spread beyond the potatoes to threaten animal and human health.

Cendrowicz explains that the EU imports GM crops, such as soybeans, but requires that they be labeled and separated to prevent contamination of other crops. For example, produce that has a GM content of over 0.9% must be labeled, and this policy has resulted in shipments being returned to the exporting country. Regardless, some observers believe the EU is softening its stance on GM foods in the face of rising food prices and increasing global demand.

Molecular Farming Harvests New Drugs

Molecular farming (also known as phytomanufacturing or biopharming) is the production of pharmaceutical proteins or other materials in GE plants and animals. It runs the gamut from tobacco plants harboring drugs to treat acquired immunodeficiency syndrome (AIDS) to plants intended to yield fruit-based hepatitis vaccines. One example of a plant-based expression system with the potential to produce a GM pharmaceutical protein is corn grown by Epicyte Pharmaceutical in San Diego, California. According to David Liénard et al., in "Pharming and Transgenic Plants" (*Biotechnology Annual Review*, vol. 13, 2007), a human gene that codes for an antibody to genital herpes (a sexually transmitted disease that affects about 60 million Americans) is being grown in the corn plants. By January 2011 a wide range of plant-derived biotech drugs, such as Enbrel (an anti-inflammatory agent that is used to treat arthritis), were being prescribed in the United States.

Opponents call molecular farming "Pharmageddon," and environmentalists fear that the artificially combined genes will have unintended, untoward consequences for the environment. Consumer advocates, wary about the proliferation of GM foods, dread the possibility that plant-grown drugs and industrial chemicals will end up in food crops. To prevent such a scenario, the FDA issued new regulations to safeguard the food supply in 2003. Suzanne White Junod notes in "Celebrating a Milestone: FDA's Approval of First Genetically-Engineered Product" (April 10, 2009, http://www.fda.gov/AboutFDA/WhatWeDo/History/ProductRegulation/SelectionsFromFDLIUpdateSeriesonFDAHistory/ucm081964.htm) that in 2007 the FDA celebrated the 25th anniversary of the agency's first approval of a recombinant DNA drug product: human insulin. This celebration was not only intended to highlight the beneficial uses of GM products but also to help counter the fears expressed by consumer advocates and dispel any existing consumer discomfort with them.

In "Drawing Bright Lines: Food and the Futures of Biopharming" (*Sociological Review*, vol. 58, no. S1, May 2010), Richard Milne describes biopharming as "re-imagining [a] biotechnological promise." Milne observes that this technology "represents a coming together of medical and agricultural applications of biotechnology" and that for its supporters it "recombines not only plant and human biologies, but also the promise of past applications."

THE PROMISE AND PROGRESS OF GENOMIC MEDICINE

Progress in understanding the genetic basis of disease has arrived at a rapid-fire pace. Genetic and genomic information gained from the Human Genome Project promises to revolutionize prevention and treatment of disease in the 21st century. Physicians are increasingly able to accurately predict patients' risks of acquiring specific diseases and advise them of actions they may take to reduce their risks, prevent disease, and protect their health. There are promising therapeutic applications of genetic research including custom-tailored treatment that relies on knowledge of the patient's genetic profile and development of highly specific and effective medications to combat diseases.

According to Zsofia K. Stadler et al., in "Genome-wide Association Studies of Cancer: Principles and Potential Utility" (*Oncology*, vol. 24, no. 7, June 2010), the genomics revolution has already arrived in terms of investigating the genetic basis of complex diseases. The researchers observe that genome-wide studies of nearly all common malignancies have been performed and that more than 100 genetic variants associated with increased risk for cancer have been identified as well as common genetic variations that contribute to susceptibility to more than 40 other diseases, including heart disease, diabetes, asthma, psychiatric disorders, and inflammatory bowel disease. Identification of new genetic risk variants has raised the expectation that these markers may prove useful for disease prevention, increase scientists' understanding of disease processes, and improve treatment. Even though industry observers hope that genomic research will catalyze a shift from reactive patient care to predictive, preventive, and personalized care, by 2010 this goal had not yet been realized. Stadler et al. point out that there are still scientific obstacles to overcome before these genomic data can be effectively translated into improved patient care.

Stadler et al. also caution against the commercialization and direct-to-consumer marketing of genetic tests with uncertain clinical utility, such as genomic risk assessments. They worry that increased risk scores with unproven clinical relevance may produce unwarranted anxiety and that reduced risk scores may offer false reassurance about disease risk, which in turn may prompt patients to ignore medical advice, such as age-appropriate screening recommendations, because their test results appear to place them at lower risk. Stadler et al. assert that genetic tests should be performed in environments where pre- and posttest counseling by experienced health care professionals is available.

Preclinical Disease Detection

By the close of the 20th century technological advances offered opportunities to identify diseases at stages before they were visible, in terms of biochemical or symptomatic expression. Called preclinical detection, this ability to predict and as a result intervene to prevent or avert serious disease requires an understanding of three levels of disease detection involving genomes, transcriptomes (the transcribed messenger ribonucleic acid [RNA] complement), and proteomes (the full range of translated proteins).

In "Molecular Markers for Early Detection" (*Seminars in Oncology*, vol. 37, no. 3, June 2010), Barbara K. Dunn et al. report on the identification of molecular markers that improve the ability to predict and detect

cancer before it develops and at the earliest signs of impending carcinogenic transformation. Of the many molecular markers in development, an early detection marker with acceptable sensitivity (the ability to identify people who have a disease) and specificity (the ability to identify people who do not have a disease) is available for bladder cancer. Dunn et al. opine that there will likely be comparable markers for other cancers that will become available in the coming decade.

Genes Explain Disease, Drug Resistance, and Treatment Failures and Successes

The discovery of predictive genes and proteins that reflect sensitivity and/or resistance to treatment and genes that predict the risk of relapse has personalized the treatment of certain cancers, such as colorectal cancer. Jeffrey S. Rosset et al. note in "Biomarker-Based Prediction of Response to Therapy for Colorectal Cancer" (*American Journal of Clinical Pathology*, vol. 134, no. 3, September 2010) that new biomarkers have been proposed as specific predictors of how patients will respond to cancer chemotherapy and radiotherapy. In addition, other markers can successfully predict toxic effects and other responses to standard therapies.

In "Resistance May Not Be Futile: MicroRNA Biomarkers for Chemoresistance and Potential Therapeutics" (*Molecular Cancer Therapeutics*, vol. 9, no. 12, December 2010), Kristi E. Allen and Glen J. Weiss observe a set of microRNAs (small RNA molecules in the genome that regulate the expression of genes by binding to specific regions of messenger RNAs) that regulate key genes, such as EGFR, MDR1, PTEN, Bak1, and PDCD4, that are linked to resistance and response to commonly used anticancer drugs. Allen and Weiss assert that because microRNAs act as both regulators of key target genes for chemoresistance and biomarkers for treatment response, they "hold much promise for the development of targeted therapies and personalized medicine."

Responses to treatment for depression, especially with antidepressant medication, vary widely. In "Biomarkers to Predict Antidepressant Response" (*Current Psychiatry Reports*, vol. 12, no. 6, December 2010), Andrew F. Leuchter et al. explain that predicting which antidepressant drug and specific form of psychotherapy would be most effective for each individual would prove to be very valuable for people suffering from depression. They observe that researchers have identified genetic and genomic biomarkers and measures that, once validated, may help clinicians and patients to choose between medication and psychotherapy, guide the selection of specific medication or the choice of a specific type of psychotherapy, and predict the outcome of treatment.

Similarly, George R. Uhl et al. list in "Genome-wide Association for Smoking Cessation Success in a Trial of

Precessation Nicotine Replacement" (*Molecular Medicine*, vol. 16, nos. 11–12, November–December 2010) several genetic variations that appear to distinguish successful quitters from people who are unable to quit smoking and that also predict the likelihood of success or failure of two different types of smoking cessation treatments: nicotine replacement therapy and drug treatment with bupropion. Uhl et al. assert that their findings support the premise that smoking cessation treatment may be personalized based on genetic predictors of outcome.

Gene Therapy

Gene therapy aims to correct defective or faulty genes with one of several techniques. Most gene therapy involves the insertion of functioning genes into the genome to replace nonfunctioning genes. Other techniques entail swapping an abnormal gene for a normal one in a process known as homologous recombination, restoring an abnormal gene to normal function through selective reverse mutation, or changing the regulation of a gene by influencing the extent to which the gene is "turned on" or "turned off."

According to the National Cancer Institute, in "Gene Therapy for Cancer: Questions and Answers" (August 31, 2006, http://www.cancer.gov/cancertopics/factsheet/Therapy/gene), the first disease approved for gene therapy was adenosine deaminase (ADA) deficiency, a rare genetic disease in which children do not produce ADA, an enzyme that is necessary for immune function. ADA deficiency was chosen because, if left untreated, it is fatal in the first year of life, it is caused by a defect in a single gene that is relatively simply regulated, and the amount of ADA present is not critical—small quantities of the enzyme are helpful and larger quantities are tolerated with no ill effects. In "Bone Marrow Transplantation and Alternatives for Adenosine Deaminase Deficiency" (*Immunology and Allergy Clinics of North America*, vol. 30, no. 2, May 2010), H. Bobby Gaspar of the UCL Institute of Child Health in London, England, explains that there are three accepted courses of treatment. Hematopoietic stem cell transplantation is the treatment of choice, but it generally requires a fully matched sibling or family donor. ADA deficiency may also be treated by enzyme replacement therapy, which is a costly, life-long treatment by injection, and by stem cell gene therapy. The advantage of gene therapy over transplantation when there is no fully matched donor is that it avoids the potential complications of graft-versus-host disease (a common condition that results when donated cells or tissue attack the tissue of the transplant patient; the symptoms of graft-versus-host disease include a skin rash or blisters, jaundice, dry mouth, or dry eyes). Gaspar opines that "as further developments in transplant and GT [gene therapy] occur, it is likely that safer and more effective therapies will become available."

Early work in gene therapy focused on replacing a gene that was defective in a specific well-defined genetic disease such as cystic fibrosis. According to Clinicaltrials.gov (October 2010, http://www.clinicaltrials.gov/ct2/results?term=gene+therapy), a service of the National Institutes of Health (NIH), at the close of 2010 gene therapy clinical trials were under way for a variety of cancers (including liver and prostate cancer), for thyroid and brain tumors, as well as for Alzheimer's disease, severe angina pectoris (chest pain arising from heart disease), and Parkinson's disease.

To treat each of these disorders, the appropriate therapeutic genes are inserted by using in vivo (in a living organism, rather than in the laboratory) gene therapy with adenoviral vectors. (Adenoviruses have double-stranded DNA genomes and cause respiratory, intestinal, and eye infections in humans.) A vector is a carrier molecule that is used to deliver the therapeutic gene to the target cells. The most frequently used vectors are viruses that have been genetically altered to contain normal human DNA. Researchers exploit viruses' natural abilities to encapsulate and deliver their genes to human cells. An unanticipated benefit of this type of gene therapy is that highly specific immunity can be induced by injecting into skeletal muscle DNA that is manipulated to carry a gene encoding a specific antigen. Because this form of therapy does not lead to integration of the donor gene into the host DNA, it eliminates potential problems that result from the disruption of the host's DNA, a phenomenon that was observed by investigators using ex vivo (outside a living organism, usually in the laboratory) gene therapy. Figure 10.3 shows how ex vivo cells are used to create a gene therapy product.

FIGURE 10.3

Complex manufacture of a gene therapy product

Ex vivo transduced CD34+ cells expressing gammaC-R for X-SCID

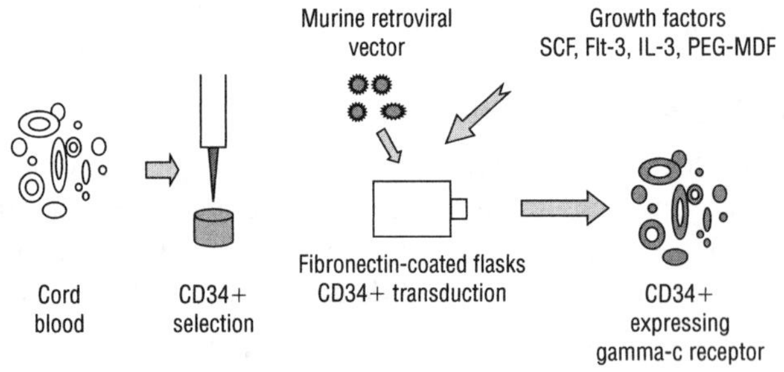

Murine retroviral vector

Growth factors SCF, Flt-3, IL-3, PEG-MDF

Cord blood

CD34+ selection

Fibronectin-coated flasks CD34+ transduction

CD34+ expressing gamma-c receptor

Note: X-SCID = X-linked severe combined immunodeficiency syndrome. This syndrome can cause death within the first year of life from severe recurrent infections.

SOURCE: Adapted from Philip D. Noguchi, "Slide 7. Complexity of (a) Gene Therapy Product," in *Simple Complexity in an Evolving World: Rising to the Challenge*, U.S. Food and Drug Administration Division of Dockets Management, Office of Cellular, Tissue, and Gene Therapies, October 2002, http://www.fda.gov/ohrms/dockets/ac/02/slides/3902s1-07-noguchi/sld007.htm (accessed September 30, 2008)

Figure 10.4 shows the steps that are involved in gene therapy using a retrovirus (a virus that has RNA rather than DNA as its genetic material and is able to become part of the host cell's DNA using an enzyme called reverse transcriptase) as the vector. In this process the vector discharges its DNA into the affected cells, which then begin to produce the missing or absent protein and are restored to their normal state. In this example the patient's own bone marrow cells are used as the vector to deliver severe combined immune deficiency–repaired genes to restore the function of the immune system.

Besides virus-mediated gene delivery, there are non-viral techniques for gene delivery. Nonviral carriers were developed to overcome concerns related to viral carriers' potential to trigger immune responses or to give rise to tumors. Direct introduction of therapeutic DNA into target cells requires large amounts of DNA and can be used only with certain tissues. Genes can also be delivered via an artificial lipid sphere, called a liposome, with a liquid core. This liposome, which contains the DNA, passes the DNA through the target cell's membrane. DNA can also enter target cells when it is chemically bound to a molecule

FIGURE 10.4

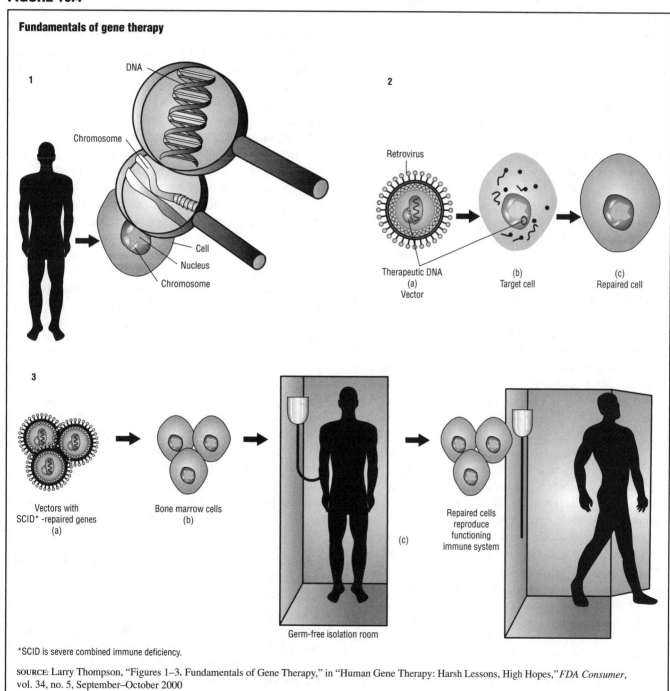

Fundamentals of gene therapy

1

DNA

Chromosome

Cell

Nucleus

Chromosome

2

Retrovirus

Therapeutic DNA
(a)
Vector

(b)
Target cell

(c)
Repaired cell

3

Vectors with
SCID* -repaired genes
(a)

Bone marrow cells
(b)

Germ-free isolation room

(c)

Repaired cells
reproduce
functioning
immune system

*SCID is severe combined immune deficiency.

SOURCE: Larry Thompson, "Figures 1–3, Fundamentals of Gene Therapy," in "Human Gene Therapy: Harsh Lessons, High Hopes," *FDA Consumer*, vol. 34, no. 5, September–October 2000

that in turn binds to special cell receptors. Once united with these receptors, the therapeutic DNA is engulfed by the cell membrane and enters the target cell.

In "Nonviral Gene Delivery: Principle, Limitations, and Recent Progress" (*American Association of Pharmaceutical Scientists Journal*, vol. 11, no. 4, December 2009), Mohammed S. Al-Dosari and Xiang Gao describe some additional advantages of nonviral vectors such as their ease of production and the potential for repeat administration. The researchers note that even though nonviral vectors overcome issues of immunogenicity or insertion mutations often resulting from viral vectors, they vary widely in their ability to effectively and efficiently convey genes. Physical nonviral gene delivery methods involve creating holes in the cell membrane using methods such as rapid injection, particle impact, electric pulse, or ultrasound. The holes or defects in the membrane are temporary but enable the transfer of genes from extracellular space to the nucleus. Chemical vectors, such as cationic lipids and cationic polymers, form complexes with negatively charged DNA. These complexes protect the DNA and facilitate its uptake and intracellular delivery. Other very tiny inorganic particles composed of metals or ceramics have been used to target gene delivery. These tiny particles, known as nanoparticles, are readily transported through cell membranes to the nucleus. Al-Dosari and Gao opine that a more effective approach might be to combine nonviral and viral vectors to produce highly effective nontoxic gene delivery systems.

In 1999 the death of the American teenager Jesse Gelsinger after he participated in a clinical gene therapy trial for ornithine transcarbamylase deficiency shocked and saddened the scientific community and diminished enthusiasm for technology that promises that healthy genes can replace faulty ones. However, gene therapy trials continued, some with successful outcomes. In "Scientists Use Gene Therapy to Cure Immune Deficient Child" (*British Medical Journal*, vol. 325, no. 7354, July 6, 2002), Judy Siegel-Itzkovich notes that in 2002 an international team of scientists reported curing a child with severe combined immunodeficiency using gene therapy. The child spent the first seven months of her life inside a plastic bubble to protect her from all disease-causing agents because she totally lacked an immune system. After suppressing her defective bone marrow cells, the researchers introduced, using a GE virus, a healthy copy of the gene she was missing into her purified bone marrow stem cells. The baby recovered quickly; within a few weeks she was no longer in isolation and went home healthy. Researchers credited the gene alteration treatment with the cure.

Human Genome Project Information notes in "Gene Therapy" (June 11, 2009, http://www.ornl.gov/sci/techre sources/Human_Genome/medicine/genetherapy.shtml) that since then other adverse outcomes have tempered hopes raised by such success. Two other children with similar conditions who received comparable gene therapy subsequently developed conditions resembling leukemia. As a result, in January 2003 the FDA temporarily stopped all gene therapy trials using retroviral vectors in blood stem cells. The following month the FDA reconvened its Biological Response Modifiers Advisory Committee to determine whether to permit retroviral gene therapy trials for the treatment of life-threatening diseases. The committee eventually eased the ban on gene therapy trials using viral vectors. However, as of January 2011 gene therapy in the United States remained limited to clinical trials and research.

In 2004 the Biological Response Modifiers Advisory Committee was renamed the Cellular, Tissue, and Gene Therapies Advisory Committee to describe more accurately the areas for which the committee is responsible. The FDA (October 26, 2010, http://www.fda.gov/Advisory Committees/CommitteesMeetingMaterials/BloodVaccines andOtherBiologics/CellularTissueandGeneTherapiesAdvisory Committee/default.htm) describes the function of the renamed committee as evaluating "data relating to the safety, effectiveness, and appropriate use of human cells, human tissues, gene transfer therapies and xenotransplantation products." Xenotransplantation is any procedure that involves transplantation, implantation, or infusion into a human recipient of live cells, tissues, or organs from a nonhuman animal source; or human body fluids, cells, tissues, or organs that have had any contact with live nonhuman animal cells, tissues, or organs. It has been used experimentally to treat certain diseases such as liver failure and diabetes, when there are insufficient human donor organs and tissues to meet demand.

The FDA and the NIH jointly oversee regulation of human gene therapy in the United States. The FDA focuses on ensuring that manufacturers produce quality, safe gene therapy products and that these products are adequately studied in human subjects. The NIH evaluates the quality of the science involved in human gene therapy research and funds the laboratory scientists who are involved in the development and refinement of gene transfer technology and clinical studies.

RECENT ADVANCES IN GENE THERAPY. In "Effect of Gene Therapy on Visual Function in Leber's Congenital Amaurosis" (*New England Journal of Medicine*, vol. 358, no. 21, May 22, 2008), James W. B. Bainbridge et al. report the promising results of the first clinical trial to test a new gene therapy treatment for Leber's congenital amaurosis (LCA), a form of inherited blindness that is caused by an abnormality in a gene called RPE65. The clinical trial involved inserting healthy copies of the RPE65 gene into the cells of the retina to help them regain normal functioning. A virus served as a vector to deliver the RPE65 gene to the retina and carry the gene into the cells. The clinical trial was designed to determine whether this new gene therapy for retinal disease was safe and to find out if it could improve the vision of young adults who already have advanced retinal disease. Bainbridge et al. indicate that the treatment proved safe and effective; it produced modest improvements in vision even in patients with advanced disease.

By 2010 gene therapy had been successfully used to treat eye diseases in three clinical trials. Kamolika Roy, Linda Stein, and Shalesh Kaushal note in "Ocular Gene Therapy: An Evaluation of Recombinant Adeno-Associated Virus-Mediated Gene Therapy Interventions for the Treatment of Ocular Disease" (*Human Gene Therapy*, vol. 21, no. 8, August 2010) that gene therapy provides an alternative to current treatments for eye diseases that are associated with high rates of complications, offer only short-term relief of symptoms, or do not effectively treat the loss of vision. In the first human trials, a recombinant adeno-associated viral (rAAV) vector was shown to be a safe, effective, and enduring treatment for LCA, which was previously considered untreatable. Roy, Stein, and Kaushal aver that this type of gene therapy may prove effective for a wide range of eye disorders.

MORE APPLICATIONS OF GENETIC RESEARCH

Most applications of genetic biotechnology are agricultural, medical, and scientific. However, geneticists are also engaged in product research and development of related technology and in legal determinations. Involvement with legal matters and the criminal justice system often takes the form of DNA profiling, also known as DNA fingerprinting. Because every organism has its own unique DNA, genetic testing can definitively determine whether individuals are related to one another and whether DNA evidence at a crime scene belongs to a suspect. It can also accurately identify a specific strain of a bacterium.

One example of researchers' use of genetic profiling to identify disease-causing bacteria is reported by Chayapa Techathuvanan, Frances Ann Draughon, and Doris Helen D'Souza in "Real-Time Reverse Transcriptase PCR for the Rapid and Sensitive Detection of Salmonella Typhimurium from Pork" (*Journal of Food Protection*, vol. 73, no. 3, March 2010). The researchers used reverse transcriptase polymerase chain reaction (PCR) to improve the detection of *Salmonella Typhimurium* (a bacterium that causes the second-most commonly occurring bacterial foodborne illness in the United States) in samples of pork chops and sausage.

Forensics

The 18th edition of *Merck Manual of Diagnosis and Therapy* (2006) describes forensic genetics as using molecular genetic techniques to identify an individual's genetic makeup. Forensic genetics relies on the measurement of many different genetic markers, each of which normally varies from individual to individual, and may be used to determine whether two people are genetically related. For example, because half of a person's genetic markers come from the father and half from the mother, analyses of these DNA markers enable laboratory technologists to establish that one person is the offspring of another. Analysis of DNA derived from blood samples can determine whether the supposed parents of a particular child are actually the biological parents. DNA markers may also be used to identify a specimen and establish its origin—that is, definitively determine the individual from whom it came. Biopsies, pathologic specimens, and blood and semen samples can all be used to measure DNA markers.

Forensic investigations often involve analyses of evidence left at a crime scene such as trace amounts of blood, a single hair, or skin cells. Using PCR, DNA from a single cell can be amplified to provide a sample quantity that is large enough to determine the source of the DNA. In PCR the double strand of DNA is denatured into single strands, which are placed in a medium with the chemicals needed for DNA replication. Each single strand becomes a double strand, yielding twice the amount of the initial DNA sample. The double strands are once again denatured to form single strands, and the process is repeated until there is a sufficient quantity of DNA for analysis. Analysis is usually performed using gel electrophoresis, in which DNA is loaded onto a gel and an electrical current is passed through the DNA. Investigators are then able to observe larger molecules migrating more slowly than smaller ones. Examining variable-number tandem repeats (VNTRs) is a procedure that identifies the length of tandem repeats in an individual's DNA. VNTR is an especially useful technique in forensic genetics, because when it is performed in careful laboratory conditions the probability of two individuals having the exact same VNTR results is less than one in a million.

DNA evidence is preferred by forensic specialists because fingerprints can often be erased or eliminated and hair color and appearance may be altered, but DNA is unalterable. It can be used to identify individuals with extremely high probability and is more stable than other biological samples such as proteins or blood groups. Forensic genetics can be a powerful tool and has been used successfully to eliminate suspects and clear their names. Even though it is not as useful in proving guilt, forensic genetics provides solid and often convincing evidence when many alleles match.

The first admission of DNA evidence in criminal court occurred in *Florida v. Tommy Lee Andrews* (1987), when the state of Florida used it as part of the prosecution's case to convict a suspect of a series of sexual assaults. In 1989 the Federal Bureau of Investigation (FBI) began accepting work from state forensic laboratories, and in 1996 the National Research Council published *The Evaluation of Forensic DNA Evidence*, which cited the FBI statistic that approximately one-third of primary suspects in rape cases are excluded by using DNA evidence.

Many Americans first became aware of the use and importance of DNA evidence during the sensational and widely publicized 1994–95 trial of O. J. Simpson (1947–) for the alleged murder of Nicole Brown Simpson (1959–1994) and Ron Goldman (1968–1994). Nicole Simpson's blood was found in O. J. Simpson's vehicle and house, but the defense discredited the DNA results by claiming the police had conducted a sloppy investigation that caused the samples to be contaminated and, alternatively, that the blood was planted in an attempt to frame Simpson.

In the 17 years since this sensational case informed Americans about the role of DNA evidence in confirming or refuting innocence or guilt, the use of such evidence has become routine and has been used to successfully exonerate wrongfully convicted people. According to the Innocence Project, in the fact sheet "Facts on Post-conviction DNA Exonerations" (http://www.innocenceproject.org/Content/Facts_on_PostConviction_DNA_Exonerations.php), as of 2010, 266 people had been exonerated by DNA testing, including 17 on death row. In the fact sheet "Access to Post-conviction DNA Testing" (2010, http://www.innocenceproject.org/Content/Access_To_PostConviction_DNA_Testing.php), the Innocence Project also reports that even though 48 states have postconviction DNA testing access statutes, many of these testing laws are limited. (As of 2010, Massachusetts and Oklahoma were the only states that did not have any DNA access statutes incorporated into their state laws.) For example, some state legislation does require adequate safeguards for the preservation of DNA evidence, whereas other states fail to require "full, fair and prompt proceedings" once a DNA testing petition has been filed, which can result in an innocent victim spending long periods being wrongfully imprisoned. Furthermore, despite its ability to prove innocence, some courts still refuse to consider newly discovered DNA evidence after trial.

The editorial "Death, DNA, and the Supreme Court" (*New York Times*, October 17, 2010) observes that in the 21st century, when sophisticated analysis of DNA evidence can not only help identify criminals but can also prevent the innocent from suffering the consequences of false accusations, the states should not be permitted to block the analysis of biological evidence before putting someone to death. The editorial cites the case of a death row inmate in Texas who was granted a stay of execution by the U.S. Supreme Court in March 2010 to consider the testing of untested biological evidence. The editorial asserts that such testing should not be left to the discretion of the courts; instead, it should be the standard operating procedure in all serious criminal investigations.

Nanotechnology

A nanometer (the width of 10 hydrogen atoms laid side by side) is among the smallest units of measure. It is one-billionth of a meter, one-millionth the size of a pinhead, or one-thousandth the length of a typical bacterium. Figure 10.5 shows the incredibly tiny scale of nanometers. According to the National Nanotechnology Initiative (May 21, 2007, http://www.nano.gov/html/about/docs/20070521NNI_Industrial_Nano_Impact_NSTI_Carim.pdf), "Nanotechnology is the understanding and control at dimensions of roughly 1 to 100 nanometers, where unique phenomena enable novel applications. ... Nanotechnology involves imaging, measuring, modeling, and manipulating matter at this length scale. ... The term is meant to broadly encompass nanoscale science, engineering, and technology."

Nanotechnology is sometimes called molecular manufacturing because it draws from many disciplines (including physics, engineering, molecular biology, and chemistry) and considers the design and manufacture of extremely small electronic circuits and mechanical devices built at the molecular level of matter. K. Eric Drexler, the chief technical adviser of Nanorex, a company that develops software for the design and simulation of molecular machine systems, coined the term *nanotechnology* during the 1980s to describe atomically precise molecular manufacturing systems and their products. He explains in *Engines of Creation* (1986) that it is an emerging technology with the potential to fulfill many scientific, engineering, and medical objectives.

Researchers anticipate many applications combining genetic engineering and nanotechnology such as home food-growing machines that could produce virtually unlimited food supplies and chip-sized diagnostic devices that would revolutionize the detection and management of illness by identifying the genetic variants that are associated with diseases and delivering genetically targeted treatments. The field of genetic nanomedicine encompasses myriad approaches to gene detection and delivery. Researchers envision computer-controlled molecular tools much smaller than a human cell and constructed with the accuracy and precision of drug molecules. Such tools would enable medicine to intervene in a sophisticated and controlled way at the cellular, molecular, and genetic level by removing obstructions in the circulatory system, destroying cancer cells, or assuming the function of organelles such as the mitochondria. Other potential uses of nanotechnology in

FIGURE 10.5

Visual representation of the size of a nanometer

Less than a nanometer
Individual atoms are up to a few angstroms, or up to a few tenths of a nanometer, in diameter.

Nanometer
Ten shoulder-to-shoulder hydrogen atoms span 1 nanometer. DNA molecules are about 2.5 nanometers wide.

Thousands of nanometers
Biological cells, like these red blood cells, have diameters in the range of thousands of nanometers.

A million nanometers
The pinhead sized patch of this thumb (circled in black) is a million nanometers across.

Billions of nanometers
A two meter tall male is two billion nanometers tall.

SOURCE: Mildred S. Dresselhaus, "The Incredible Tininess of Nano," in *Science Challenges in Nanomaterials*, National Transmission Electron Achromatic Microscopy Project, July 2002, http://ncem.lbl.gov/team/presentations/Dresselhaus/sld004.htm (accessed October 22, 2010)

---

medicine include the early detection and treatment of disease via exquisitely precise sensors for use in the laboratory, clinic, and human body, plus new formulations and delivery systems for pharmaceutical drugs.

Some of these nanoscale drug delivery structures are called dendrimers. (Figure 10.6 is an image of the many branches of a dendrimer.) Because dendrimers can carry different materials on their branches, they can perform various functions simultaneously, such as searching for and detecting diseased cells and apoptotic cells (apoptosis is the programmed death of a cell), diagnosing and locating gene variants that are associated with specific diseases, delivering drugs, and reporting the responses to treatment.

The technology is used to develop immediately compatible, rejection-resistant implants made of high-performance materials that respond as the body's needs change. Nanotechnology and genetic engineering will lead to new biomedical therapies as well as to prosthetic devices and medical implants. Some of these will help attract and assemble raw materials in bodily fluids to regenerate bone, skin, or other missing or damaged tissues. Nanotubes that act like tiny straws can circulate in a person's bloodstream and deliver medicines slowly over time or to highly specific locations in the body.

NEW DEVELOPMENTS IN NANOMEDICINE. The article "Nano Device 'Times Drug Release'" (BBC News, January 4, 2009) describes a promising recent application of nanotechnology in medicine. Massachusetts Institute of Technology researchers have developed devices and systems that use nanoparticles to deliver timed drug release to specific body parts at predetermined intervals. Site-

FIGURE 10.6

Illustration of a dendrimer

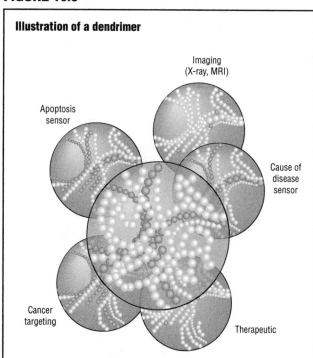

Imaging (X-ray, MRI)

Apoptosis sensor

Cause of disease sensor

Cancer targeting

Therapeutic

SOURCE: Nakissa Sadrieh, "Illustration of Dendrimer Agents Attached," in *Nanotechnology: Regulatory Perspective for Drug Development*, Food and Drug Administration, Center for Drug Evaluation and Research, Office of Pharmaceutical Science, undated, http://satyen.baindur.org/FDA-Regulatory_Perspective.pdf (accessed October 22, 2010)

specific release is important when toxic drugs, such as many chemotherapeutic agents used in cancer treatment, are targeting a specific organ or tissue and when genes must be delivered to specific organs or sites in the body.

Targeted nanotechnology can be used to control gene delivery in viral and nonviral gene-delivery systems. Rama Mallipeddi and Lisa Cencia Rohan explain in "Progress in Antiretroviral Drug Delivery Using Nanotechnology" (*International Journal of Nanomedicine*, vol. 5, August 9, 2010) that besides their size, which enables them to penetrate through cells and deliver drugs intracellularly, the advantages of nanocarriers (nanoscale particulate drug carriers) include protecting drugs from degrading and targeting drugs to specific sites of action. There also are challenges associated with nanocarrier drug delivery. For example, the smaller the size of the nanocarrier, the harder it is to control the rate of drug release from the nanocarrier. Drugs carried by smaller nanocarriers are released faster, and in bursts, which makes it harder to sustain or control the rate of drug delivery over time.

Mallipeddi and Rohan report that nanocarriers may improve the efficiency of the delivery of antiretroviral drugs that are used to prevent and treat human immunodeficiency virus (HIV is the virus that causes AIDS). Various types of nanocarriers, such as nanoparticles that are made of polymers, inorganic nanocarriers, liposomes, dendrimers, and cyclodextrins (compounds that are composed of sugar molecules in a ring formation) are currently being studied for the delivery of drugs that are intended for HIV prevention or treatment.

THE PROMISE OF NANOMEDICINE. According to Marco A. Zarbin et al., in "Nanomedicine in Ophthalmology: The New Frontier" (*American Journal of Ophthalmology*, vol. 150, no. 2, August 2010), the goal of nanomedicine is the "comprehensive monitoring, control, construction, repair, defense, and improvement of human biological systems at the molecular level, using engineered nanodevices and nanostructures, operating massively in parallel at the single cell level, ultimately to achieve medical benefit." Even though the most immediate impact of nanomedicine is likely in the area of drug delivery and implantable medical devices, such as intraocular pressure monitors that measure pressure in the eye, nanomedicine is also being used to advance biological imaging and diagnostics. For example, gold nanoparticles have been used to enhance tumor identification with computed tomography.

Nanomedicine is especially relevant to the fields of surgery and regenerative medicine, which is concerned with the improvement and replacement of diseased, aging, or damaged cells, tissues, and organs. Tools such as nanotubes and nanotweezers have already been developed for nanosurgery and even though the idea of "swallowing the surgeon" may seem like the stuff of science fiction, in the not-too-distant future nanosurgical devices may be able travel in blood vessels and identify and repair damaged tissue. Regenerative nanomedicine, a new subspecialty of nanomedicine, uses nanoparticles containing gene transcription factors and other modulating molecules that facilitate the reprogramming of cells. Zarbin et al. opine that the "application of nano-scale technologies to the practice of medicine will alter profoundly our approach to diagnosis, treatment, and prevention of disease."

CHAPTER 11
ETHICAL ISSUES AND PUBLIC OPINION

The institutions of genetic scientific and technical research, and the industries of genetic application, are relatively well organized and generously funded. Their imperatives are clear: push toward new knowledge and its applications. By contrast, our ethical, social discussion is unfocused, episodic, and scattered. We need to harness moral thinking to genetic technique. The need for organized, intelligent debate involving an active public and committed scientists has never been clearer.

—Everett Mendelsohn, "The Eugenic Temptation: When Ethics Lag behind Technology" (*Harvard Magazine*, March–April 2000)

Rapid advances in genetics and its applications pose new and complicated ethical, legal, regulatory, and policy issues for individuals and society. The issues society must consider include the responsible use of genetic engineering technology, the consequences of knowledge about personal genetic information for individuals, and the repercussions of genomic information for groups such as ethnic and racial minorities. For example, African-Americans were among the first to call for the inclusion of African-American genetic sequences in the human genome's template, and their concerns are vitally important because historically they have been victimized by "genetic inquiries" and research.

Even though the potential for discrimination based on genetic information has been effectively eliminated by the provisions of the Genetic Information Nondiscrimination Act of 2008, a host of ethical issues related to health and medical care remain. These include the use of genetic information to guide reproductive decision making and the application of genetic engineering to reproductive technology. Reproductive applications are an especially emotionally charged topic as many antiabortion advocates staunchly oppose any action to alter the development or course of a naturally occurring pregnancy, such as in vitro fertilization (fertilization that takes place outside the body) and preimplantation intervention as well as the use of stem cells that are obtained from human embryos. Scientists, ethicists, and other observers also fear that genetic engineering technology will rapidly progress from enabling prenatal diagnosis and intervention to prevent serious disease to preconception selection of the traits and attributes of offspring—the creation of so-called designer babies.

Genetic testing challenges health care professionals to rapidly acquire new knowledge and skills to effectively use new technology and to assist their patients in making informed choices. The quality, reliability, and utility of genetic testing must be continuously reevaluated and regulated. Along with quality-control measures, health care professionals and consumers must determine whether it is appropriate to perform tests for conditions for which no treatment exists and how to interpret and act on test results, such as an individual's increased susceptibility to a disease that is associated with several genes and environmental triggers.

Philosophical, psychological, and spiritual considerations invite mental health professionals, ethicists, members of the clergy, and society overall to redefine concepts of human responsibility, free will, and genetic determinism. Behavioral genetics looks at the extent to which genes influence behavior and the human capacity to control behavior. It also attempts to describe the biological basis and heritability of traits ranging from intelligence and risk-taking behaviors to sexual orientation and alcoholism.

There are also environmental issues associated with the applications of genetic research, such as weighing the benefits and risks to the environment by creating genetically modified animals, crops, and other products for human consumption. Legal and financial issues center on the ownership of genes, deoxyribonucleic acid (DNA), and related data; property rights; patents and copyrights; and public access to research data and other genetic information.

PUBLIC OPINION ABOUT GENETIC RESEARCH, ETHICAL ISSUES, AND PROTECTING THE PUBLIC INTEREST

The Virginia Commonwealth University's Center for Public Policy notes in *VCU Life Sciences Survey 2010* (2010, http://www.vcu.edu/lifesci/images2/survey2010.pdf), a nationwide survey of 1,001 adults, that awareness and understanding of developments in therapeutic cloning and stem cell research are not high. More than two-thirds (67%) of survey respondents said they had not heard or read about the technique that enables adult human stem cells to take on the characteristics of human embryonic stem cells. (See Table 11.1.) Only 20% of survey respondents said they have a "very clear" understanding of the differences between human embryonic stem cells and adult stem cells, whereas 34% described themselves as "somewhat clear" and 44% said they are "not too clear/not clear at all" about the distinction between the different sources of stem cells. Even though understanding of the types of stem cells is low, support for stem cell research that does not use human embryonic stem cells remains relatively high—the majority (71%) of the 2010 survey respondents said they favor such research. (See Figure 11.1.) This proportion had not changed significantly between 2007 and 2010.

The American public appears to be somewhat more favorably inclined toward embryonic stem cell research than in past years. Over six out of 10 (62%) respondents were in favor of such research in 2010, up eight percentage points from 2007. (See Figure 11.2.) Consistent with

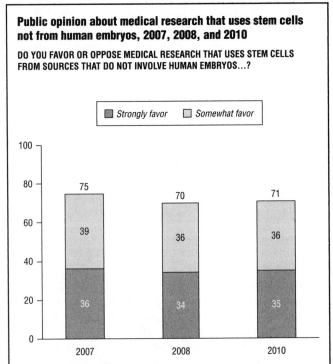

FIGURE 11.1

Public opinion about medical research that uses stem cells not from human embryos, 2007, 2008, and 2010

DO YOU FAVOR OR OPPOSE MEDICAL RESEARCH THAT USES STEM CELLS FROM SOURCES THAT DO NOT INVOLVE HUMAN EMBRYOS...?

■ Strongly favor   ☐ Somewhat favor

	2007	2008	2010
Total	75	70	71
Somewhat favor	39	36	36
Strongly favor	36	34	35

SOURCE: "Do you favor or oppose medical research that uses stem cells from sources that do NOT involve human embryos...?" in "Non-embryonic Stem Cell Research," in *VCU Life Sciences Survey 2010*, Virginia Commonwealth University (VCU), http://www.vcu.edu/lifesci/images2/survey2010.pdf (accessed October 25, 2010)

this finding, there is also a slight increase in the percentage of Americans favoring therapeutic cloning—in 2010, 55% of respondents supported therapeutic cloning, up eight percentage points from 2007. (See Figure 11.3.)

In contrast, the overwhelming majority of Americans are opposed to genetic modifications of humans. When asked, "The technology now exists to clone or genetically alter animals. How much do you favor or oppose allowing the same thing to be done in humans?" 80% of adults (22% somewhat opposed and 58% strongly opposed) said they oppose such genetic alteration of humans. (See Table 11.2.) Interestingly, unfavorable feelings about using genetic technology to alter humans is shared by people of all ages, levels of educational attainment, religious views, political party affiliation, and the extent to which Americans feel they are knowledgeable about scientific and medical discoveries.

The Impact of Genetic Testing on Behavior

Many health professionals hope that genetic testing for hereditary cancers will enable early disease detection, offer targeted surveillance, result in effective prevention strategies, and improve health outcomes (how affected individuals fare as a result of treatment). In "The Beliefs, and Reported and Intended Behaviors of Unaffected Men in Response to Their Family History of Prostate Cancer"

TABLE 11.1

Public awareness of developments in stem cell research, 2010

Awareness of iPS cells?

HAVE YOU HEARD OR READ ANYTHING ABOUT RECENT DISCOVERIES THAT ALLOW STEM CELLS WITH EQUAL ABILITIES TO HUMAN EMBRYONIC STEM CELLS TO BE CREATED FROM NON-EMBRYONIC HUMAN ADULT CELLS, OR NOT?

	%
Yes, heard	28
Not heard	67
Don't know/refused	6
	100

HOW CLEAR ARE YOU PERSONALLY ON THE DIFFERENCE BETWEEN STEM CELLS THAT COME FROM HUMAN EMBRYOS, STEM CELLS THAT COME FROM ADULTS, AND STEM CELLS THAT COME FROM OTHER SOURCES?

Very clear	20
Somewhat clear	34
Not too/not at all clear	44
Don't know/refused	3
	100

Figures may add to 99 or 101 due to rounding

SOURCE: "Awareness of iPS Cells?" in "Stem Cell Research and Human Cloning," in *VCU Life Sciences Survey 2010*, Virginia Commonwealth University (VCU), http://www.vcu.edu/lifesci/images2/survey2010.pdf (accessed October 25, 2010)

FIGURE 11.2

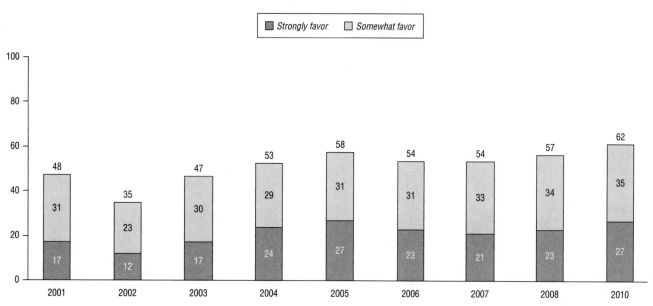

Public opinion about embryonic stem cell research, 2001–10

ON THE WHOLE, HOW MUCH DO YOU FAVOR OR OPPOSE MEDICAL RESEARCH THAT USES STEM CELLS FROM HUMAN EMBRYOS—DO YOU STRONGLY FAVOR, SOMEWHAT FAVOR, SOMEWHAT OPPOSE, OR STRONGLY OPPOSE THIS?

SOURCE: "On the whole, how much do you favor or oppose medical research that uses stem cells from human embryos—Do you strongly favor, somewhat favor, somewhat oppose, or strongly oppose this?" in "Embryonic Stem Cell Research," in *VCU Life Sciences Survey 2010*, Virginia Commonwealth University (VCU), http://www.vcu.edu/lifesci/images2/survey2010.pdf (accessed October 25, 2010)

FIGURE 11.3

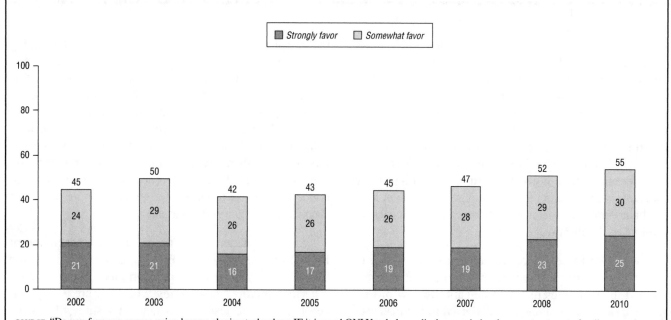

Public opinion about therapeutic cloning, 2002–10

DO YOU FAVOR OR OPPOSE USING HUMAN CLONING TECHNOLOGY IF IT IS USED ONLY TO HELP MEDICAL RESEARCH DEVELOP NEW TREATMENTS FOR DISEASE—DO YOU STRONGLY FAVOR, SOMEWHAT FAVOR, SOMEWHAT OPPOSE OR STRONGLY OPPOSE THIS?

SOURCE: "Do you favor or oppose using human cloning technology IF it is used ONLY to help medical research develop new treatments for disease—do you strongly favor, somewhat favor, somewhat oppose or strongly oppose this?" in "Therapeutic Cloning," in *VCU Life Sciences Survey 2010*, Virginia Commonwealth University (VCU), http://www.vcu.edu/lifesci/images2/survey2010.pdf (accessed October 25, 2010)

TABLE 11.2

Public opinion about genetic alterations of humans, 2010

	The technology now exists to clone or genetically alter animals. How much do you favor or oppose allowing the same thing to be done in humans—do you...?					
	Strongly favor	Somewhat favor	Somewhat oppose	Strongly oppose	Don't know/ no answer (vol.)	Number of cases
All adults	3%	12%	22%	58%	4%	1,001
Gender						
Men	4%	14%	23%	54%	4%	489
Women	3%	10%	21%	62%	4%	512
Age						
18–29	6%	17%	33%	42%	2%	151
30–44	2%	14%	18%	59%	7%	184
45–64	3%	8%	18%	69%	3%	407
65 and older	4%	12%	20%	58%	6%	244
Education						
High school or less	5%	10%	20%	61%	4%	330
Some college	2%	13%	26%	55%	4%	265
College grad or more	2%	15%	21%	58%	4%	401
Religious guidance in daily life						
Great deal	3%	11%	18%	64%	4%	424
Quite a bit	3%	15%	26%	53%	3%	210
Some guidance	3%	13%	21%	57%	7%	166
Not important	6%	14%	24%	53%	3%	182
Views about the Bible						
Actual word of God	1%	7%	18%	71%	3%	378
Word of God, not everything to be taken literally	4%	15%	24%	53%	4%	366
Written by men	6%	16%	27%	47%	3%	205
Scientific and medical discoveries						
More informed	4%	15%	18%	59%	4%	195
Somewhat informed	3%	13%	22%	59%	3%	487
Less informed	4%	10%	25%	58%	4%	314
Party identification						
Democrat	5%	12%	30%	51%	2%	349
Republican	1%	11%	16%	69%	3%	252
Independent	3%	14%	18%	60%	4%	325

SOURCE: "Q20. The technology now exists to clone or genetically alter animals. How much do you favor or oppose allowing the same thing to be done in humans?" in *VCU Life Sciences Survey 2010*, Virginia Commonwealth University (VCU), http://www.vcu.edu/lifesci/images2/survey2010.pdf (accessed October 25, 2010)

(*Genetics in Medicine*, vol. 10, no. 6, June 2008), Ruth Cowan et al. note the results of a survey of men with a family history of prostate cancer that sought to explore their intention to adopt preventive behaviors in response to test results. Even though genetic testing for prostate cancer is not yet available, clinical trials are under way to investigate the role of specific preventive measures for at-risk individuals. The survey of 280 men found that 92% were interested in genetic testing and that a positive test result would prompt 93% of the men to make at least one behavioral change intended to reduce their risk of developing the disease.

Jodi T. Heshka et al. looked at whether knowledge of carrier status might stimulate behavioral change and reported their findings in "A Systematic Review of Perceived Risks, Psychological and Behavioral Impacts of Genetic Testing" (*Genetics in Medicine*, vol. 10, no. 1, January 2008). The researchers reviewed 30 studies that measured the perceived risk (the extent to which an individual believes that he or she is at risk of developing a disease or disorder) as well as the psychological and behavioral effects of genetic testing on individuals. The studies considered hereditary colorectal carcinoma, hereditary breast and ovarian cancer, and Alzheimer's disease. Heshka et al. observe an increase in screening behavior among carriers but the change in other behaviors was less than expected. In terms of perceived risk, there were generally no differences between carriers and noncarriers by 12 months after genetic testing and, over time, risk perception decreased.

Susan M. Domchek et al. report in "Association of Risk-Reducing Surgery in BRCA1 or BRCA2 Mutation Carriers with Cancer Risk and Mortality" (*Journal of the American Medical Association*, vol. 304, no. 9, September 1, 2010) the results of a landmark study of women who carry BRCA mutations. The surgical removal of the ovaries, known as oophorectomy, has proven effective at preventing ovarian and breast cancer in women with these mutations. Women who carry the harmful BRCA mutations have a significantly increased cancer risk—15% to 40% will develop ovarian cancer during their lifetime (compared with about 1% of the general population) and 60% will develop breast cancer (compared with 12% of the general population of women).

Domchek et al. followed 2,482 women with BRCA1 and BRCA2 mutations for a mean (average) duration of three to six years and compared the cancer and mortality outcomes between those who had risk-reducing surgery (172 had a mastectomy and 993 had an oophorectomy) and those who chose not to have surgery. The women who chose not to have the prophylactic (preventive) surgery (about half of the subjects) were offered intensified surveillance for breast cancer—annual mammography and breast magnetic resonance imaging (MRI is a radiological technique that provides detailed images of organs and tissues)—and for ovarian cancer—ultrasound and Ca125 testing every four to 12 months.

The risk of developing ovarian cancer was higher in BRCA1 carriers (7% to 8%) than in BRCA2 carriers (3%) who opted against oophorectomy. No BRCA2 carriers who had an oophorectomy developed peritoneal cancer (cancer of the tissue lining the abdominal and pelvic cavities) during the follow-up. Cancer was diagnosed in the ovaries of nine women who underwent prophylactic oophorectomy.

Prophylactic oophorectomy also protected against breast cancer, reducing the risk by half—from 22% to 11%. Prophylactic salpingo-oophorectomy (removal of the fallopian tubes and ovaries) appears to be more effective in BRCA2 carriers than in BRCA1 carriers—there were no breast or ovarian cancer-related deaths in BRCA2 carriers who had a salpingo-oophorectomy. Interestingly, prophylactic mastectomy reduced the risk of developing breast cancer but did not reduce the mortality from breast cancer. This finding may reflect the fact that breast cancer screening and treatment are effective at reducing mortality.

Intensified screening in the form of more frequent mammograms, sonograms, and MRIs offer BRCA mutation carriers ways to ensure early detection of breast cancer. There are not yet comparably effective screening tests for ovarian cancer—ultrasound and Ca125 tests have not been shown to reduce the mortality from ovarian cancer. Domchek et al. find that surveillance does not confer the same protection against ovarian cancer as does oophorectomy.

Domchek et al. conclude that among women with BRCA1 and BRCA2 mutations, "the use of risk-reducing mastectomy was associated with a lower risk of breast cancer; [and] risk-reducing salpingo-oophorectomy was associated with a lower risk of ovarian cancer, first diag-nosis of breast cancer, all-cause mortality, breast cancer–specific mortality, and ovarian cancer–specific mortality."

Even in view of these findings, the decision to have prophylactic surgery is often not straightforward. Variables such as the woman's age, marital status, fertility, and overall quality of life must be considered along with the risk of cancer. In "Risk-Reducing Strategies for Women Carrying BRCA1/2 Mutations with a Focus on Prophylactic Surgery" (*BMC Women's Health*, vol. 10, no. 1, October 2010), Mohamed Salhab, Selina Bismohun, and Kefah Mokbel observe that nonsurgical precautions such as intensified surveillance and chemoprevention (prescription drug therapy that is intended to reduce risk) offer the benefits of unimpaired body image and a good quality of life, but may be associated with an increased risk of advanced-stage cancer and mortality. Even though prophylactic surgery confers a very high degree of protection from cancer, Salhab, Bismohun, and Mokbel believe that its benefits must be weighed against "an array of potential costs, including surgical complications and the potential impacts of mastectomy or oophorectomy on a woman's self-image, as well as on their sexual and reproductive function."

The Impact of Disclosing the Results of Genetic Testing

In an effort to better understand the frequency with which genetic test results are disclosed and the personal impact of disclosing genetic test results, Angela R. Bradbury et al. looked at parents' decisions to communicate BRCA test results to children under the age of 25 years and reported their findings in "How Often Do BRCA Mutation Carriers Tell Their Young Children of the Family's Risk for Cancer? A Study of Parental Disclosure of BRCA Mutations to Minors and Young Adults" (*Journal of Clinical Oncology*, vol. 25, no. 24, August 20, 2007). The researchers find that 55% of the parents disclosed genetic test results to their children years before preventive or therapeutic interventions were appropriate. The parents reported that some of the children appeared unable to understand the significance of the test results and that about half of the children learning the test results reacted negatively by expressing anxiety or crying.

Because interventions such as increased surveillance, chemoprevention, or prophylactic surgery generally do not need to begin before the age of 25 years, most health care professionals advise against genetic testing of minors for conditions that are unlikely to arise before adulthood and for which preventive or therapeutic measures are not indicated until adulthood. Bradbury et al. also find that parents who chose to disclose the test results to their children generally did so without consultation from a health care professional. When parents were asked if they thought it would be helpful to have their offspring meet with a genetics counselor, 62% said no, 31% responded yes, and 7% were unsure. The researchers conclude that the consequences of communicating genetic test results and the long-term psychological effects on children are important areas for future research. They assert that "a better understanding of how children conceptualize genetic risk can help define a role for health care professionals and genetic counseling."

In "Disclosing the Disclosure: Factors Associated with Communicating the Results of Genetic Susceptibility Testing for Alzheimer's Disease" (*Journal of Health Communication*, vol. 14, no. 8, December 2009), Sato Ashida et al. examine the extent to which people who had genetic susceptibility testing for Alzheimer's disease (AD) communicated their results to others. The researchers find that the majority (82%) of subjects who tested for the apoE genotype shared their results with someone. Communicating test results was associated with greater optimism about AD treatment, lower perceived risk of developing AD, and higher perceived importance of genetics in the development of AD.

CONCERNS ABOUT GENETIC RESEARCH AND ENGINEERING

Genetic research and engineering technologies promise to address some of the most pressing problems of the 21st century, such as cleaning the environment, feeding the world's hungry, and preventing serious diseases. However, like all new technologies, they pose risks as well as benefits. In general, Americans support advances in genetic research and technology, and they are optimistic that the outcomes will ultimately be used to reduce sickness and suffering, as opposed to generating legal and ethical controversies.

Some of the fears about genetic engineering presume that the use of such technologies will be alien, impersonal, and technologically difficult. Scientists and advocates of applying genetic engineering to improve human life foresee a future in which parenting is guided by genetic testing and new reproductive technologies are routine aspects of medical care. Bioethicists and others concerned about the implications of the routine use of this biotechnology contend that it is possible and advantageous to fund, discuss, and regulate genetics in the same way that society currently considers environmental medicine, nutrition, and public health.

ETHICAL CONSIDERATIONS AND CONCERNS ABOUT NANOTECHNOLOGY AND NANOMEDICINE

Nanotechnology and nanomedicine have generated ethical concern and scrutiny in recent years. In "How Should We Do Nanoethics? A Network Approach for Discerning Ethical Issues in Nanotechnology" (*Nano-Ethics*, vol. 2, no. 1, April 2008), Ibo van de Poel of the Delft University of Technology outlines some of the ethical considerations, including "the environmental and health opportunities and risks of nanotechnology, in particular the potential risks of nanoparticles, privacy and control threats raised by the use of nanodevices, the possibilities for human disease treatment and enhancement and the ethical consequences thereof" and observes that there is no agreement on how nanoethics should proceed. He proposes the use of an approach that is able to discern ethical issues in nanotechnology as they emerge and that acknowledges the complexity of the issues rather than assuming that the issues are clear-cut. Van de Poel asserts that an optimal approach to nanoethics should make it possible to "use ethical insights to inform and steer nanotechnological R&D and should not say simply yes or no to certain nanotechnological developments." He also exhorts scientists and medical ethicists to consider the ethical issues in their real-world context rather than viewing them as abstract issues that may be resolved using existing ethical concepts and theories.

Todd Kulken of the Woodrow Wilson International Center for Scholars in Washington, D.C., opines in "Nanomedicine and Ethics: Is There Anything New or Unique?" (Wiley Interdisciplinary Reviews Nanomedicine and Nanobiotechnology, April 20, 2010) that the ethical questions surrounding nanomedicine are the same questions that would be raised about any new medical device or drug that is being evaluated. Kulken asserts that the key issue that will determine the widespread acceptance of nanomedicine is determining "how much risk society is willing to accept with a new technology before it is proven effective and safe." Kulken observes that ethics may be unable to evaluate a technology that has yet to fulfill its promise.

NOTED AUTHORS ADDRESS THE RISKS AND BENEFITS OF GENETIC ENGINEERING

In *Our Posthuman Future: Consequences of the Biotechnology Revolution* (2002), Francis Fukuyama of Johns Hopkins University cautions about the use and misuse of science and biotechnology. He observes that genetic engineering has gained popular acceptance as it is used to prevent or correct selected medical conditions, but he shares the widely held concerns about the unforeseeable results of gene manipulation and the potential for altering the complexion of society through the use of "enhancement technology" to customize the attributes of offspring. Fukuyama asserts his fear that those able to afford the genetic interventions to produce offspring who are smarter, stronger, more athletic, talented, and better looking will further widen the chasm between the economic classes. He states that the ways in which society chooses to employ, regulate, and restrict genetic engineering may challenge traditional concepts of human equality, and as a result will change the existing understanding of human personality, identity, and the capacity for moral choice. Furthermore, he contends that genetic engineering technologies may afford societies new techniques for controlling the behavior of their citizens and could potentially overturn existing social hierarchies and affect the rate of intellectual, material, and political progress. Fukuyama predicts that genetic engineering and other applications of biotechnology have the potential to sharply alter the nature of global politics.

Despite his fears and cautions about the relatively recent ability to intentionally modify the human organism as opposed to waiting for evolution, Fukuyama does not consider it necessary or advisable to prohibit any actions that alter genetic codes. He would, however, start by banning reproductive cloning outright, to establish a precedent for political control over biotechnology. If society does not institute effective regulation, Fukuyama warns, "We may be about to enter into a posthuman future, in which technology will give us the capacity gradually to alter [human] essence over time. ... We do not have to regard ourselves as slaves to inevitable technological progress when that progress does not serve human ends. True freedom means the freedom of political communities to protect the values they hold most dear, and it is that freedom that we need to exercise with regard to the biotechnology revolution today."

Bill McKibben, the author of *The End of Nature* (1989), which describes the consequences of damage done to the environment by overpopulation and global warming, also published *Enough: Staying Human in an Engineered Age* (2003), in which he warns that genetic engineering may not fulfill its promises. McKibben worries that society has been oversold on the potential benefits of genetic manipulation and that promises that it will make future generations healthier, smarter, happier, taller, thinner, better-looking, stronger, and saner may instead rob future generations of free will and freedom of choice. Like Fukuyama, McKibben is concerned that enabling the wealthy to custom-equip their offspring with good looks, high intelligence quotients, and athletic prowess will reinforce and deepen existing social class distinctions.

Even though advocates claim that genetic engineering will prevent debilitating and fatal diseases and forestall death, McKibben believes that even the genetically enhanced will suffer. He fears they will be beset by self-doubt, wondering if their achievements are their own or simply attributable to the geneticist. He wonders whether children whose parents have chosen to enhance them in the lab before birth will be able to make choices about their own lives, or will they just be making choices according to their prescripted genetic plans? McKibben contends that "the person left without any choice at all is the one you've engineered."

McKibben also argues against using germline engineering technologies to prevent diseases or germline gene therapy to treat diseases. Such therapies involve more than simply altering the affected individuals' genes; they embed the genetic changes in their reproductive cells (sperm and eggs) so that the genetic alteration is heritable by all future generations. Furthermore, McKibben notes that society is as yet unable to distinguish between preventing disease and simply enhancing natural characteristics. Citing efforts to prevent dwarfism (short stature) and genetic enhancements to increase the height of a child who is naturally of short stature, he observes that "there's no obvious line between repair and improvement." He also expresses some skepticism about the feasibility of effectively regulating the distinction between repair and improvement in medical offices and clinics throughout the United States. McKibben asserts that the pace of change has accelerated such that decisions about how to use nanotechnology and other genetic engineering techniques must be made before their use is widely available and accepted. He fears that society may accept the view that humans are an "endlessly improvable species" and exhorts readers to conclude that human beings as currently constituted are good enough.

The bioethicist Arthur L. Caplan thinks that the ethical implications of new knowledge in genetics are not at the forefront of the minds of professionals and health care consumers because of the uncertainty about how to use the new knowledge. In "If Gene Therapy Is the Cure, What Is the Disease?" (November 8, 2002, http://www .bioethics.net/articles.php?viewCat=6&articleId=58), Caplan, a strong supporter of the Human Genome Project, asserts that the greatest challenge to securing funding and support for genomic research is public fear of germline engineering (manipulating the human genome to improve the human species), and he acknowledges that this fear is based on the historical reality of horrendous eugenics (hereditary improvement of a race by genetic control) practices in Germany during World War II (1939–1945) and in other countries.

Caplan cites examples of genetic engineering in the United States such as the Repository for Germinal Choice in Escondido, California, also known as the "Nobel Prize sperm bank," which solicits and stores sperm from men who are selected for their scientific, athletic, or entrepreneurial acumen. The banked sperm is available for use by women of high intelligence for the express purpose of creating genetically superior children. Caplan observes that there have been relatively few critics of this practice, whereas the mere suggestion of the possibility of directly modifying the genetic blueprint of gametes (sperm and eggs) has generated fiery debate in professional and lay communities. He contends that the history of eugenically driven social policy is reason enough to question and even protest the actions of the Repository for Germinal Choice but that it does not argue against allowing voluntary, therapeutic efforts using germline manipulations to prevent certain serious or fatal disorders from besetting future generations.

According to Caplan, the decision to forgo germline engineering does not make ethical sense. He laments that "some genetic diseases are so miserable and awful that at least some genetic interventions with the germline seem justifiable. ... It is at best cruel to argue that some people must bear the burden of genetic disease in order to allow benefits to accrue to the group or species. At best, genetic diversity is an argument for creating a gamete bank to preserve diversity. It is hard to see why an unborn child has any obligation to preserve the genetic diversity of the species at the price of grave harm or certain death."

Caplan fears that choosing to refrain from efforts to modify the germline will result in lives sacrificed—that is, important benefits will be delayed or lost for people with disorders that might be effectively treated with germline engineering. He recommends responding to justifiable concerns about the dangers and potential for abuse of new knowledge that is generated by the genome with frank, objective assessments of the appropriate goals of this application of biotechnology.

In the September 2, 2006, presentation "Biobanking, Genomics, and Genetic Engineering: Where Are We Headed, and What Rules Should Take Us There?" Caplan reexamined the question: "Should society limit how far we push genetic manipulation?" He also discussed concerns about biobanking—physical stores of human tissues and DNA, which often include the donor's personal information. He opined that biobanks offer tremendous potential benefits for selected patients, including those receiving organ transplants, but questioned whether there are controls in place to effectively prevent black market sales of human organs and other misuse of biobank tissues and confidential donor information.

According to Susan Martinuk, in "IVF Is Nobel-Worthy, but Ethical Issues Remain" (*Calgary [Canada] Herald*, October 8, 2010), Caplan reiterated his concerns about genetic engineering in 2010 when the Nobel Prize in Physiology or Medicine was awarded to Robert Geoffrey Edwards (1925–), the developer of in vitro fertilization technology. Edwards's groundbreaking work contributed to preimplantation genetic diagnostic techniques. Many of the ethical considerations raised by his work more than three decades ago remain unresolved in the 21st century. For example, what is the disposition of unused fertilized eggs? Should they be used as a source of embryonic stem cells or donated to infertile couples? In the quest for ideal children, should society accept the buying and selling of sperm and eggs? Caplan opined, "Edwards's discoveries will make the issue of designing our descendants ... trying to create children who are stronger, faster, live longer ... the biggest issue in the first half of the 21st century."

THE ETHICS OF SYNTHETIC BIOLOGY

Katharine Sanderson reports in "Synthetic Biology Gets Ethical" (*Nature*, May 12, 2009) that the Centre for Synthetic Biology and Innovation, the first publicly funded institute in the United Kingdom that is dedicated to synthetic biology (creation of complex, biological systems that have functions that do not occur in nature), officially opened in May 2009. The center trains students and investigators to examine the social and ethical ramifications of their work and assists industry and government to develop appropriate regulatory measures. It also focuses on engaging the public in the discussion of advances in synthetic biology, in the hope of promoting understanding and appreciation of this technology. Furthermore, the center is creating a database of engineered biological components, one that is comparable to the Massachusetts Institute of Technology's Registry of Standard Biological Parts.

In "Researchers Say They Created a 'Synthetic Cell'" (*New York Times*, May 20, 2010), Nicholas Wade reports that in May 2010 the geneticist J. Craig Venter (1946–) and his colleagues announced that they had created a "synthetic cell." The group used the genome of a bacterium that infects goats and then extracted the DNA from it. The researchers created a synthetic genome that was much like the original genome; however, they added a few sequences that would not affect the genome's activity but would enable them to identify the synthetic code. After synthesizing the genome, they transplanted it into a different type of bacterium to create a self-replicating bacterium. Venter described the bacterium as "the first self-replicating species we've had on the planet whose parent is a computer." He also opined that "this is a philosophical advance as much as a technical advance" and suggested that it will reinvigorate the discussion about the nature of life.

That same day *Nature* published in "Sizing up the 'Synthetic Cell'" (May 20, 2010) the responses of leading scientists and bioethicists to Venter's announcement. Mark Bedau of Reed College called for "precautionary thinking and risk analysis," but also observed that the announcement will "revitalize perennial questions about the significance of life." George Church of the Harvard Medical School warned that society must take measures to prevent erroneous applications of this technology as well as its potential use to create bioterrorism threats. In contrast, Caplan compared the accomplishment to "the discoveries of Galileo, Copernicus, Darwin and Einstein" in terms of its ability to fundamentally change deeply held beliefs about the nature of life.

IMPORTANT NAMES
AND ADDRESSES

Alzheimer's Association
225 N. Michigan Ave., 17th Floor
Chicago, IL 60601-7633
(312) 335-8700
1-800-272-3900
E-mail: info@alz.org
URL: http://www.alz.org/

American Cancer Society
1599 Clifton Rd. NE
Atlanta, GA 30329
(404) 320-3333
1-800-227-2345
URL: http://www.cancer.org/

American Diabetes Association
1701 N. Beauregard St.
Alexandria, VA 22311
1-800-342-2383
E-mail: askada@diabetes.org
URL: http://www.diabetes.org/

American Heart Association
7272 Greenville Ave.
Dallas, TX 75231
1-800-242-8721
URL: http://www.americanheart.org/

American Parkinson Disease Association
135 Parkinson Ave.
Staten Island, NY 10305
(718) 981-8001
1-800-223-2732
FAX: (718) 981-4399
E-mail: apda@apdaparkinson.org
URL: http://www.apdaparkinson.org/

American Society of Gene and Cell Therapy
555 E. Wells St., Ste. 1100
Milwaukee, WI 53202
(414) 278-1341
FAX: (414) 276-3349
E-mail: info@asgt.org
URL: http://www.asgt.org/

American Society of Human Genetics
9650 Rockville Pike
Bethesda, MD 20814-3998
(301) 634-7300
FAX: (301) 634-7079
E-mail: society@ashg.org
URL: http://www.ashg.org/

Center for Food Safety
660 Pennsylvania Ave. SE, Ste. 302
Washington, DC 20003
(202) 547-9359
FAX: (202) 547-9429
E-mail: office@centerforfoodsafety.org
URL: http://www.centerforfoodsafety.org/

Cold Spring Harbor Laboratory
One Bungtown Rd.
Cold Spring Harbor, NY 11724
(516) 367-8800
URL: http://www.cshl.org/

Council for Biotechnology Information
1201 Maryland Ave. SW, Ste. 900
Washington, DC 20024
(202) 962-9200
URL: http://www.whybiotech.com/

Cystic Fibrosis Foundation
6931 Arlington Rd., Second Floor
Bethesda, MD 20814
(301) 951-4422
1-800-344-4823
FAX: (301) 951-6378
E-mail: info@cff.org
URL: http://www.cff.org/

Genetics Society of America
9650 Rockville Pike
Bethesda, MD 20814-3998
(301) 634-7300
1-866-486-4363
FAX: (301) 634-7079
URL: http://www.genetics-gsa.org/

Genome Programs of the U.S. Department of Energy Office of Science
1000 Independence Ave. SW
Washington, DC 20585
URL: http://genomics.energy.gov/

Huntington's Disease Society of America
505 Eighth Ave., Ste. 902
New York, NY 10018
(212) 242-1968
1-800-345-4372
FAX: (212) 239-3430
E-mail: hdsainfo@hdsa.org
URL: http://www.hdsa.org/

International Genetic Epidemiology Society
Division of Biostatistics
Washington University School of Medicine
Box 8067, 660 S. Euclid Ave.
St. Louis, MO 63108
(314) 362-3616
FAX: (314) 362-2693
URL: http://www.geneticepi.org/

International Society for Forensic Genetics
c/o Institut für Rechtsmedizin,
Röntgenstrasse 23
D-48149 Münster, Germany
FAX: (011-49) 251-83-55635
URL: http://www.isfg.org/

J. Craig Venter Institute
9704 Medical Center Dr.
Rockville, MD 20850
(301) 795-7000
URL: http://www.jcvi.org/

Muscular Dystrophy Association
National Headquarters
3300 E. Sunrise Dr.
Tucson, AZ 85718
1-800-572-1717
E-mail: mda@mdausa.org
URL: http://www.mdausa.org/

National Down Syndrome Society
666 Broadway, Eighth Floor
New York, NY 10012
1-800-221-4602
E-mail: info@ndss.org
URL: http://www.ndss.org/

National Human Genome Research Institute
National Institutes of Health
Bldg. 31, Rm. 4B09
31 Center Dr., MSC 2152
9000 Rockville Pike
Bethesda, MD 20892-2152
(301) 402-0911
FAX: (301) 402-2218
URL: http://www.genome.gov/

National Multiple Sclerosis Society
733 Third Ave.
New York, NY 10017
(212) 986-3240
1-800-344-4867
URL: http://www.nationalmssociety.org/

National Tay-Sachs and Allied Diseases Association
2001 Beacon St., Ste. 204
Boston, MA 02135
1-800-906-8723
FAX: (617) 277-0134
E-mail: info@ntsad.org
URL: http://www.ntsad.org/

Presidential Commission for the Study of Bioethical Issues
1425 New York Ave. NW, Ste. C-100
Washington, DC 20005
(202) 233-3960
FAX: (202) 233-3990
E-mail: info@bioethics.gov
URL: http://www.bioethics.gov/

Wellcome Images
183 Euston Rd.
London, NW1 2BE United Kingdom
(011-44) (0)20-7611-8348
FAX: (011-44) (0)20-7611-8577
E-mail: images@wellcome.ac.uk
URL: http://medphoto.wellcome.ac.uk/

RESOURCES

There are many published accounts of the history of genetics, but some of the most exciting versions were written by the pioneering researchers themselves. Even though many sources were used to construct the historical overview and highlights contained in this book, James D. Watson's *The Double Helix: A Personal Account of the Discovery of the Structure of DNA* (1968), Francis Crick's memoir *What Mad Pursuit: A Personal View of Scientific Discovery* (1988), and Alfred H. Sturtevant's *A History of Genetics* (1965) provided especially useful insights. Thomas Hunt Morgan's *The Mechanism of Mendelian Heredity* (1915) and *The Theory of the Gene* (1926) offer detailed descriptions of groundbreaking genetic research. Ricki Lewis and Bernard Possidente offer more recent history in *A Short History of Genetics and Genomics* (2003, http://www.dna50.com/dna50.swf).

Charles Darwin's *On the Origin of Species by Means of Natural Selection, or the Preservation of Favoured Races in the Struggle for Life* (1859) and *The Descent of Man, and Selection in Relation to Sex* (1871) provide historical perspectives on evolution. The discussion of nature versus nurture is drawn from Neil Whitehead and Briar Whitehead's *My Genes Made Me Do It* (1999) and from Richard J. Herrnstein and Charles Murray's *The Bell Curve: Intelligence and Class Structure in American Life* (1994). Ethical issues arising from genetic research and engineering are analyzed in *Our Posthuman Future: Consequences of the Biotechnology Revolution* (2002) by Francis Fukuyama and in *Enough: Staying Human in an Engineered Age* (2003) by Bill McKibben.

The journals *Nature* and *Science* have reported every significant finding and development in genetics, and articles dating from 1953 from both publications are cited in this text, as are articles from *Cell Research, e-biomed: The Journal of Regenerative Medicine, Nature Biotechnology,* and *Nature Medicine.* Research describing genetic testing, disorders, and genetic predisposition to disease is reported in professional medical journals. Studies cited in this book were published in peer-reviewed journals including *American Journal of Ophthalmology, Archives of Disease in Childhood, Archives of Internal Medicine, Archives of Neurology, Blood, British Medical Journal, Cell Stem Cell, Current Opinion in Allergy and Clinical Immunology, Current Psychiatry Reports, Fetal Diagnosis and Therapy, Genetics in Medicine, Gynecologic Oncology, Human Gene Therapy, International Journal of Nanomedicine, Journal of Allergy and Clinical Immunology, Journal of the American Medical Association, Journal of Clinical Oncology, Journal of Food Protection, Journal of the National Cancer Institute, Nature Genetics, New England Journal of Medicine, PLoS Biology, Proceedings of the National Academy of Science, Seminars in Respiratory and Critical Care Medicine,* and *Trends in Genetics.*

Ethical and psychological issues and the contributions of genetics to personality and behavior are examined in articles published in the *Archives of General Psychiatry, Archives of Sexual Behavior, Behavior Genetics, Biological Psychiatry, BMC Genetics, Human Genetics, Journal of the American Academy of Child and Adolescent Psychiatry, Journal of Educational Psychology, Journal of Neuroscience, Journal of Personality and Social Psychology, PLoS One,* and *Psychological Science.*

Besides the U.S. daily newspapers and electronic media, accounts of genetics research and milestones in the Human Genome Project were reported in the magazines *Newsweek* and *Time.*

The Human Genome Project Information website, which is operated by the U.S. Department of Energy, describes the ambitious goals and accomplishments of the Human Genome Project since its inception in 1990. The Environmental Genome Project was launched by the National Institute of Environmental Health Sciences. The National Institutes of Health provides definitions, epidemiological data, and research findings about a comprehensive range of genetic tests and genetic disorders.

Public opinion data from the following organizations was also very helpful: the Gallup Organization, the Virginia Commonwealth University's Center for Public Policy, and the Genetics and Public Policy Center of the Phoebe R. Berman Bioethics Institute at the Johns Hopkins University.

Additionally, many colleges, universities, medical centers, professional associations, and foundations dedicated to research, education, and advocacy about genetic disorders and diseases provided up-to-date information included in this edition.

INDEX

human genome sequencing groundwork, 107–110

Mendel, Gregor, research of, 3–7

model organisms, genome sequencing of, 111–112, 111*t*

modern genetics milestones, 12–14

post-Mendel era, 7

reproductive cloning, 128, 131–135

therapeutic cloning, 138–140

transgenic crops, adoption of, 150*f*

Watson and Crick, research of, 9–12

HLA gene, 79, 79*f*

Holley, Robert W., 107

Home genetic tests, 101–102

Homosexuality, 56–57

Hooke, Robert, 2

Hormones, 139

Horses, cloning of, 133–134

HUGO (Human Genome Organization), 111

Human chorionic gonadotropin (hCG), 89

Human cloning

ethical issues, 135–136, 175

public opinion, 146, 147*t*, 170

public policy, 142–145

Human Cloning and Human Dignity: An Ethical Inquiry (President's Council on Bioethics), 135–136

Human Epigenome Project, 15

Human Genome Organization (HUGO), 111

Human Genome Project

cloning types, 127

competition with private enterprises, 113–114

completion, 117

funding, 110–111, 110*t*

goals, 111, 117

history, 13–14

impact, 114–115, 116*f*, 119–121, 122*f*, 123

landmarks, 118*f*–121*f*

technology, 112–113

Human migration, 120

Humans

adaptive radiation, 46

brain development, 59

cells, 24

genetic makeup, 38

International HapMap Project, 123–124

mutation rates, 48–49

sex distribution, 34, 36*t*

shared genes, 116–117

therapeutic cloning of stem cells, 138–140

See also Sequencing, human genome

Huntington's disease, 12, 79–81, 80*f*, 108

Hydroxyurea, 84

Hypertension, 102*t*

I

Idaho Gem (mule), 133

IDDM genes, 78–79

IHEC (International Human Epigenome Consortium), 15

Immunodeficiency, 13, 164

In vitro fertilization, 176

Inborn Errors of Metabolism (Garrod), 7

Incomplete dominance, 29

Independent assortment, 5, 6*f*, 26

Induced pluripotent stem cells (IPSCs), 142

Information disclosure, 101

Information dissemination, 115, 115*f*, 124–127, 125*f*

Ingram, Vernon M., 12

Inheritance, genetic, 28–30, 31*f*, 32, 32*f*, 33*f*, 34*f*

Inheritance patterns, 34*f*, 63–64, 89(*f*6.1)

Innocence Project, 166

Institute for Genomic Research, 14, 113

Institute of Medicine (IOM), 155–156

Insulin, 128, 160

Intelligence, 57–59

Intelligent design, 45

International HapMap Project, 123–124

International Human Epigenome Consortium (IHEC), 15

International issues

Cartagena Protocol on Biosafety, 158–159

evolution, beliefs about, 44

genomic research, 111

International Service for the Acquisition of Agri-biotech Applications (ISAAA), 151

IOM (Institute of Medicine), 155–156

IPSCs (induced pluripotent stem cells), 142

Iron-overload disease, 65(*f*5.3)

ISAAA (International Service for the Acquisition of Agri-biotech Applications), 151

J

Jeffreys, Alec, 12

Johanns et al., Center for Food Safety et al. v., 157

Joint Genome Institute, 115–116

Juvenile diabetes. *See* Diabetes

K

Karyotypes, 26*f*

Khorana, Har Gobind, 107

King, Mary-Claire, 109

Kornberg, Roger D., 14

L

La Via Campesina, 153

Labeling, 158, 159–160

Laboratory of Reproductive Technology, 133–134

Laboratory techniques, 91–96, 92*f*

Lamarck, Jean Baptiste, 40–41

Lamberth, Royce C., 144

Lander, Eric, 14

Lawsuits, 153–154

LCA (Leber's congenital amaurosis), 95, 164–165

Leber's congenital amaurosis (LCA), 95, 164–165

Leder, Philip, 107

Legislation

cloning and stem cell research, 144–145

Genetic Information Nondiscrimination Act, 101

See also Government regulation and oversight

Lesch-Nyhan syndrome, 66

Levene, Phoebus A., 9

Life expectancy, 34, 37*t*

Ligase detection reaction/capillary electrophoresis, 94

Livestock, 133

Livestock breeding, 120

Louisiana, 45

M

Macleod, Colin Munro, 8

Maize, 152

Male gamete formation, 26

Malthus, Thomas Robert, 40

Mammals, 128, 131

Mammaprint, 98

MAOA (monoamine oxidase A) gene, 61

Mapping

epigenetics, 14–15

haplotype, 123–124

human genomes, 14

Mastectomies, 173

Maternal health, 34

Mathematics, 7, 8

McCarty, Maclyn, 8

McClintock, Barbara, 8

McKibben, Bill, 175

Media, 115

Medicine. *See* Health and medicine

Meiosis, 3, 7, 25–28, 27*f*, 28*f*, 29*f*, 140–141

Mencken, H. L., 41

Mendel, Gregor, 3–7, 41

Mental illness, 56, 66

Mental retardation, 66

Meselson, Matthew, 12

Messenger RNA (mRNA), 21–23, 22(*f*2.7)

Metastasis, 70–71

Mice studies

behavior and attitudes, 60

classical genetics, 8

cloning, 131, 134

embryonic stem cell research, 14, 139, 141

epigenetics, 56